Ron S. Kenett • Shelemyahu Zacks • Peter Gedeck

T0235818

Modern Statistics

A Computer-Based Approach with Python

 Birkhäuser

Ron S. Kenett
KPA Ltd. Raanana and Samuel Neaman
Institute, Technion
Haifa, Israel

Shelemyahu Zacks
Mathematical Sciences
Binghamton University
Mc Lean, VA, USA

Peter Gedeck
Data Science
University of Virginia
Falls Church, VA, USA

ISSN 2662-5555   ISSN 2662-5563   (electronic)
Statistics for Industry, Technology, and Engineering
ISBN 978-3-031-07568-1   ISBN 978-3-031-07566-7   (eBook)
https://doi.org/10.1007/978-3-031-07566-7

Mathematics Subject Classification: 62E15, 62G30, 62M10, 62P30, 62P10, 97K40, 97K70, 97K80

This book is published under the imprint Birkhäuser, www.birkhauser-science.com by the registered
company Springer Nature Switzerland AG
The registered company address is: Gewerbestrasse 11, 6330 Cham, Switzerland

To my wife Sima, our children and their children: Yonatan, Alma, Tomer, Yadin, Aviv, Gili, Matan, Eden, and Ethan. RSK

To my wife Hanna, our sons Yuval and David, and their families with love. SZ

To Janet with love. PG

Preface

Statistics has developed by combining the needs of science, business, industry, and government. More recent development is connected with methods for generating insights from data, using statistical theory and delivery platforms. This integration is at the core of applied statistics and most of theoretical statistics.

Before the beginning of the twentieth century, statistics meant observed data and descriptive summary figures, such as means, variances, indices, etc., computed from data. With the introduction of the χ^2-test for goodness of fit by Karl Pearson (1900) and the t-test by Gosset (Student, 1908) for drawing inference on the mean of a normal population, statistics became a methodology of analyzing sample data to determine the validity of hypotheses about the source of the data (the population). Fisher (1922) laid the foundations for statistics as a discipline. He considered the object of statistical methods to be reducing data into the essential statistics, and he identified three problems that arise in doing so:

1. Specification-choosing the right mathematical model for a population
2. Estimation-methods to calculate, from a sample, estimates of the parameters of the hypothetical population
3. Distribution-properties of statistics derived from samples

Forty years later, Tukey (1962) envisioned a data-centric development of statistics, sketching the pathway to data science. Forty years after that, we entered the age of big data, data science, artificial intelligence, and machine learning. These new developments are built on the methods, applications, and experience of statisticians around the world.

The first two authors started collaborating on a book in the early 1990s. In 1998, we published with Duxbury Wadsworth *Modern Industrial Statistics: Design and Control of Quality and Reliability*. The book appeared in a Spanish edition (Estadística Industrial Moderna: Diseño y Control de Calidad y la Confiabilidad, Thomson International, 2000). An abbreviated edition was published as *Modern Statistics: A Computer based Approach* (Thomson Learning, 2001); this was followed by a Chinese edition (China Statistics Press, 2003) and a softcover edition, (Brooks/Cole, 2004). The book used QuickBasic, S-Plus, and MINITAB. In 2014 we published, with Wiley, an extended second edition titled *Modern Industrial Statistics: With Applications in R, MINITAB and JMP*. That book was translated into Vietnamese by the Vietnam Institute for Advanced Studies in Mathematics (VIASM, 2016). A third, expanded edition, was published by Wiley in 2021.

This book is about modern statistics with Python. It reflects many years of experience of the authors in doing research, teaching and applying statistics in science, healthcare, business, defense, and industry domains. The book invokes over 40 case studies and provides comprehensive Python applications. In 2019, there were 8.2 million developers in the world who code using Python which is considered the fastest-growing programming language. A special Python package, mistat, is available for download https://gedeck.github.io/mistat-code-solutions/ ModernStatistics/. Everything in the book can be reproduced with mistat. We therefore provide, in this book, an integration of needs, methods, and delivery platform for a large audience and a wide range of applications.

Modern Statistics: A Computer-Based Approach with Python is a companion text to another book published by Springer titled: *Industrial Statistics: A Computer Based Approach with Python*. Both books include mutual cross references, but both books are stand-alone publications. This book can be used as textbook in a one semester or two semester course on modern statistics. The technical level of the presentation in both books can serve both undergraduate and graduate students. The example and case studies provide access to hands on teaching and learning. Every chapter includes exercises, data sets, and Python applications. These can be used in regular classroom setups, flipped classroom setups, and online or hybrid education programs. The companion text is focused on industrial statistics with special chapters on advanced process monitoring methods, cybermanufacturing, computer experiments, and Bayesian reliability. *Modern Statistics* is a foundational text and can be combined with any program requiring data analysis in its curriculum. This, for example, can be courses in data science, industrial statistics, physics, biology, chemistry, economics, psychology, social sciences, or any engineering discipline.

Modern Statistics: A Computer-Based Approach with Python includes eight chapters. Chapter 1 is on analyzing variability with descriptive statistics. Chapter 2 is on probability models and distribution functions. Chapter 3 introduces statistical inference and bootstrapping. Chapter 4 is on variability in several dimensions and regression models. Chapter 5 covers sampling for estimation of finite population quantities, a common situation when one wants to infer on a population from a sample. Chapter 6 is dedicated to time series analysis and prediction. Chapters 7 and 8 are about modern data analytic methods.

Industrial Statistics: A Computer-Based Approach with Python contains 11 chapters: Chapter 1— Introduction to Industrial Statistics, Chapter 2—Basic Tools and Principles of Process Control, Chapter 3—Advanced Methods of Statistical Process Control, Chapter 4—Multivariate Statistical Process Control, Chapter 5— Classical Design and Analysis of Experiments, Chapter 6—Quality by Design, Chapter 7—Computer Experiments, Chapter 8—Cybermanufacturing and Digital Twins, Chapter 9—Reliability Analysis, Chapter 10—Bayesian Reliability Estimation and Prediction, and Chapter 11—Sampling Plans for Batch and Sequential Inspection. This second book is focused on industrial statistics with applications to monitoring, diagnostics, prognostic, and prescriptive analytics. It can be used as a stand-alone book, or in conjunction with *Modern Statistics*. Both books include solution manuals to exercises listed at the end of each chapter. This was designed to support self-learning as well as instructor led courses.

We made every possible effort to ensure the calculations are correct and the text is clear. However, should errors have skipped to the printed version, we would appreciate feedback from readers noticing these. In general, any feedback will be much appreciated.

Finally, we would like to thank the team at Springer Birkhäuser, including Dana Knowles and Christopher Tominich. They made everything in the publication process look easy.

Ra'anana, Israel   Ron S. Kenett
McLean, VA, USA   Shelemyahu Zacks
Falls Church, VA, USA   Peter Gedeck
April 2022

Contents

1 **Analyzing Variability: Descriptive Statistics** 1
   1.1   Random Phenomena and the Structure of Observations 1
   1.2   Accuracy and Precision of Measurements 6
   1.3   The Population and the Sample 8
   1.4   Descriptive Analysis of Sample Values 9
        1.4.1   Frequency Distributions of Discrete Random Variables ... 9
        1.4.2   Frequency Distributions of Continuous Random
                  Variables 14
        1.4.3   Statistics of the Ordered Sample 17
        1.4.4   Statistics of Location and Dispersion 19
   1.5   Prediction Intervals 23
   1.6   Additional Techniques of Exploratory Data Analysis 25
        1.6.1   Density Plots 25
        1.6.2   Box and Whiskers Plots 27
        1.6.3   Quantile Plots 29
        1.6.4   Stem-and-Leaf Diagrams 30
        1.6.5   Robust Statistics for Location and Dispersion 31
   1.7   Chapter Highlights 34
   1.8   Exercises 34

2 **Probability Models and Distribution Functions** 39
   2.1   Basic Probability 39
        2.1.1   Events and Sample Spaces: Formal Presentation
                  of Random Measurements 39
        2.1.2   Basic Rules of Operations with Events: Unions
                  and Intersections 41
        2.1.3   Probabilities of Events 44
        2.1.4   Probability Functions for Random Sampling 46
        2.1.5   Conditional Probabilities and Independence of Events 49
        2.1.6   Bayes' Theorem and Its Application 51
   2.2   Random Variables and Their Distributions 54

	2.2.1	Discrete and Continuous Distributions	55
		2.2.1.1 Discrete Random Variables	55
		2.2.1.2 Continuous Random Variables	56
	2.2.2	Expected Values and Moments of Distributions	59
	2.2.3	The Standard Deviation, Quantiles, Measures of Skewness, and Kurtosis	62
	2.2.4	Moment Generating Functions	65
2.3		Families of Discrete Distribution	66
	2.3.1	The Binomial Distribution	66
	2.3.2	The Hypergeometric Distribution	69
	2.3.3	The Poisson Distribution	72
	2.3.4	The Geometric and Negative Binomial Distributions	74
2.4		Continuous Distributions	78
	2.4.1	The Uniform Distribution on the Interval (a, b), $a < b$	78
	2.4.2	The Normal and Log-Normal Distributions	79
		2.4.2.1 The Normal Distribution	79
		2.4.2.2 The Log-Normal Distribution	84
	2.4.3	The Exponential Distribution	85
	2.4.4	The Gamma and Weibull Distributions	88
	2.4.5	The Beta Distributions	92
2.5		Joint, Marginal, and Conditional Distributions	93
	2.5.1	Joint and Marginal Distributions	93
	2.5.2	Covariance and Correlation	96
	2.5.3	Conditional Distributions	99
2.6		Some Multivariate Distributions	102
	2.6.1	The Multinomial Distribution	102
	2.6.2	The Multi-Hypergeometric Distribution	104
	2.6.3	The Bivariate Normal Distribution	105
2.7		Distribution of Order Statistics	108
2.8		Linear Combinations of Random Variables	111
2.9		Large Sample Approximations	117
	2.9.1	The Law of Large Numbers	117
	2.9.2	The Central Limit Theorem	117
	2.9.3	Some Normal Approximations	119
2.10		Additional Distributions of Statistics of Normal Samples	120
	2.10.1	Distribution of the Sample Variance	121
	2.10.2	The "Student" t-Statistic	122
	2.10.3	Distribution of the Variance Ratio	123
2.11		Chapter Highlights	125
2.12		Exercises	126
3		**Statistical Inference and Bootstrapping**	139
3.1		Sampling Characteristics of Estimators	139
3.2		Some Methods of Point Estimation	141
	3.2.1	Moment Equation Estimators	142

		3.2.2	The Method of Least Squares	144
		3.2.3	Maximum Likelihood Estimators	146
3.3	Comparison of Sample Estimates			149
	3.3.1	Basic Concepts		149
	3.3.2	Some Common One-Sample Tests of Hypotheses		152

3.2.2   The Method of Least Squares 144
3.2.3   Maximum Likelihood Estimators 146
3.3 Comparison of Sample Estimates 149
3.3.1   Basic Concepts 149
3.3.2   Some Common One-Sample Tests of Hypotheses 152
         3.3.2.1   The Z-Test: Testing the Mean of a
                   Normal Distribution, σ^2 Known 152
         3.3.2.2   The t-Test: Testing the Mean of a
                   Normal Distribution, σ^2 Unknown 155
         3.3.2.3   The Chi-Squared Test: Testing the
                   Variance of a Normal Distribution 156
         3.3.2.4   Testing Hypotheses About the Success
                   Probability, p, in Binomial Trials 158
3.4 Confidence Intervals 160
3.4.1   Confidence Intervals for μ; σ Known 161
3.4.2   Confidence Intervals for μ; σ Unknown 162
3.4.3   Confidence Intervals for σ^2 162
3.4.4   Confidence Intervals for p 163
3.5 Tolerance Intervals 166
3.5.1   Tolerance Intervals for the Normal Distributions 166
3.6 Testing for Normality with Probability Plots 169
3.7 Tests of Goodness of Fit 173
3.7.1   The Chi-Square Test (Large Samples) 173
3.7.2   The Kolmogorov-Smirnov Test 175
3.8 Bayesian Decision Procedures 176
3.8.1   Prior and Posterior Distributions 177
3.8.2   Bayesian Testing and Estimation 181
         3.8.2.1   Bayesian Testing 181
         3.8.2.2   Bayesian Estimation 184
3.8.3   Credibility Intervals for Real Parameters 185
3.9 Random Sampling from Reference Distributions 186
3.10 Bootstrap Sampling 189
3.10.1   The Bootstrap Method 189
3.10.2   Examining the Bootstrap Method 190
3.10.3   Harnessing the Bootstrap Method 192
3.11 Bootstrap Testing of Hypotheses 192
3.11.1   Bootstrap Testing and Confidence Intervals
         for the Mean 192
3.11.2   Studentized Test for the Mean 193
3.11.3   Studentized Test for the Difference of Two Means 195
3.11.4   Bootstrap Tests and Confidence Intervals
         for the Variance 197
3.11.5   Comparing Statistics of Several Samples 199
         3.11.5.1 Comparing Variances of Several Samples 200

3.11.5.2 Comparing Several Means: The
One-Way Analysis of Variance 201
3.12 Bootstrap Tolerance Intervals 204
3.12.1 Bootstrap Tolerance Intervals for Bernoulli Samples 204
3.12.2 Tolerance Interval for Continuous Variables................ 205
3.12.3 Distribution-Free Tolerance Intervals 206
3.13 Non-Parametric Tests.. 208
3.13.1 The Sign Test.. 208
3.13.2 The Randomization Test...................................... 210
3.13.3 The Wilcoxon Signed-Rank Test 211
3.14 Chapter Highlights.. 214
3.15 Exercises 215

4 Variability in Several Dimensions and Regression Models 225
4.1 Graphical Display and Analysis 226
4.1.1 Scatterplots 226
4.1.2 Multiple Boxplots 229
4.2 Frequency Distributions in Several Dimensions 230
4.2.1 Bivariate Joint Frequency Distributions..................... 231
4.2.2 Conditional Distributions 234
4.3 Correlation and Regression Analysis 235
4.3.1 Covariances and Correlations 236
4.3.2 Fitting Simple Regression Lines to Data.................... 237
4.3.2.1 The Least Squares Method...................... 239
4.3.2.2 Regression and Prediction Intervals 243
4.4 Multiple Regression 245
4.4.1 Regression on Two Variables................................ 246
4.4.2 Partial Regression and Correlation 251
4.4.3 Multiple Linear Regression.................................. 254
4.4.4 Partial-F Tests and the Sequential SS...................... 260
4.4.5 Model Construction: Step-Wise Regression 263
4.4.6 Regression Diagnostics 265
4.5 Quantal Response Analysis: Logistic Regression 268
4.6 The Analysis of Variance: The Comparison of Means............... 271
4.6.1 The Statistical Model 271
4.6.2 The One-Way Analysis of Variance (ANOVA)............. 271
4.7 Simultaneous Confidence Intervals: Multiple Comparisons 275
4.8 Contingency Tables... 279
4.8.1 The Structure of Contingency Tables 279
4.8.2 Indices of association for contingency tables............... 282
4.8.2.1 Two Interval-Scaled Variables 282
4.8.2.2 Indices of Association for Categorical
Variables 284
4.9 Categorical Data Analysis.. 288
4.9.1 Comparison of Binomial Experiments...................... 288

	4.10	Chapter Highlights	290
	4.11	Exercises	291

5	**Sampling for Estimation of Finite Population Quantities**		299
	5.1	Sampling and the Estimation Problem	299
		5.1.1 Basic Definitions	299
		5.1.2 Drawing a Random Sample from a Finite Population	301
		5.1.3 Sample Estimates of Population Quantities and Their Sampling Distribution	302
	5.2	Estimation with Simple Random Samples	305
		5.2.1 Properties of $\bar{X}_n$ and S_n^2 Under RSWR	306
		5.2.2 Properties of $\bar{X}_n$ and S_n^2 Under RSWOR	310
	5.3	Estimating the Mean with Stratified RSWOR	314
	5.4	Proportional and Optimal Allocation	316
	5.5	Prediction Models with Known Covariates	320
	5.6	Chapter Highlights	324
	5.7	Exercises	325

6	**Time Series Analysis and Prediction**		329
	6.1	The Components of a Time Series	330
		6.1.1 The Trend and Covariances	330
		6.1.2 Analyzing Time Series with Python	331
	6.2	Covariance Stationary Time Series	336
		6.2.1 Moving Averages	337
		6.2.2 Auto-Regressive Time Series	338
		6.2.3 Auto-Regressive Moving Average Time Series	343
		6.2.4 Integrated Auto-Regressive Moving Average Time Series	344
		6.2.5 Applications with Python	345
	6.3	Linear Predictors for Covariance Stationary Time Series	346
		6.3.1 Optimal Linear Predictors	346
	6.4	Predictors for Non-stationary Time Series	349
		6.4.1 Quadratic LSE Predictors	349
		6.4.2 Moving Average Smoothing Predictors	351
	6.5	Dynamic Linear Models	352
		6.5.1 Some Special Cases	353
		6.5.1.1 The Normal Random Walk	353
		6.5.1.2 Dynamic Linear Model With Linear Growth	354
		6.5.1.3 Dynamic Linear Model for ARMA(p,q)	355
	6.6	Chapter Highlights	358
	6.7	Exercises	359

7	**Modern Analytic Methods: Part I**		361
	7.1	Introduction to Computer Age Statistics	361
	7.2	Data Preparation	362
	7.3	The Information Quality Framework	363

7.4   Determining Model Performance 364
7.5   Decision Trees 368
7.6   Ensemble Models 376
7.7   Naïve Bayes Classifier.. 378
7.8   Neural Networks.. 381
7.9   Clustering Methods... 386
        7.9.1   Hierarchical Clustering 386
        7.9.2 K-Means Clustering 389
        7.9.3   Cluster Number Selection 390
7.10   Chapter Highlights... 392
7.11   Exercises 392

8 **Modern Analytic Methods: Part II** 395
8.1   Functional Data Analysis... 395
8.2   Text Analytics... 401
8.3   Bayesian Networks 405
8.4   Causality Models 411
8.5   Chapter Highlights... 416
8.6   Exercises 417

A **Introduction to Python** 421
A.1   List, Set, and Dictionary Comprehensions 421
A.2   Pandas Data Frames 422
A.3   Data Visualization Using `Pandas` and `Matplotlib` 423

B **List of Python Packages** 425

C **Code Repository and Solution Manual** 427

Bibliography 429

Index 433

Industrial Statistics: A Computer-Based Approach with Python (Companion volume)

1 **The Role of Statistical Methods in Modern Industry**
   1.1   The Evolution of Industry
   1.2   The Evolution of Quality
   1.3   Industry 4.0 Characteristics
   1.4   The Digital Twin
   1.5   Chapter Highlights
   1.6   Exercises

2 **Basic Tools and Principles of Process Control**
   2.1   Basic Concepts of Statistical Process Control
   2.2   Driving a Process with Control Charts
   2.3   Setting Up a Control Chart: Process Capability Studies
   2.4   Process Capability Indices
   2.5   Seven Tools for Process Control and Process Improvement
   2.6   Statistical Analysis of Pareto Charts
   2.7   The Shewhart Control Charts
   2.8   Process analysis with data segments
   2.9   Chapter Highlights
   2.10   Exercises

3 **Advanced Methods of Statistical Process Control**
   3.1   Tests of Randomness
   3.2   Modified Shewhart Control Charts for $\bar{X}$
   3.3   The Size and Frequency of Sampling for Shewhart Control Charts
   3.4   Cumulative Sum Control Charts
   3.5   Bayesian Detection
   3.6   Process Tracking
   3.7   Automatic Process Control
   3.8   Chapter Highlights
   3.9   Exercises

4   Multivariate Statistical Process Control
   4.1   Introduction
   4.2   A Review Multivariate Data Analysis
   4.3   Multivariate Process Capability Indices
   4.4   Advanced Applications of Multivariate Control Charts
   4.5   Multivariate Tolerance Specifications
   4.6   Tracking structural changes
   4.7   Chapter Highlights
   4.8   Exercises

5   Classical Design and Analysis of Experiments
   5.1   Basic Steps and Guiding Principles
   5.2   Blocking and Randomization
   5.3   Additive and Non-Additive Linear Models
   5.4   The Analysis of Randomized Complete Block Designs
   5.5   Balanced Incomplete Block Designs
   5.6   Latin Square Design
   5.7   Full Factorial Experiments
   5.8   Blocking and Fractional Replications of 2^m Factorial Designs
   5.9   Exploration of Response Surfaces
   5.10 Evaluating Designed Experiments
   5.11 Chapter Highlights
   5.12 Exercises

6   Quality by Design
   6.1   Off-Line Quality Control, Parameter Design and The Taguchi Method
   6.2   The Effects of Non-Linearity
   6.3   Taguchi's Designs
   6.4   Quality by Design in the Pharmaceutical Industry
   6.5   Tolerance Designs
   6.6   Case Studies
   6.7   Chapter Highlights
   6.8   Exercises

7   Computer Experiments
   7.1   Introduction to Computer Experiments
   7.2   Designing Computer Experiments
   7.3   Analyzing Computer Experiments
   7.4   Stochastic Emulators
   7.5   Integrating Physical and Computer Experiments
   7.6   Simulation of Random Variables
   7.7   Chapter Highlights
   7.8   Exercises

8   Cybermanufacturing and Digital Twins
   8.1   Introduction to Cybermanufacturing
   8.2   Cybermanufacturing Analytics
   8.3   Information Quality in Cybermanufacturing
   8.4   Modeling in Cybermanufacturing
   8.5   Computational pipelines
   8.6   Digital Twins
   8.7   Chapter Highlights
   8.8   Exercises

9   Reliability Analysis
   9.1   Basic Notions
   9.2   System Reliability
   9.3   Availability of Repairable Systems
   9.4   Types of Observations on TTF
   9.5   Graphical Analysis of Life Data
   9.6   Non-Parametric Estimation of Reliability
   9.7   Estimation of Life Characteristics
   9.8   Reliability Demonstration
   9.9   Accelerated Life Testing
   9.10   Burn-In Procedures
   9.11   Chapter Highlights
   9.12   Exercises

10 Bayesian Reliability Estimation and Prediction
   10.1   Prior and Posterior Distributions
   10.2   Loss Functions and Bayes Estimators
   10.3   Bayesian Credibility and Prediction Intervals
   10.4   Credibility Intervals for the Asymptotic Availability of
         Repairable Systems: The Exponential Case
   10.5   Empirical Bayes Method
   10.6   Chapter Highlights
   10.7   Exercises

11 Sampling Plans for Batch and Sequential Inspection
   11.1   General Discussion
   11.2   Single-Stage Sampling Plans for Attributes
   11.3   Approximate Determination of the Sampling Plan
   11.4   Double-Sampling Plans for Attributes
   11.5   Sequential Sampling and A/B testing
   11.6   Acceptance Sampling Plans for Variables
   11.7   Rectifying Inspection of Lots
   11.8   National and International Standards
   11.9   Skip-Lot Sampling Plans for Attributes
   11.10 The Deming Inspection Criterion
   11.11 Published Tables for Acceptance Sampling

11.12 Sequential Reliability Testing
11.13 Chapter Highlights
11.14 Exercises
References

A   List of Python packages

List of Abbreviations

AIC	Akaike information criteria
ANOVA	Analysis of variance
ANSI	American National Standard Institute
AOQ	Average outgoing quality
AOQL	Average outgoing quality limit
AQL	Acceptable quality level
ARIMA	Autoregressive integrated moving average
ARL	Average run length
ASN	Average sample number
ASQ	American Society for Quality
ATE	Average treatment effect
ATI	Average total inspection
BECM	Bayes estimation of the current mean
BI	Business intelligence
BIBD	Balanced incomplete block design
BIC	Bayesian information criteria
BN	Bayesian network
BP	Bootstrap population
c.d.f.	Cumulative distribution function
CAD	Computer-aided design
CADD	Computer-aided drawing and drafting
CAM	Computer-aided manufacturing
CART	Classification and regression trees
CBD	Complete block design
CED	Conditional expected delay
cGMP	Current good manufacturing practices
CHAID	Chi-square automatic interaction detector
CIM	Computer integrated manufacturing
CLT	Central limit theorem
CMM	Coordinate measurement machines
CMMI	Capability maturity model integrated

CNC	Computerized numerically controlled
CPA	Circuit pack assemblies
CQA	Critical quality attribute
CUSUM	Cumulative sum
DACE	Design and analysis of computer experiments
DAG	Directed acyclic graph
DFIT	Difference in fits distance
DLM	Dynamic linear model
DoE	Design of experiments
DTM	Document term matrix
EBD	Empirical bootstrap distribution
ETL	Extract-transform-load
EWMA	Exponentially weighted moving average
FDA	Food and Drug Administration
FDA	Functional data analysis
FPCA	Functional principal component analysis
FPM	Failures per million
GFS	Google file system
GRR	Gage repeatability and reproducibility
HPD	Highest posterior density
HPLC	High-performance liquid chromatography
i.i.d.	Independent and identically distributed
IDF	Inverse document frequency
InfoQ	Information quality
IPO	Initial public offering
IPS	Inline process control
IQR	Inter quartile range
ISC	Short circuit current of solar cells (in Ampere)
KS	Kolmogorov-Smirnov test
LCL	Lower control limit
LLN	Law of large numbers
LQL	Limiting quality level
LSA	Latent semantic analysis
LSL	Lower specification limit
LTPD	Lot tolerance percent defective
LWL	Lower warning limit
MAE	Mean absolute error
m.g.f.	Moment generating function
MLE	Maximum likelihood estimator
MSD	Mean squared deviation
MSE	Mean squared error
MTBF	Mean time between failures
MTTF	Mean time to failure
NID	Normal independently distributed
OAB	One-armed bandit

OC	Operating characteristic
p.d.f.	Probability density function
PCA	Principal component analysis
PERT	Project evaluation and review technique
PFA	Probability of false alarm
PL	Product limit estimator
PPM	Defects in parts per million
PSE	Practical statistical efficiency
QbD	Quality by design
QMP	Quality measurement plan
QQ-Plot	Quantile vs. quantile plot
RCBD	Randomized complete block design
Regex	Regularized expression
RMSE	Root mean squared error
RSWOR	Random sample without replacement
RSWR	Random sample with replacement
SE	Standard error
SL	Skip lot
SLOC	Source lines of code
SLSP	Skip lot sampling plans
SPC	Statistical process control
SPRT	Sequential probability ratio test
SR	Shiryaev Roberts
SSE	Sum of squares of errors
SSR	Sum of squares around the regression model
SST	Total sum of squares
STD	Standard deviation
SVD	Singular value decomposition
TAB	Two-armed bandit
TF	Term frequency
TTC	Time till censoring
TTF	Time till failure
TTR	Time till repair
TTT	Total time on test
UCL	Upper control limit
USL	Upper specification limit
UWL	Upper warning limit
WSP	Wave soldering process

Chapter 1
Analyzing Variability: Descriptive Statistics

Preview The chapter focuses on statistical variability and various methods of analyzing random data. Random results of experiments are illustrated with distinction between deterministic and random components of variability. The difference between accuracy and precision is explained. Frequency distributions are defined to represent random phenomena. Various characteristics of location and dispersion of frequency distributions are defined. The elements of exploratory data analysis are presented.

1.1 Random Phenomena and the Structure of Observations

Many phenomena which we encounter are only partially predictable. It is difficult to predict the weather or the behavior of the stock market. In this book we focus on industrial phenomena, like performance measurements from a product which is being manufactured, or the sales volume in a specified period of a given product model. Such phenomena are characterized by the fact that measurements performed on them are often not constant but reveal a certain degree of variability. Variability is also a reflection of uncertainty. For a comprehensive treatment of uncertainty in engineering applications, see del Rosario and Iaccarino (2022). The objective of this chapter is to present methods for analyzing this variability, in order to understand the variability structure and enhance our ability to control, improve, and predict future behavior of such phenomena. We start with a few simple examples. The data and code used throughout the book is available from https://gedeck.github.io/mistat-code-solutions/ModernStatistics/

Example 1.1 A piston is a mechanical device that is present in most types of engines. One measure of the performance of a piston is the time it takes to complete

Supplementary Information The online version contains supplementary material available at https://doi.org/10.1007/978-3-031-07566-7_1.

1

one cycle. We call this measure **cycle time**. In Table 1.1 we present 50 cycle times of a piston operating under fixed operating conditions (a sample dataset is stored in file **CYCLT.csv**). We provide with this book code in Python for running a piston software simulation. If you installed Python, install the `mistat` package using the Python package installer `pip`.

```
pip install mistat
```

You will get access to a piston simulator with seven factors that you can change. We will also use this simulator when we discuss statistical process control (Chaps. 2, 3, and 4 in the Industrial Statistics book) and the design of experiments (Chaps. 5, 6, and 7 in the Industrial Statistics book). We continue at a pedestrian pace by recreating Table 1.1 using Python. All the Python applications referred to in this book are contained in a package called `mistat` available from the Python package index and on GitHub at https://github.com/gedeck/mistat. The following Python commands will import the `mistat` package, read the cycle time data, and print the first five values on your monitor:

```
import mistat
data = mistat.load_data('CYCLT')
print(data.head())
```

```
0    1.008
1    1.098
2    1.120
3    0.423
4    1.021
Name: CYCLT, dtype: float64
```

Notice that functions in Python have parenthesis. The `import` statement imports the `mistat` package and makes its functionality available. `mistat.load_data` is a function that loads the CYCLT dataset as a *Pandas* data series. A *Pandas* data series is a simple vector of values.

The differences in cycle times values are quite apparent and we can make the statement "cycle times are varying." Such a statement, in spite of being true, is not very useful. We have only established the existence of variability—we have not yet characterized it and are unable to predict and control future behavior of the piston. ∎

Example 1.2 Consider an experiment in which a coin is flipped once. Suppose the coin is fair in the sense that it is equally likely to fall on either one of its faces. Furthermore, assume that the two faces of the coin are labeled with the numbers "0" and "1". In general, we cannot predict with certainty on which face the coin will fall. If the coin falls on the face labeled "0", we assign to a variable X the value 0; if the coin falls on the face labeled "1", we assign to X the value 1. Since the values which X will obtain in a sequence of such trials cannot be predicted with certainty, we call X a **random variable**. A typical random sequence of 0, 1 values that can be generated in this manner might look like the following:

Table 1.1 Cycle times of piston (in seconds) with control factors set at minimum levels

1.008	1.098	1.120	0.423	1.021
1.069	0.271	0.431	1.095	1.088
1.117	1.080	0.206	0.330	0.314
1.132	0.586	1.118	0.319	0.664
1.141	0.662	0.531	0.280	0.489
1.080	0.628	0.302	0.179	1.056
0.449	1.057	0.437	0.175	0.482
0.275	1.084	0.287	1.068	1.069
0.215	1.107	0.348	0.213	0.200
0.187	0.339	0.224	1.009	0.560

0, 1, 0, 0, 0, 0, 0, 0, 0, 1, 0, 1, 0, 1, 0, 1, 0, 1, 0, 0,
1, 1, 0, 1, 1, 1, 0, 0, 0, 1, 0, 0, 1, 1, 1, 0, 1, 1, 0, 1.

In this sequence of 40 random numbers, there are 22 0s and 18 1s. We expect in a large number of trials, since the coin is unbiased, that 50% of the random numbers will be 0s and 50% of them will be 1s. In any particular short sequence, the actual percentage of 0s will fluctuate around the expected number of 50%.

At this point we can use the computer to "**simulate**" a coin tossing experiment. There are special routines for generating random numbers on the computer. We will illustrate this by using Python. The following commands generate a sequence of 50 random binary numbers (0 and 1).

```
from scipy.stats import binom
import numpy as np

X = binom.rvs(1, 0.5, size=50)

print(X)
```

```
[1 1 0 1 0 0 1 0 0 0 0 1 0 0 0 0 1 0 1 1 0 0 1 0 0 1 1 1 1
 1 1 0 0 1 0 0 1 0 1 1 1 1 1 0 0 1 0 1 0 1 1 1]
```

The command uses a binomial function (`binom.rvs`). The first argument defines the number of trials, here set to 1, and the second argument the probability of 1s. The `size` argument specifies the number of observations. Execute this command to see another random sequence of 50 0s and 1s. Compare this sequence to the one given earlier.   ■

Example 1.3 Another example of a random phenomenon is illustrated in Fig. 1.1 where 50 measurements of the length of steel rods are presented. These data are stored in file **STEELROD.csv**. To generate Fig. 1.1 in Python, type at the prompt the following commands:

```
import matplotlib.pyplot as plt

steelrod = mistat.load_data('STEELROD')
```

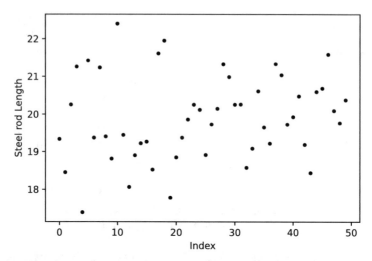

Fig. 1.1 Length of 50 steel rods (in cm)

```
# create a scatterplot
ax = steelrod.plot(y='STEELROD', style='.', color='black')
ax.set_xlabel('Index')          # set the x axis title
ax.set_ylabel('Steel rod Length') # set the y axis title
plt.show()
```

Steel rods are used in the car and truck industry to strengthen vehicle structures. Automation of assembly lines has created stringent requirements on the physical dimensions of parts. Steel rods supplied by Urdon Industries for Peugeot car plants are produced by a process adjusted to obtain rods with a length of 20 cm. However, due to natural fluctuations in the production process, the actual length of the rods varies around the nominal value of 20 cm. Examination of this sequence of 50 values does not reveal any systematic fluctuations. We conclude that the deviations from the nominal values are random. It is impossible to predict with certainty what the values of additional measurements of rod length will be. However, we shall learn later that with further analysis of this data, we can determine that there is a high likelihood that new observations will fall close to 20 cm.

It is possible for a situation to arise in which, at some time, the process will start to malfunction, causing a **shift** to occur in the average value of the process. The pattern of variability might then look like the one in Fig. 1.2. An examination of Fig. 1.2 shows that a significant shift has occurred in the level of the process after the 25th observation and that the systematic deviation from the average value of the process has persisted constantly. The deviations from the nominal level of 20 cm are first just random and later systematic and random. The steel rods obviously became shorter. A quick investigation revealed that the process got accidentally misadjusted by a manager who played with machine knobs while showing the plant to important guests.   ■

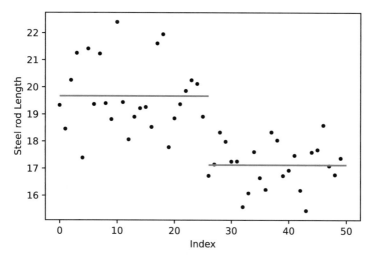

Fig. 1.2 Level shift after the first 25 observations

In formal notation, if X_i is the value of the i-th observation, then

$$X_i = \begin{cases} O + E_i & i = 1, \cdots, 25 \\ N + E_i & i = 26, \cdots, 50, \end{cases}$$

where $O = 20$ is the original level of the process, $N = 17$ is its new level after the shift, and E_i is a random component. Note that O and N are fixed and, in this case, constant non-random levels. Thus, a random sequence can consist of values which have two components: a **fixed** component and a **random** component. A fixed-nonrandom pattern is called a **deterministic** pattern. As another example, in Fig. 1.3 we present a sequence of 50 values of

$$X_i = D_i + E_i, \quad i = 1, \cdots, 50,$$

where the D_i's follow a sinusoidal pattern shown on Fig. 1.3 by dots and E_i's are random deviations having the same characteristics as those of Fig. 1.1. The sinusoidal pattern is $D_i = \sin(2\pi i/50)$, $i = 1, \ldots, 50$. This component can be determined exactly for each i and is therefore called deterministic while E_i is a random component. In Python we can construct such a sequence and plot it with the following commands:

```
import math
from scipy.stats import norm

# create a list of 50 values forming a sine curve
x = [math.sin(x * 2 * math.pi / 50) for x in range(1, 51)]

# Add a random normal with mean 0 and standard deviation 0.05
```

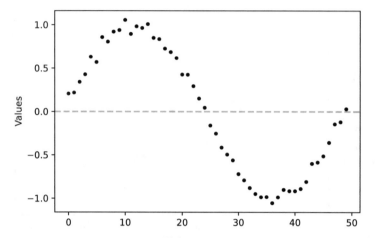

Fig. 1.3 Random variation around a systematic trend

```
x = [xi + norm.rvs(loc=0, scale=0.05) for xi in x]

ax = pd.Series(x).plot(style='.', color='black')
ax.set_ylabel('Values')
ax.axhline(y=0, linestyle='--', color='darkgray')
plt.show()
```

If the random component could be eliminated, we would be able to predict exactly the future values of X_i. For example, by following the pattern of the D_i's, we can determine that X_{100} would be equal to 0. However, due to the existence of the random component, an exact prediction is impossible. Nevertheless, we **expect** that the actual values will fall around the deterministic pattern. In fact, certain prediction limits can be assigned, using methods which will be discussed later.

1.2 Accuracy and Precision of Measurements

Different measuring instruments and gages or gauges (such as weighing scales, voltmeters, etc.) may have different characteristics. For example, we say that an instrument is **accurate** if repetitive measurements of the same object yield an average equal to its true value. An instrument is **inaccurate** if it yields values whose average is different from the true value. **Precision**, on the other hand, is related to the dispersion of the measurements around their average. In particular, small dispersion of the measurements reflects high precision, while large dispersion reflects low precision. It is possible for an instrument to be inaccurate but precise, or accurate but imprecise. Precision, sometimes called **repeatability**, is a property of the measurement technology. **Reproducibility** is assessing the impact of the measurement procedure on measurement uncertainty, including the contribution of the individuals taking the measurement. Differences between lab operators are reflected by the level of reproducibility. There are other properties of measuring devices or gages, like stability, linearity etc., which will not be discussed here. A common term for describing techniques for empirical assessment of the uncertainty of a measurement device is **gage repeatability and reproducibility** (GR&R). These involve repeated testing of a number of items, by different operators. In addition to a (GR&R) assessment, to ensure proper accuracy, measuring instruments need to be calibrated periodically relative to an external standard. In the USA, the National Institute of Standards and Technologies (NIST) is responsible for such activities.

Example 1.4 In Fig. 1.4, we present weighing measurements of an object whose true weight is 5 kg. The measurements were performed on three instruments, with ten measurements on each one. We see that instrument A is accurate (the average is 5.0 kg), but its dispersion is considerable. Instrument B is not accurate (the average is 2.0 kg) but is more precise than A. Instrument C is as accurate as A but is more precise than A.   ∎

```
np.random.seed(seed=1)

x = np.concatenate([5 + norm.rvs(loc=0, scale=0.5, size=10),
                    2 + norm.rvs(loc=0, scale=0.2, size=10),
                    5 + norm.rvs(loc=0, scale=0.1, size=10)])
ax = pd.Series(x).plot(style='.', color='black')
ax.set_ylabel('Values')
ax.set_xlabel('Index')
ax.set_ylabel('Weight')
ax.hlines(y=5, xmin=0, xmax=9, color='darkgray')
ax.hlines(y=2, xmin=10, xmax=19, color='darkgray')
ax.hlines(y=5, xmin=20, xmax=29, color='darkgray')
ax.text(4, 6.5, 'A')
ax.text(14, 3.5, 'B')
ax.text(24, 6.5, 'C')
ax.set_ylim(0, 8)
plt.show()
```

As a note it should be mentioned that repeatability and reproducibility are also relevant in the wider context of research. For a dramatic failure in reproducibility, see the article by Nobel Prize winner Paul Krugman on the research of Harvard economists, Carmen Reinhart and Kenneth Rogoff, that purported to identify a critical threshold or tipping point, for government indebtedness. Their findings were flawed because of self-selected data points and coding errors in Excel (https://www.nytimes.com/2013/04/19/opinion/krugman-the-excel-depression.html).   Another dramatic example of irreproducible research is a Duke university genomic study which proposed genomic tests that looked at the molecular traits of a cancerous tumor and recommended which chemotherapy would work best. This research proved flawed because of errors such as moving a row or a column over by one in a giant spreadsheet and other more complex reasons (https://www.nytimes.com/2011/07/08/health/research/08genes.html). Repeatability in microarray studies is related to identifying the same set of active genes in large and smaller studies. These topics

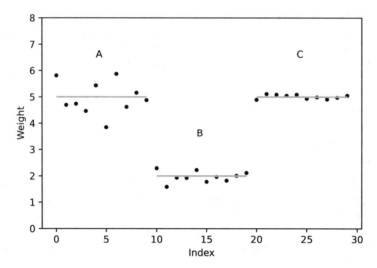

Fig. 1.4 Samples of ten measurements from three different instruments

are however beyond the scope of this book. For a discussion of reproducibility in science, with references, see Kenett and Rubinstein (2021).

1.3   The Population and the Sample

A **statistical population** is a collection of units having a certain common attribute. For example, the set of all the citizens of the USA on January 1, 2021, is a statistical population. Such a population is comprised of many subpopulations, e.g., all males in the age group of 19–25 living in Illinois, etc. Another statistical population is the collection of all concrete cubes of specified dimensions that can be produced under well-defined conditions. The first example of all the citizens of the USA on January 1, 2021, is a **finite** and **real** population, while the population of all units that can be produced by a specified manufacturing process is **infinite** and **hypothetical**.

A **sample** is a subset of the elements of a given population. A sample is usually drawn from a population for the purpose of observing its characteristics and making some statistical decisions concerning the corresponding characteristics of the whole population. For example, consider a lot of 25,000 special screws which were shipped by a vendor to factory A. Factory A must decide whether to accept and use this shipment or reject it (according to the provisions of the contract). Suppose it is agreed that, if the shipment contains no more than 4% defective items, it should be accepted and, if there are more than 6% defectives, the shipment should be rejected and returned to the supplier. Since it is impractical to test each item of this population (although it is finite and real), the decision of whether or not to accept the lot is based on the number of defective items found in a **random sample** drawn from the population. Such procedures for making statistical decisions are called **acceptance sampling** methods. Chapter 11 in the Industrial Statistics book is dedicated to these methods. Chapter 5 provides the foundations for estimation of parameters using samples from finite populations including random sample with replacement (RSWR) and random sample without replacement (RSWOR). Chapter 3 includes a description of a technique called bootstrapping, which is a special case of RSWR.

1.4 Descriptive Analysis of Sample Values

In this section we discuss the first step for analyzing data collected in a sampling process. One way of describing a distribution of sample values, which is particularly useful in large samples, is to construct a **frequency distribution** of the sample values. We distinguish between two types of frequency distributions, namely, frequency distributions of (i) **discrete** variables and (ii) **continuous** variables.

A random variable, X, is called discrete if it can assume only a finite (or at most a countable) number of different values. For example, the number of defective computer cards in a production lot is a discrete random variable. A random variable is called continuous if, theoretically, it can assume all possible values in a given interval. For example, the output voltage of a power supply is a continuous random variable.

1.4.1 Frequency Distributions of Discrete Random Variables

Consider a random variable, X, that can assume only the values $x_1, x_2, \cdots, x_k$, where $x_1 < x_2 < \cdots < x_k$. Suppose that we have made n different observations on X. The frequency of x_i $(i = 1, \cdots, k)$ is defined as the number of observations having the value x_i. We denote the frequency of x_i by f_i. Notice that

$$\sum_{i=1}^{k} f_i = f_1 + f_2 + \cdots + f_k = n.$$

The set of ordered pairs

$$\{(x_1, f_1), (x_2, f_2), \cdots, (x_k, f_k)\}$$

constitutes the frequency distribution of X. We can present a frequency distribution in a tabular form as

It is sometimes useful to present a frequency distribution in terms of the proportional or **relative frequencies** p_i, which are defined by

Value	Frequency
x_1	f_1
x_2	f_2
$\vdots$	$\vdots$
x_k	f_k
Total	n

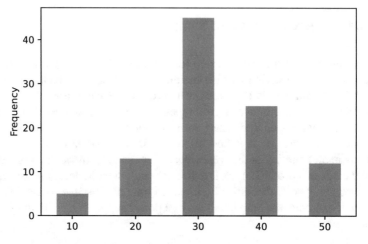

Fig. 1.5 Bar diagram of a frequency distribution

$$p_i = f_i/n \quad (i = 1, \cdots, k).$$

A frequency distribution can be presented graphically in a form which is called a **bar diagram**, as shown in Fig. 1.5. The height of the bar at x_j is proportional to the frequency of this value.

In addition to the frequency distribution, it is often useful to present the cumulative frequency distribution of a given variable. The **cumulative frequency** of x_i is defined as the sum of frequencies of values less than or equal to x_i. We denote it by F_i and the proportional cumulative frequencies or cumulative relative frequency by

$$P_i = F_i/n.$$

A table of proportional cumulative frequency distribution could be represented as follows:

The graph of the cumulative relative frequency distribution is a step function and looks typically like the graph shown in Fig. 1.6.

Example 1.5 A US manufacturer of hybrid microelectronic components purchases ceramic plates from a large Japanese supplier. The plates are visually inspected

Value	p	P
x_1	p_1	$P_1 = p_1$
x_2	p_2	$P_2 = p_1 + p_2$
$\vdots$	$\vdots$	
x_k	p_k	$P_k = p_1 + \cdots + p_k$
Total	1	

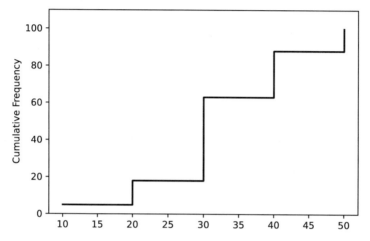

Fig. 1.6 Step function of a cumulative relative frequency distribution

Table 1.2 Frequency distribution of blemishes on ceramic plates

x	f	p	P
0	15	0.50	0.50
1	8	0.27	0.77
2	3	0.10	0.87
3	3	0.10	0.97
4	0	0.00	0.97
5	1	0.03	1.00

before screen printing. Blemishes will affect the final product's electrical performance and overall yield. In order to prepare a report for the Japanese supplier, the US manufacturer decided to characterize the variability in the number of blemishes found on the ceramic plates. The following measurements represent the number of blemishes found on each of 30 ceramic plates:

$$0, 2, 0, 0, 1, 3, 0, 3, 1, 1, 0, 0, 1, 2, 0,$$
$$0, 0, 1, 1, 3, 0, 1, 0, 0, 0, 5, 1, 0, 2, 0.$$

Here the variable X assumes the values 0, 1, 2, 3, and 5. The frequency distribution of X is displayed in Table 1.2.

We did not observe the value $x = 4$, but since it seems likely to occur in future samples, we include it in the frequency distribution, with frequency $f = 0$.

For pedagogical purposes, we show next how to calculate a frequency distribution and how to generate a bar diagram in Python:

```
blemishes = mistat.load_data('BLEMISHES')
```

The object BLEMISHES is not a simple vector like CYCLT—it is called a data frame, i.e., a matrix-like structure whose columns (variables) may be of differing types. Here are the first few rows of the data frame.

```
blemishes.head(3)
```

```
   plateID  count
0        1      0
1        2      2
2        3      0
```

We can access individual elements in a list using square brackets in Python. For a Pandas data frame, we need to use the square brackets on the .iloc[i, j] property. i defines the row and j the column.

```
print(blemishes.iloc[1, 0])
print(blemishes.iloc[2, 1])
```

```
2
0
```

Note that, like many other programming languages, Python starts the index of lists at 0. For the first example, this means we accessed row 2 (index 1) and column 1 (index 0).

It is also possible to extract a whole column from a Pandas data frame by name.

```
blemishes['count'].head(5)
```

```
0    0
1    2
2    0
3    0
4    1
Name: count, dtype: int64
```

We can create a bar diagram of the blemish count distribution as follows.

```
# use value_counts with normalize to get relative frequencies
X = pd.DataFrame(blemishes['count'].value_counts(normalize=True))
X.loc[4, 'count'] = 0  # there are no samples with 4 blemishes add a row
X = X.sort_index()  # sort by number of blemishes

ax = X['count'].plot.bar(color='grey', legend=False)
ax.set_xlabel('Number of blemishes')
ax.set_ylabel('Proportional Frequency')
plt.show()
```

```
X['Number'] = X.index  # add number of blemishes as column
X['Cumulative Frequency'] = X['count'].cumsum()
```

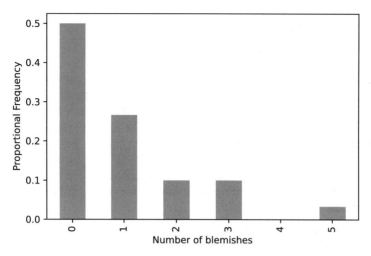

Fig. 1.7 Bar diagram for number of blemishes on ceramic plates

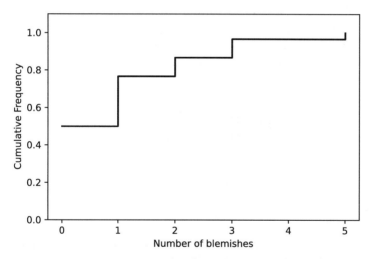

Fig. 1.8 Cumulative relative frequency distribution for number of blemishes on ceramic plates

```
ax = X.plot.line(x='Number', y='Cumulative Frequency', color='black',
                 drawstyle='steps-post', legend=False)
ax.set_xlabel('Number of blemishes')
ax.set_ylabel('Cumulative Frequency')
ax.set_ylim(0, 1.1)
plt.show()
```

The bar diagram and cumulative frequency step function are shown in Figs. 1.7 and 1.8, respectively.   ∎

1.4.2   Frequency Distributions of Continuous Random Variables

For the case of a continuous random variable, we partition the possible range of variation of the observed variable into k subintervals. Generally speaking, if the possible range of X is between L and H, we specify numbers $b_0, b_1, b_2, \cdots, b_k$ such that $L = b_0 < b_1 < b_2 < \cdots < b_{k-1} < b_k = H$. The values $b_0, b_1, \cdots, b_k$ are called the limits of the k subintervals. We then classify the X values into the interval (b_{i-1}, b_i) if $b_{i-1} < X \leq b_i$ ($i = 1, \cdots, k$). (If $X = b_0$, we assign it to the first subinterval.) Subintervals are also called **bins, classes**, or **class intervals**.

In order to construct a frequency distribution, we must consider the following two questions:

 (i) How many subintervals should we choose?
(ii) How large should the width of the subintervals be?

In general, it is difficult to give these important questions exact answers which apply in all cases. However, the general recommendation is to use between 10 and 15 subintervals in large samples and apply equal width subintervals. The frequency distribution is given then for the subintervals, where the mid-point of each subinterval provides a numerical representation for that interval. A typical frequency distribution table might look like the following:

Subintervals	Mid-point	Freq.	Cum. Freq.
$b_0 - b_1$	$\bar{b}_1$	f_1	$F_1 = f_1$
$b_1 - b_2$	$\bar{b}_2$	f_2	$F_2 = f_1 + f_2$
$\vdots$			
$b_{k-1} - b_k$	$\bar{b}_k$	f_k	$F_k = n$

Example 1.6 Nilit, a large fiber supplier to US and European textile manufacturers, has tight control over its yarn strength. This critical dimension is typically analyzed on a logarithmic scale. This logarithmic transformation produces data that is more symmetrically distributed. Consider $n = 100$ values of $Y = \ln(X)$ where X is the yarn strength [lb./22 yarns] of woolen fibers. The data is stored in file **YARNSTRG.csv** and shown in Table 1.3.

The smallest value in Table 1.3 is $Y = 1.1514$ and the largest value is $Y = 5.7978$. This represents a range of $5.7978 - 1.1514 = 4.6464$. To obtain approximately 15 subintervals, we need the width of each interval to be about $4.6464/15 = .31$. A more convenient choice for this class width might be 0.50. The first subinterval would start at $b_0 = 0.75$ and the last subinterval would end with $b_k = 6.25$. The frequency distribution for this data is presented in Table 1.4.

A graphical representation of the distribution is given by a **histogram** as shown in Fig. 1.9. Each rectangle has a height equal to the frequency (f) or relative frequency

Table 1.3 A sample of 100 log (yarn strength)

2.4016	1.1514	4.0017	2.1381	2.5364
2.5813	3.6152	2.5800	2.7243	2.4064
2.1232	2.5654	1.3436	4.3215	2.5264
3.0164	3.7043	2.2671	1.1535	2.3483
4.4382	1.4328	3.4603	3.6162	2.4822
3.3077	2.0968	2.5724	3.4217	4.4563
3.0693	2.6537	2.5000	3.1860	3.5017
1.5219	2.6745	2.3459	4.3389	4.5234
5.0904	2.5326	2.4240	4.8444	1.7837
3.0027	3.7071	3.1412	1.7902	1.5305
2.9908	2.3018	3.4002	1.6787	2.1771
3.1166	1.4570	4.0022	1.5059	3.9821
3.7782	3.3770	2.6266	3.6398	2.2762
1.8952	2.9394	2.8243	2.9382	5.7978
2.5238	1.7261	1.6438	2.2872	4.6426
3.4866	3.4743	3.5272	2.7317	3.6561
4.6315	2.5453	2.2364	3.6394	3.5886
1.8926	3.1860	3.2217	2.8418	4.1251
3.8849	2.1306	2.2163	3.2108	3.2177
2.0813	3.0722	4.0126	2.8732	2.4190

Table 1.4 Frequency distribution for log yarn strength data

$b_{i-1} - b_i$	$\bar{b}_i$	f_i	p_i	F_i	P_i
0.75–1.25	1.0	2	0.02	2	0.02
1.25–1.75	1.5	9	0.09	11	0.11
1.75–2.25	2.0	12	0.12	23	0.23
2.25–2.75	2.5	26	0.26	49	.49
2.75–3.25	3.0	17	0.17	66	0.66
3.25–3.75	3.5	17	0.17	83	0.83
3.75–4.25	4.0	7	0.07	90	0.90
4.25–4.75	4.5	7	0.07	97	0.97
4.75–5.25	5.0	2	0.02	99	0.99
5.25–5.75	5.5	0	0.00	99	0.99
5.75–6.25	6.0	1	0.01	100	1.00

(p) of the corresponding subinterval. In either case the area of the rectangle is proportional to the frequency of the interval along the base. The cumulative frequency distribution is presented in Fig. 1.10.

Computer programs select a default midpoint and width of class intervals but provide the option to change these choices. The shape of the histogram depends on the number of class intervals chosen. You can experiment with the dataset **YARNSTRG.csv**, by choosing a different number of class intervals, starting with the default value.

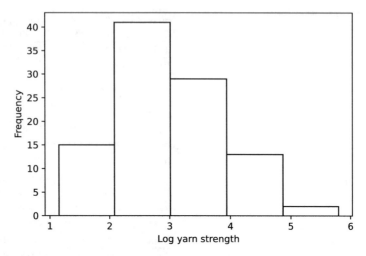

Fig. 1.9 Histogram of log yarn strength (Table 1.4)

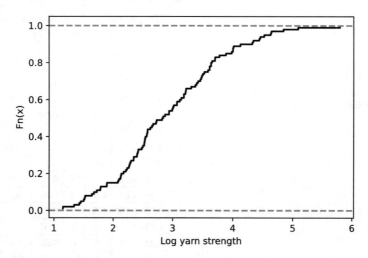

Fig. 1.10 Cumulative relative distribution of log yarn strength

Apply the following Python commands:

```
X = mistat.load_data('YARNSTRG')
ax = X.plot.hist(bins=5, color='white', edgecolor='black', legend=False)
ax.set_xlabel('Log yarn strength')
plt.show()

ecdf = pd.DataFrame({'Log yarn strength': X.sort_values(),
                     'Fn(x)': range(1, len(X) + 1)})
ecdf['Fn(x)'] = ecdf['Fn(x)'] / len(X)
ax = ecdf.plot(x='Log yarn strength', y='Fn(x)', color='black',
               drawstyle='steps-post', legend=False)
ax.axhline(y=0, color='grey', linestyle='--')
ax.axhline(y=1, color='grey', linestyle='--')
ax.set_ylabel('Fn(x)')
plt.show()
```

The commands produce a histogram with five bins (Fig. 1.9) and a cumulative relative distribution shown in Fig. 1.10.   ∎

1.4.3   Statistics of the Ordered Sample

In this section we identify some characteristic values of a sample of observations that have been sorted from smallest to largest. Such sample characteristics are called **order statistics**. In general **statistics** are computed from observations and are used to make an inference on characteristics of the population from where the sample was drawn. Statistics that do not require to sort observation are discussed in Sect. 1.4.4.

Let $X_1, X_2, \cdots, X_n$ be the observed values of some random variable, as obtained by a random sampling process. For example, consider the following ten values of the shear strength of welds of stainless steel (lb./weld): 2385, 2400, 2285, 2765, 2410, 2360, 2750, 2200, 2500, 2550. What can we do to characterize the variability and location of these values?

The first step is to sort the sample values in an increasing order, that is, we rewrite the list of sample values as 2200, 2285, 2360, 2385, 2400, 2410, 2500, 2550, 2750, 2765. These ordered values are denoted by $X_{(1)}, X_{(2)}, \cdots, X_{(n)}$, where $X_{(1)} = 2200$ is the smallest value in the sample, $X_{(2)} = 2285$ is the second smallest, and so on. We call $X_{(i)}$ the i-**the order statistic** of the sample. For convenience, we can also denote the average of consecutive order statistics by

$$X_{(i.5)} = (X_{(i)} + X_{(i+1)})/2 = X_{(i)} + .5(X_{(i+1)} - X_{(i)}). \tag{1.1}$$

For example, $X_{(2.5)} = (X_{(2)} + X_{(3)})/2$. We now identify some characteristic values that depend on these order statistics, namely, the sample minimum, the sample maximum, the sample range, the sample median, and the sample quartiles. The **sample minimum is** $X_{(1)}$ and the **sample maximum** is $X_{(n)}$. In our example $X_{(1)} = 2200$ and $X_{(n)} = X_{(10)} = 2765$. The **sample range** is the difference $R = X_{(n)} - X_{(1)} = 2765 - 2200 = 565$. The "middle" value in the ordered sample is called the **sample median**, denoted by M_e. The sample median is defined as $M_e = X_{(m)}$ where $m = (n+1)/2$. In our example, $n = 10$ so $m = (10+1)/2 = 5.5$. Thus

$$M_e = X_{(5.5)} = (X_{(5)} + X_{(6)})/2 = X_{(5)} + .5(X_{(6)} - X_{(5)})$$
$$= (2400 + 2410)/2$$
$$= 2405.$$

The median characterizes the center of dispersion of the sample values and is therefore called a **statistic of central tendency**, or **location statistic**. Approximately 50% of the sample values are smaller than the median. Finally we define the **sample quartiles** as $Q_1 = X_{(q_1)}$ and $Q_3 = X_{(q_3)}$ where

$$q_1 = \frac{(n+1)}{4}$$

and   (1.2)

$$q_3 = \frac{3(n+1)}{4}.$$

Q_1 is called the **lower quartile** and Q_3 is called the **upper quartile**. These quartiles divide the sample so that approximately one fourth of the values are smaller than Q_1, one half are between Q_1 and Q_3, and one fourth are greater than Q_3. In our example, $n = 10$ so

$$q_1 = \frac{11}{4} = 2.75$$

and

$$q_3 = \frac{33}{4} = 8.25.$$

Thus, $Q_1 = X_{(2.75)} = X_{(2)} + .75 \times (X_{(3)} - X_{(2)}) = 2341.25$ and $Q_3 = X_{(8.25)} = X_{(8)} + .25 \times (X_{(9)} - X_{(8)}) = 2600$.

These sample statistics can be obtained from a frequency distribution using the cumulative relative frequency as shown in Fig. 1.11 which is based on the log yarn strength data of Table 1.3.

Using linear interpolation within the subintervals, we obtain $Q_1 = 2.3$, $Q_3 = 3.6$ and $M_e = 2.9$. These estimates are only slightly different from the exact values $Q_1 = X_{(.25)} = 2.2789$, $Q_3 = X_{(.75)} = 3.5425$, and $M_e = X_{(.5)} = 2.8331$.

The sample median and quartiles are specific forms of a class of statistics known as sample quantiles. The **p-th sample quantile** is a number that exceeds exactly $100p\%$ of the sample values. Hence, the median is the 0.5 sample quantile, Q_1 is the 0.25-th quantile, and Q_3 is the 0.75-th sample quantile. We may be interested, for example, in the 0.9 sample quantile. Using linear interpolation in Fig. 1.11, we obtain the value 4.5, while the value of $X_{(.9)} = 4.2233$. The p-th sample quantile is also called the $100p$-th sample percentile. The following Python commands yield these statistics of the data: median, min, max, Q_1, and Q_3.

```
cyclt = mistat.load_data('CYCLT')

print(cyclt.quantile(q=[0, 0.25, 0.5, 0.75, 1.0]))
```

```
0.00     0.1750
0.25     0.3050
0.50     0.5455
0.75     1.0690
1.00     1.1410
Name: CYCLT, dtype: float64
```

The mean is calculated with the mean method.

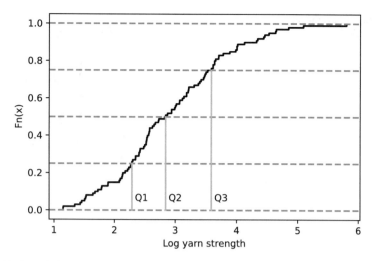

Fig. 1.11 Cumulative relative distribution function with linear interpolation lines at quartiles

```
print(cyclt.mean())
```

```
0.6524599999999999
```

The `describe` method returns several of the statistics in a concise format.

```
print(cyclt.describe())
```

```
count    50.000000
mean      0.652460
std       0.372971
min       0.175000
25%       0.305000
50%       0.545500
75%       1.069000
max       1.141000
Name: CYCLT, dtype: float64
```

Applying this command on the piston cycle time of file **CYCLT.csv**, we find $X_{(1)} = 0.1750$, $Q_1 = 0.3050$, $M_e = 0.5455$, $\bar{X} = 0.6525$, $Q_3 = 1.0690$, and $X_{(50)} = 1.1410$.

1.4.4 Statistics of Location and Dispersion

Given a sample of n measurements, $X_1, \cdots, X_n$, we can compute various statistics to describe the distribution. The **sample mean** is determined by the formula

Table 1.5 Computing the
sample variance

	X	$(X - \bar{X})$	$(X - \bar{X})^2$
	45	15	225
	60	30	900
	21	−9	81
	19	−11	121
	4	−26	676
	31	1	1
Sum	180	0	2004

$$\bar{X} = 180/6 = 30$$
$$S^2 = 2004/5 = 400.8$$

$$\bar{X} = \frac{1}{n} \sum_{i=1}^{n} X_i. \tag{1.3}$$

Like the sample median, $\bar{X}$ is a measure of central tendency. In Physics, the sample mean represents the "center of gravity" for a system consisting of n equal-mass particles located on the points X_i on the line.

As an example consider the following measurements, representing component failure times in hours since initial operation

$$45, \ 60, \ 21, \ 19, \ 4, \ 31.$$

The sample mean is

$$\bar{X} = (45 + 60 + 21 + 19 + 4 + 31)/6 = 30.$$

To measure the spread of data about the mean, we typically use the **sample variance** defined by

$$S^2 = \frac{1}{n-1} \sum_{i=1}^{n} (X_i - \bar{X})^2, \tag{1.4}$$

or the **sample standard deviation**, given by

$$S = \sqrt{S^2}.$$

The sample standard deviation is used more often since its units (cm., lb.) are the same as those of the original measurements. In the next section, we will discuss some ways of interpreting the sample standard deviation. Presently we remark only that datasets with greater dispersion about the mean will have larger standard deviations. The computation of S^2 is illustrated in Table 1.5 using the failure time data.

The sample standard deviation and sample mean provide information on the variability and central tendency of observation. For the dataset (number of blemishes on ceramic plates) in Table 1.2, one finds that $\bar{X} = 0.933$ and $S = 1.258$. Looking at the histogram in Fig. 1.7, one notes a marked asymmetry in the data. In 50% of the ceramic plates, there were no blemishes and in 3% there were five blemishes. In contrast, consider the histogram of log yarn strength which shows remarkable symmetry with $\bar{X} = 2.9238$ and $S = 0.93776$. The difference in shape is obviously not reflected by $\bar{X}$ and S. Additional information pertaining to the shape of a distribution of observations is derived from the **sample skewness** and **sample kurtosis**. The sample skewness is defined as the index

$$\beta_3 = \frac{1}{n} \sum_{i=1}^{n} (X_i - \bar{X})^3 / S^3. \tag{1.5}$$

The sample kurtosis (steepness) is defined as

$$\beta_4 = \frac{1}{n} \sum_{i=1}^{n} (X_i - \bar{X})^4 / S^4. \tag{1.6}$$

These indices can be computed in Python using the `Pandas` or the `scipy` package.

```
X = mistat.load_data('YARNSTRG')
print(f'Skewness {X.skew():.4f}')      # Computes the skewness
print(f'Kurtosis {X.kurtosis():.4f}') # Computes the kurtosis

from scipy.stats import skew, kurtosis
print(f'Skewness {skew(X):.4f}')       # Computes the skewness
print(f'Kurtosis {kurtosis(X):.4f}') # Computes the kurtosis
```

```
Skewness 0.4164
Kurtosis -0.0080
Skewness 0.4102
Kurtosis -0.0670
```

Skewness and kurtosis are provided by most statistical computer packages, and we can see from these two examples that there are subtle differences in implementation. If a distribution is symmetric (around its mean), then skewness $= 0$. If skewness > 0, we say that the distribution is positively skewed or skewed to the right. If skewness < 0, then the distribution is negatively skewed or skewed to the left. We should also comment that in distributions which are positively skewed $\bar{X} > M_e$, while in those which are negatively skewed $\bar{X} < M_e$. In symmetric distributions $\bar{X} = M_e$.

The steepness of a distribution is determined relative to that of the normal (Gaussian) distribution, which is described in the next section and specified in Sect. 2.4.2. In a normal distribution, kurtosis $= 3$. Thus, if kurtosis > 3, the distribution is called steep. If kurtosis < 3, the distribution is called flat. A schematic representation of shapes is given in Figs. 1.12 and 1.13.

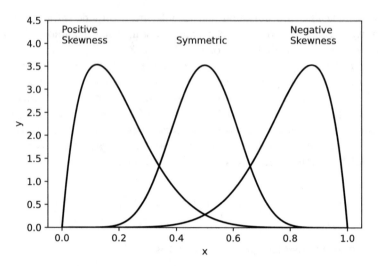

Fig. 1.12 Symmetric and asymmetric distributions

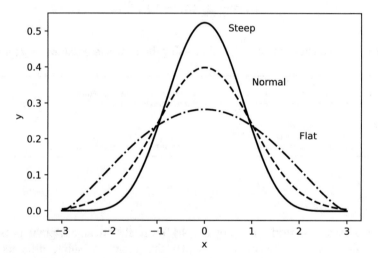

Fig. 1.13 Normal, steep, and flat distributions

To illustrate these statistics, we compute $\bar{X}$, S^2, S, skewness and kurtosis for the log yarn strength data of Table 1.3, We obtain

$$\bar{X} = 2.9238$$

$$S^2 = 0.8794 \quad S = 0.9378$$

$$\text{Skewness} = 0.4164 \quad \text{Kurtosis} = -0.0080.$$

The sample mean is $\bar{X} = 2.9238$, for values on a logarithmic scale. To return to the original scale [lb/22 yarns], we can use the measure

$$G = \exp\{\bar{X}\} \tag{1.7}$$

$$= \left(\prod_{i=1}^{n} Y_i\right)^{1/n} = 18.6127, \tag{1.8}$$

where $Y_i = \exp(X_i), i = 1, \ldots, n$. The measure G is called the **geometric mean** of Y. The geometric mean, G, is defined only for positive valued variables. It is used as a measure of central tendency for rates of change and index numbers such as the desirability function. One can prove the following general result:

$$G \leq \bar{X}.$$

Equality holds only if all values in the sample are the same.

Additional statistics to measure the dispersion are as follows.

 (i) The **interquartile range**

$$\text{IQR} = Q_3 - Q_1, \text{ and} \tag{1.9}$$

(ii) The **coefficient of variation**

$$\gamma = \frac{S}{|\bar{X}|}. \tag{1.10}$$

The interquartile range, IQR, is a useful measure of dispersion when there are extreme values (outliers) in the sample. It is easy to compute and can yield an estimate of S; for more details, see Sect. 1.6.5. The coefficient of variation is a dimensionless index, used to compare the variability of different datasets, when the standard deviation tends to grow with the mean. The coefficient of variation of the log yarn strength data is $\gamma = \dfrac{0.9378}{2.9238} = 0.3207$.

1.5   Prediction Intervals

When the data $X_1, \cdots, X_n$ represents a sample of observations from some population, we can use the sample statistics discussed in the previous sections to predict how future measurements will behave. Of course, our ability to predict accurately depends on the size of the sample.

Prediction using order statistics is very simple and is valid for any type of distribution. Since the ordered measurements partition the real line into $n + 1$ subintervals,

$$(-\infty, X_{(1)}), (X_{(1)}, X_{(2)}), \cdots, (X_{(n)}, \infty),$$

we can predict that $100/(n + 1)\%$ of all future observations will fall in any one of these subintervals; hence, $100\,i/(n + 1)\%$ of future sample values are expected to be less than the i-th order statistic $X_{(i)}$. It is interesting to note that the sample minimum, $X_{(1)}$, is **not** the smallest possible value. Instead, we expect to see one out of every $n + 1$ future measurements to be less than $X_{(1)}$. Similarly, one out of every $n + 1$ future measurements is expected to be greater than $X_{(n)}$.

Predicting future measurements using sample skewness and kurtosis is a bit more difficult because it depends on the type of distribution that the data follow. If the distribution is symmetric (skewness ≈ 0) and somewhat "bell-shaped" or "normal"[1] (kurtosis ≈ 3) as in Fig. 1.9, for the log yarn strength data, we can make the following statements:

1. Approximately **68%** of all future measurements will lie within **one standard deviation** of the mean
2. Approximately **95%** of all future measurements will lie within **two standard deviations** of the mean
3. Approximately **99.7%** of all future measurements will lie within **three standard deviations** of the mean.

The sample mean and standard deviation for the log yarn strength measurement are $\bar{X} = 2.92$ and $S = 0.94$. Hence, we predict that 68% of all future measurements will lie between $\bar{X} - S = 1.98$ and $\bar{X} + S = 3.86$, 95% of all future observations will be between $\bar{X} - 2S = 1.04$ and $\bar{X} + 2S = 4.80$, and 99.7% of all future observations will be between $\bar{X} - 3S = 0.10$ and $\bar{X} + 3S = 5.74$. For the data in Table 1.4, there are exactly 69, 97, and 99 of the 100 values in the above intervals, respectively.

When the data does not follow a normal distribution, we may use the following result:

Chebyshev's Inequality
For any number $k > 1$ **the percentage of future measurements within k standard deviations of the mean will be at least** $100(1 - 1/k^2)\%$.

This means that at least 75% of all future measurements will fall within two standard deviations ($k = 2$). Similarly, at least 89% will fall within three standard deviations ($k = 3$). These statements are true for any distribution; however,

---

[1] The normal or Gaussian distribution will be defined in Chap. 2.

the actual percentages may be considerably larger. Notice that for data which is normally distributed, 95% of the values fall in the interval $[\bar{X} - 2S, \bar{X} + 2S]$. The Chebyshev inequality gives only the lower bound of 75% and is therefore very conservative.

Any prediction statements, using the order statistics or the sample mean and standard deviation, can only be made with the understanding that they are based on a sample of data. They are accurate only to the degree that the sample is representative of the entire population. When the sample size is small, we cannot be very confident in our prediction. For example, if based on a sample of size $n = 10$, we find $\bar{X} = 20$ and $S = 0.1$, then we might make the statement that 95% of all future values will be between $19.8 = 20 - 2(.1)$ and $20.2 = 20 + 2(.1)$. However, it would not be too unlikely to find that a second sample produced $\bar{X} = 20.1$ and $S = .15$. The new prediction interval would be wider than 19.8 to 20.4, a considerable change. Also, a sample of size 10 does not provide sufficient evidence that the data has a "normal" distribution. With larger samples, say $n > 100$, we may be able to draw this conclusion with greater confidence.

In Chap. 3 we will discuss theoretical and computerized statistical inference whereby we assign a "confidence level" to such statements. This confidence level will depend on the sample size. Prediction intervals which are correct with high confidence are called **tolerance intervals**.

1.6   Additional Techniques of Exploratory Data Analysis

In the present section, we present additional modern graphical techniques, which are quite common today in exploratory data analysis. These techniques are the **density plot**, the **box and whiskers plot**, the **quantile plot**, and **stem-and-leaf Diagram**. We also discuss the problem of sensitivity of the sample mean and standard deviation to outlying observations and introduce some robust statistics.

1.6.1   Density Plots

Similar to histograms, density plots represent the distribution of a variable. In contrast to histograms, they create a continuous representation that can give a more detailed insight into the distribution.

Density plots can, for example, be created using pandas. Figure 1.14 compares histogram and density plot for the yarn strength dataset.

```
X = mistat.load_data('YARNSTRG')
X.plot.density()
```

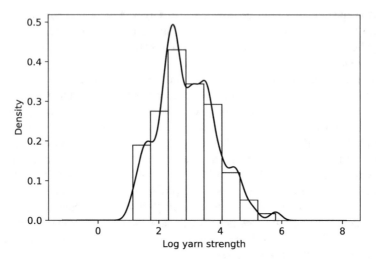

Fig. 1.14 Comparison of histogram and density plot for the log yarn strength datasets

The density plots are also known as kernel density plots, a name that reflects how they are calculated. The kernel density $f(x)$ is calculated from the n data points $x_1, \ldots, x_n$ as follows:

$$f(x) = \frac{1}{nh} \sum_{i=1}^{n} K\left(\frac{x - x_i}{h}\right)$$

K is the kernel and $h > 0$ is the smoothing parameter. There are a variety of functions for K described in the literature; however, a frequently used one is the gaussian or normal distribution $N(0, 1)$ (see Sect. 2.4.2).

The smoothing parameter, often referred to as bandwidth, controls the level of detail in the density estimate. A small value shows many details and may however lead to an undersmoothing; a large value on the other hand can hide interesting details. Figure 1.15 demonstrates the effect of changing the bandwidth for the log yarn strength dataset.

Finding a good value for h is therefore crucial. There are various approaches to derive h based on the data: rule-of-thumb methods, plugin methods, or cross-validation methods. A rule of thumb is Scott's rule:

$$h_{\text{Scott}} = 1.06\bar{S}n^{\frac{-1}{d+4}}$$

where d is the number of dimensions, here $d = 1$, n the number of data points, and $\bar{S}$ the estimate of the standard deviation. While Scott's rule may be a reasonable starting point, it is known to fail for multi-modal data. In this case plugin methods, e.g., Sheather-Jones, or cross-validation methods give better density estimates.

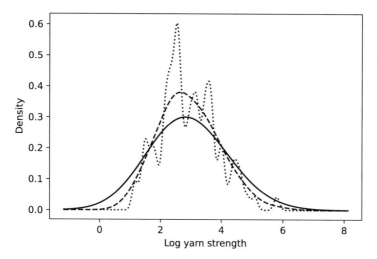

Fig. 1.15 Effect of setting the bandwidth on a density plot to 0.1 (dotted), 0.5 (dashed), and 1.0 (solid line)

Figure 1.16 shows the resulting density estimate using the Sheather-Jones plugin method.

```
from KDEpy.bw_selection import improved_sheather_jones
h = improved_sheather_jones(X.values.reshape(-1, 1))
ax = X.plot.density(color='grey')
X.plot.density(bw_method=h, color='black', ax=ax)
ax.set_xlabel('Log yarn strength')
ax.set_ylabel(f'Density (h={h:.2f})')
plt.show()
```

If data abruptly change, a single bandwidth may not be suitable to describe the distribution across the full data range. Adaptive bandwidth density estimate methods were developed to address cases like this. Implementations in Python can be found online. It is also possible to extend kernel density estimates to two or more dimensions.

1.6.2   Box and Whiskers Plots

The box and whiskers plot is a graphical presentation of the data, which provides an efficient display of various features, like location, dispersion, and skewness. A box is plotted, with its lower hinge at the first quartile $Q_1 = X_{(q_1)}$ and its upper hinge at the third quartile $Q_3 = X_{(q_3)}$. Inside the box a line is drawn at the median, M_e, and a cross is marked at the sample mean, $\bar{X}_n$, to mark the statistics of central location. The interquartile range, $Q_3 - Q_1$, which is the length of the box, is a measure of dispersion. Two whiskers are extended from the box. The lower whisker is extended

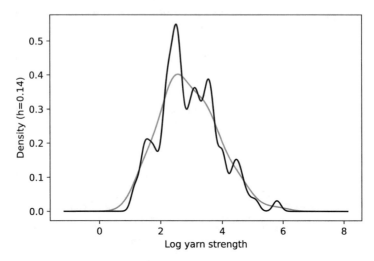

Fig. 1.16 Density estimate using Scott's rule of thumb (gray) and the improved Sheather-Jones plugin method (black) for bandwidth selection

toward the minimum $X_{(1)}$, but not lower than one and half of the interquartile range, i.e.,

$$\text{Lower whisker starts} = \max\{X_{(1)}, Q_1 - 1.5(Q_3 - Q_1)\}. \tag{1.11}$$

Similarly,

$$\text{Upper whisker ends} = \min\{X_{(n)}, Q_3 + 1.5(Q_3 - Q_1)\}. \tag{1.12}$$

Data points beyond the lower or upper whiskers are considered **outliers**. The commands below generate the box and whiskers plot shown in Fig. 1.17.

```
X = mistat.load_data('YARNSTRG')
ax = X.plot.box(color='black')
ax.set_ylabel('Log Yarn Strength')
plt.show()
```

Example 1.7 In Fig. 1.17 we present the box whiskers plot of the log yarn strength data, of Table 1.3. For this data we find the following summarizing statistics:

$$X_{(1)} = 1.1514$$

$$Q_1 = 2.2844$$

$$M_e = 2.8331, \quad \bar{X}_{100} = 2.9238$$

$$Q_3 = 3.5426$$

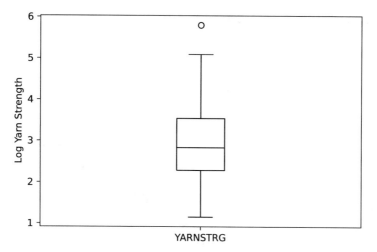

Fig. 1.17 Box whiskers plot of log yarn strength data

$$X_{(100)} = 5.7978$$

$$Q_3 - Q_1 = 1.2581, \quad S_{(100)} = 0.9378.$$

In the box whiskers plot, the end point of the lower whisker is at $\max\{X_{(1)}, 0.3973\} = X_{(1)}$. The upper whisker ends at $\min\{X_{(100)}, 5.4297\} = 5.4297$. Thus $X_{(100)}$ is an outlier. We conclude that the one measurement of yarn strength, which seems to be exceedingly large, is an outlier (could have been an error of measurement).   ∎

1.6.3   Quantile Plots

The quantile plot is a plot of the sample quantiles x_p against p, $0 < p < 1$ where $x_p = X_{(p(n+1))}$. In Fig. 1.18 we see the quantile plot of the log yarn strength. From such a plot, one can obtain graphical estimates of the quantiles of the distribution. For example, from Fig. 1.17 we immediately obtain the estimate 2.8 for the median, 2.28 for the first quartile, and 3.54 for the third quartile. These are close to the values presented earlier. We see also in Fig. 1.18 that the maximal point of this dataset is an outlier. Tracing a straight line, beginning at the median, we can also see that from $x_{.4}$ to $x_{.9}$, 50% of the data points are almost uniformly distributed, while the data between $x_{.1}$ to $x_{.4}$ tend to be larger (closer to the M_e) than those of a uniform distribution, while the largest 10% of the data values tend to be again larger (further away from the M_e) than those of a uniform distribution. This explains the slight positive skewness of the data, as seen in Fig. 1.17. For further discussion of quantile plots, see Sect. 3.6.

Fig. 1.18 Quantile plot of
log yarn strength data

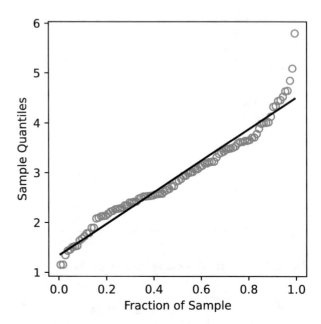

1.6.4   Stem-and-Leaf Diagrams

Table 1.6a is a stem-and-leaf display of the log yarn strength data.

Table 1.6a: Stem-and-Leaf Display
Stem-and-leaf of log yarn strength, $N = 100$, leaf unit $= 0.10$

5	1	11344
15	1	5556677788
34	2	0011112222233344444
(21)	2	555555555566677888999
45	3	000011112223344444
27	3	5556666677789
14	4	00013344
6	4	5668
2	5	0
1	5	7

In order to construct the stem-and-leaf diagram, the data is classified into class
intervals, like in the histogram. The classes are of equal length. The 100 values in
Table 1.3 start $X_{(1)} = 1.151$ and at $X_{(100)} = 5.798$. The stem-and-leaf diagram
presents only the first two digits to the left, without rounding. All values between
1.0 and 1.499 are represented in the first class as 1.1, 1.1, 1.3, 1.4, and 1.4. There are
five such values, and this frequency is written on the left-hand side. The second class

Table 1.6b Stem-and-leaf of Elec_Out $N = 99$ leaf unit $= 1.0$

5	21	01111
10	21	22333
19	21	444445555
37	21	666666667777777777
(22)	21	8888888888889999999999
40	22	0000000001111111111
21	22	22233333
13	22	44455555
5	22	6777
1	22	8

consists of all values between 1.5 and 1.999. There are ten such values, namely, 1.5, 1.5, 1.5, 1.6, 1.6, 1.7, 1.7, 1.7, 1.8, and 1.8. In a similar manner all other classes are represented. The frequency of the class to which the median, M_e, belongs is written on the left in round brackets. In this way one can immediately indicate where the median is located. The frequencies below or above the class of the median are cumulative. Since the cumulative frequency (from above) of the class right that of the median is 45, we know that the median is located right after the fifth largest value from the top of that class, namely, $M_e = 2.8$, as we have seen before. Similarly, to find Q_1, we see that $X_{(q_1)}$ is located at the third class from the top. It is the tenth value in that class, from the left. Thus, we find $Q_1 = 2.2$. Similarly we find that $X_{(q_3)} = 4.5$. This information cannot be directly obtained from the histogram. Thus, the stem-and-leaf diagram is an important additional tool for data analysis.

In Table 1.6b we present the stem-and-leaf diagram of the electric output data (**OELECT.csv**).

1.6.5 Robust Statistics for Location and Dispersion

The sample mean $\bar{X}_n$ and the sample standard deviation are both sensitive statistics to extreme deviations. Let us illustrate this point. Suppose we make three observations on the shear weld strength of steel and obtain the values 2350, 2400, and 2500. The sample mean is $\bar{X}_3 = 2416.67$. What happens if the technician by mistake punches into the computer the value 25000, instead of 2500? The sample mean would come out as 9916.67. If the result is checked on the spot, the mistake would likely be discovered and corrected. However, if there is no immediate checking, that absurd result would have remained and cause all kinds of difficulties later. Also, the standard deviations would have recorded wrongly as 13063 rather than the correct value of 76.376. This simple example shows how sensitive the mean and the standard deviation are to extreme deviations (outliers) in the data.

To avoid such complexities, a more **robust** statistic can be used, instead of the sample mean, $\bar{X}_n$. This statistic is the α-**trimmed mean**. A proportion α of the data is trimmed from the lower and the upper end of the ordered sample. The mean is then computed on the remaining $(1 - 2\alpha)$ proportion of the data. Let us denote by $\bar{T}_\alpha$ the α-trimmed mean. The formula of this statistic is

$$\bar{T}_\alpha = \frac{1}{N_\alpha} \sum_{j=[n\alpha]+1}^{[n(1-\alpha)]} X_j, \tag{1.13}$$

where $[\cdot]$ denotes the integer part of the number in brackets, e.g., $[7.3] = 7$, and $N_\alpha = [n(1 - \alpha)] - [n\alpha]$. For example, if $n = 100$ and $\alpha = 0.05$, we compute the mean of the 90 ordered values $X_{(6)}, \cdots, X_{(95)}$.

Example 1.8 Let us now examine the **robustness** of the trimmed mean.

We import the data file **OELECT.csv** using the method `mistat.load_data`. The function `trim_mean` yields different results given specific values of the trim parameter. We use this example to show how to set up a function in Python.

```
from scipy.stats import trim_mean

Oelect = mistat.load_data('OELECT')

def mySummary(x, trim=0):
    """ Returns summary information for list x

    The optional argument trim can be used to calculate a trimmed mean
    """
    x = pd.Series(x)  # convert to pandas series

    quantiles = list(x.quantile(q=[0, 0.25, 0.5, 0.75, 1.0]))
    trimmed_mean = trim_mean(x, trim)

    # return the summary information as pandas Series
    return pd.Series({
        'Min': quantiles[0],
        'Q1': quantiles[1],
        'Median': quantiles[2],
        'Mean': trimmed_mean,
        'Q3': quantiles[3],
        'Max': quantiles[4],
        'SD': x.std(),
        'IQR': quantiles[3] - quantiles[1],
    })
```

Note that we define the function mySummary with an optional argument `trim` that has a default value of 0. While it is not enforced, it is good practice to use the name of the argument in the function call.

```
print(pd.DataFrame({
    'untrimmed': mySummary(Oelect),
    'trimmed': mySummary(Oelect, trim=0.05),
}))
```

```
           untrimmed     trimmed
Min        210.896000  210.896000
```

```
Q1        216.846000   216.846000
Median    219.096000   219.096000
Mean      219.248020   219.218198
Q3        221.686000   221.686000
Max       228.986000   228.986000
SD          4.003992     4.003992
IQR         4.840000     4.840000
```

We see that $\bar{X}_{99} = 219.25$ and $\bar{T}_{.05} = 219.22$. Let us make a sorted copy of the series using the function `sort_values()` and replace the largest values from 228.986 to be $V(99) = 2289.86$ (an error in punching the data) and look at results when we apply the same commands

```
# sort and reset the index
OutVolt = Oelect.sort_values(ignore_index=True)
# in Python index starts at 0, so the 99-th value is at position 98
OutVolt[98] = 2289.86
print(pd.DataFrame({
  'untrimmed': mySummary(OutVolt),
  'trimmed': mySummary(OutVolt, trim=0.05),
}))
```

```
           untrimmed       trimmed
Min        210.896000   210.896000
Q1         216.846000   216.846000
Median     219.096000   219.096000
Mean       240.064929   219.218198
Q3         221.686000   221.686000
Max       2289.860000  2289.860000
SD         208.150486   208.150486
IQR          4.840000     4.840000
```

We see by comparing the two outputs that $\bar{X}_{99}$ changed from 219.25 to 240.1, S_{99} (STDEV) changed dramatically from 4.00 to 208.2.

On the other hand, M_e, $\bar{T}_\alpha$, Q_1, and Q_3 did not change at all. These statistics are called **robust** (non-sensitive) against extreme deviations (outliers). ∎

We have seen that the standard deviation S is very sensitive to deviations in the extremes. A robust statistic for dispersion is

$$\tilde{\sigma} = \frac{Q_3 - Q_1}{1.3490}. \tag{1.14}$$

The denominator 1.3490 is the distance between Q_3 and Q_1 in the theoretical normal distribution (see Chap. 2). Indeed, Q_3 and Q_1 are robust against outliers. Hence, $\tilde{\sigma}$, which is about 3/4 of the IQR, is often a good statistic to replace S.

Another statistic is the **α-trimmed standard deviation**

$$S_\alpha = \left(\frac{1}{N_\alpha - 1} \sum_{j=[n\alpha]+1}^{[n(1-\alpha)]} (X_j - \bar{T}_\alpha)^2 \right)^{1/2}. \tag{1.15}$$

For the OELECT data, S_α equals 3.5897. The command below calculates a robust statistic for dispersion $\tilde{\sigma}$ from the OELECT data.

```
from scipy.stats import iqr

def trim_std(data, alpha):
    data = np.array(data)
    data.sort()
    n = len(data)
    low = int(n * alpha) + 1
    high = int(n * (1 - alpha))
    return data[low:(high + 1)].std()

print('S_alpha', trim_std(OutVolt, 0.025))
print('sigma:', iqr(OutVolt) / 1.349)
```

```
S_alpha 3.5896765570264395
sigma: 3.587842846552984
```

We see these two robust statistics, $\tilde{\sigma}$ and S_α, yield close results. The sample standard deviation of OELECT is $S = 4.00399$.

1.7 Chapter Highlights

The main concepts and definitions introduced in this chapter include:

- Random variable
- Fixed and random components
- Accuracy and precision
- The population and the sample
- Random sampling with replacement (RSWR)
- Random sampling without replacement (RSWOR)
- Frequency distributions
- Discrete and continuous random variables
- Quantiles
- Sample mean and sample variance
- Skewness
- Kurtosis
- Prediction intervals
- Box and whiskers plots
- Quantile plots
- Stem-and-leaf diagrams
- Robust statistics

1.8 Exercises

Exercise 1.1 In the present problem, we are required to generate at random 50 integers from the set $\{1, 2, 3, 4, 5, 6\}$. To do this we can use the `random.choices` method from the `random` package.

Use this method of simulation and count the number of times the different integers have been repeated. This counting can be done by using the Counter class from the `collections` package.

How many times you expect each integer to appear if the process generates the numbers at random?

Exercise 1.2 Construct a sequence of 50 numbers having a linear trend for deterministic components with random deviations around it. This can be done by using these Python commands. We use Python's list comprehension to modify the elements of the list

```
random.seed(1)
x = list(range(50))
y = [5 + 2.5 * xi for xi in x]
y = [yi + random.uniform(-10, 10) for yi in y]
```

By plotting y versus x, one sees the random variability around the linear trend.

Exercise 1.3 Generate a sequence of 50 random binary numbers $(0, 1)$, when the likelihood of 1 is p using the command `binom.rvs(1, p, size=50)`.

Do this for the values $p = 0.1, 0.3, 0.7, 0.9$. Count the number of 1s in these random sequences, by summing up the result sequence.

Exercise 1.4 The following are two sets of measurements of the weight of an object, which correspond to two different weighing instruments. The object has a true weight of 10 kg.
Instrument 1:

9.490950	10.436813	9.681357	10.996083	10.226101	10.253741
10.458926	9.247097	8.287045	10.145414	11.373981	10.144389
11.265351	7.956107	10.166610	10.800805	9.372905	10.199018
9.742579	10.428091				

Instrument 2:

11.771486	10.697693	10.687212	11.097567	11.676099	10.583907
10.505690	9.958557	10.938350	11.718334	11.308556	10.957640
11.250546	10.195894	11.804038	11.825099	10.677206	10.249831
10.729174	11.027622				

Which instrument seems to be more accurate? Which instrument seems to be more precise?

Exercise 1.5 The quality control department of a candy factory uses a scale to verify compliance of the weight of packages. What could be the consequences of problems with the scale accuracy, precision, and stability?

Exercise 1.6 Draw a random sample with replacement (RSWR) of size $n = 20$ from the set of integers $\{1, 2, \cdots , 100\}$.

Exercise 1.7 Draw a random sample without replacement (RSWOR) of size $n = 10$ from the set of integers $\{11, 12, \cdots , 30\}$.

Exercise 1.8

  (i) How many words of five letters can be composed ($N = 26, n = 5$)?
 (ii) How many words of five letters can be composed, if all letters are different?
(iii) How many words of five letters can be written if the first and the last letters are x?
(iv) An electronic signal is a binary sequence of ten 0s or 1s. How many different signals are available?
 (v) How many electronic signals in a binary sequence of size 10 are there in which the number 1 appears exactly five times?

Exercise 1.9 For each of the following variable, state whether it is discrete or continuous:

  (i) The number of "heads" among the results of ten flippings of a coin;
 (ii) The number of blemishes on a ceramic plate;
(iii) The thickness of ceramic plates;
(iv) The weight of an object.

Exercise 1.10 Data file **FILMSP.csv** contains data gathered from 217 rolls of film. The data consists of the film speed as measured in a special lab. Prepare a histogram of the data.

Exercise 1.11 Data file **COAL.csv** contains data on the number of yearly disasters in coal mines in England. Prepare a table of frequency distributions of the number of coal mine disasters.

Exercise 1.12 Data file **CAR.csv** contains information on 109 different car models. For each car there are values of five variables:

1. Number of cylinders (4, 6, 8)
2. Origin (1, 2, 3)
3. Turn diameter [m]
4. Horsepower [HP]
5. Number of miles/gallon in city driving [mpg].

Prepare frequency distributions of variables 1, 2, 3, 4, 5.

Exercise 1.13 Compute the following five quantities for the data in file **FILMSP.csv**:

  (i) Sample minimum, $X_{(1)}$;
  (ii) Sample first quartile, Q_1;
  (iii) Sample median, M_e;
  (iv) Sample third quartile, Q_3;
  (v) Sample maximum, $X_{(217)}$.
  (vi) The .8-quantile.
 (vii) The .9-quantile.
(viii) The .99-quantile.

Show how you get these statistics by using the formulae. The order statistics of the sample can be obtained by first ordering the values of the sample.

Exercise 1.14 Compute with Python the indices of skewness and kurtosis of the **FILMSP.csv**, using the given formulas.

Interpret the skewness and kurtosis of this sample in terms of the shape of the distribution of film speed.

Exercise 1.15 Compare the means and standard deviations of the number of miles per gallon/city of cars by origin (1 = USA; 2 = Europe; 3 = Asia) according to the data of file **CAR.csv**.

Exercise 1.16 Compute the coefficient of variation of the turn diameter of US-made cars (Origin = 1) in file **CAR.csv**.

Exercise 1.17 Compare the mean $\bar{X}$ and the geometric mean G of the turn diameter of US- and Japanese-made cars in **CAR.csv**.

Exercise 1.18 Compare the prediction proportions to the actual frequencies of the intervals

$$\bar{X} \pm kS, \quad k = 1, 2, 3$$

for the film speed data, given in **FILMSP.csv** file.

Exercise 1.19 Present side by side the box plots of miles per gallon/city for cars by origin. Use data file **CAR.csv**.

Exercise 1.20 Prepare a stem-leaf diagram of the piston cycle time in file **OTURB.csv**. Compute the five summary statistics $(X_{(1)}, Q_1, M_e, Q_3, X_{(n)})$ from the stem-leaf.

Exercise 1.21 Compute the trimmed mean $\bar{T}_{.10}$ and trimmed standard deviation, $S_{.10}$, of the piston cycle time of file **OTURB.csv**.

Exercise 1.22 The following data is the time (in sec.) to get from 0 to 60 mph for a sample of 15 German- and 20 Japanese-made cars

German-made cars			Japanese-made cars			
10.0	10.9	4.8	9.4	9.5	7.1	8.0
6.4	7.9	8.9	8.9	7.7	10.5	6.5
8.5	6.9	7.1	6.7	9.3	5.7	12.5
5.5	6.4	8.7	7.2	9.1	8.3	8.2
5.1	6.0	7.5	8.5	6.8	9.5	9.7

Compare and contrast the acceleration times of German- and Japanese-made cars, in terms of their five summary statistics.

Exercise 1.23 Summarize variables Res 3 and Res 7 in dataset HADPAS.csv by computing sample statistics, histograms, and stem-and-leaf diagrams.

Exercise 1.24 Are there outliers in the Res 3 data of **HADPAS.csv**? Show your calculations.

Chapter 2
Probability Models and Distribution Functions

Preview The chapter provides the basics of probability theory and theory of distribution functions. The probability model for random sampling is discussed. This is fundamental for statistical inference discussed in Chap. 3 and sampling procedures in Chap. 5. Bayes' theorem also presented here has important ramifications in statistical inference, including Bayesian decision making presented in Chap. 3. The Industrial Statistics book treats, in Chap. 3, Bayesian tracking and detection methods, and has a chapter on Bayesian reliability analysis (Chap. 10).

2.1 Basic Probability

2.1.1 Events and Sample Spaces: Formal Presentation of Random Measurements

Experiments or trials of interest are those which may yield different results with outcomes that are not known ahead of time with certainty. We have seen in the previous chapter a number of examples in which outcomes of measurements vary. It is of interest to find, before conducting a particular experiment, what are the chances of obtaining results in a certain range. In order to provide a quantitative answer to such a question, we have to formalize the framework of the discussion so that no ambiguity is left.

When we say a "trial" or "experiment," in the general sense, we mean a well-defined process of measuring certain characteristic(s), or variable(s). For example, if the experiment is to measure the compressive strength of concrete cubes, we must specify exactly how the concrete mixture was prepared, i.e., proportions of cement, sand, aggregates, and water in the batch, length of mixing time, dimensions of

Supplementary Information The online version contains supplementary material available at https://doi.org/10.1007/978-3-031-07566-7_2.

mold, number of days during which the concrete has hardened, the temperature and humidity during preparation and storage of the concrete cubes, etc. All these factors influence the resulting compressive strength. Well-documented protocol of an experiment enables us to replicate it as many times as needed. In a well-controlled experiment, we can assume that the variability in the measured variables is due to **randomness**. We can think of the random experimental results as sample values from a hypothetical population. The set of all possible sample values is called the **sample space**. In other words, the sample space is the set of all possible outcomes of a specified experiment. The outcomes do not have to be numerical. They could be names, categorical values, functions, or collection of items. The individual outcome of an experiment will be called an **elementary event** or a **sample point** (element). We provide a few examples.

Example 2.1 The experiment consists of choosing ten names (without replacement) from a list of 400 undergraduate students at a given university. The outcome of such an experiment is a list of ten names. The sample space is the collection of **all** possible such sublists that can be drawn from the original list of 400 students. ∎

Example 2.2 The experiment is to produce 20 concrete cubes, under identical manufacturing conditions, and count the number of cubes with compressive strength above 200 [kg/cm^2]. The sample space is the set $S = \{0, 1, 2, \cdots, 20\}$. The elementary events, or sample points, are the elements of S. ∎

Example 2.3 The experiment is to choose a steel bar from a specific production process and measure its weight. The sample space S is the interval (ω_0, ω_1) of possible weights. The weight of a particular bar is a sample point. ∎

Thus, sample spaces could be finite sets of sample points, or countable or noncountable infinite sets.

Any subset of the sample space, S, is called an **event**. S itself is called the **sure event**. The empty set, $\emptyset$, is called the **null event**. We will denote events by the letters $A, B, C, \cdots$ or $E_1, E_2, \cdots$. All events under consideration are subsets of the same sample space S. Thus, events are sets of sample points.

For any event $A \subseteq S$, we denote by A^c the **complementary event**, i.e., the set of all points of S which are not in A.

An event A is said to **imply** an event B, if all elements of A are elements of B. We denote this **inclusion relationship** by $A \subset B$. If $A \subset B$ **and** $B \subset A$, then the two events are **equivalent**, $A \equiv B$.

Example 2.4 The experiment is to select a sequence of five letters for transmission of a code in a money transfer operation. Let $A_1, A_2, \ldots, A_5$ denote the first, second,..., fifth letter chosen. The sample space is the set of all possible sequences of five letters. Formally,

$$S = \{(A_1 A_2 A_3 A_4 A_5) : A_i \in \{a, b, c, \cdots, z\}, \ i = 1, \cdots, 5\}$$

This is a finite sample space containing 26^5 possible sequences of five letters. Any such sequence is a sample point.

Let E be the event that all the five letters in the sequence are the same. Thus

$$E = \{aaaaa, bbbbb, \cdots, zzzzz\}.$$

This event contains 26 sample points. The complement of E, E^c is the event that at least one letter in the sequence is different from the other ones. ∎

2.1.2 Basic Rules of Operations with Events: Unions and Intersections

Given events $A, B, \cdots$ of a sample space S, we can generate new events, by the operations of union, intersection, and complementation.

The **union** of two events A and B, denoted $A \cup B$, is an event having elements which belong **either** to A **or** to B.

The intersection of two events, $A \cap B$, is an event whose elements belong both to A **and** to B. By pairwise union or intersection, we immediately extend the definition to finite number of events $A_1, A_2, \cdots, A_n$, i.e.,

$$\bigcup_{i=1}^{n} A_i = A_1 \cup A_2 \cup \cdots \cup A_n$$

and

$$\bigcap_{i=1}^{n} A_i = A_1 \cap A_2 \cap \cdots \cap A_n.$$

The finite union $\bigcup_{i=1}^{n} A_i$ is an event whose elements belong to **at least one** of the n events. The finite intersection $\bigcap_{i=1}^{n} A_i$ is an event whose elements belong to **all** the n events.

Any two events, A and B, are said to be mutually **exclusive** or **disjoint** if $A \cap B = \emptyset$, i.e., they do not contain common elements. Obviously, by definition, any event is disjoint of its complement, i.e., $A \cap A^c = \emptyset$. The operations of union and intersection are:

1. **Commutative**:

$$A \cup B = B \cup A,$$
$$A \cap B = B \cap A;$$

2. **Associative**:

$$(A \cup B) \cup C = A \cup (B \cup C)$$
$$= A \cup B \cup C$$
$$(A \cap B) \cap C = A \cap (B \cap C)$$
$$= A \cap B \cap C$$

(2.1)

3. **Distributive**:

$$A \cap (B \cup C) = (A \cap B) \cup (A \cap C)$$
$$A \cup (B \cap C) = (A \cup B) \cap (A \cup C).$$

(2.2)

The intersection of events is sometimes denoted as a product, i.e.,

$$A_1 \cap A_2 \cap \cdots \cap A_n \equiv A_1 A_2 A_3 \cdots A_n.$$

The following law, called **De Morgan's law**, is fundamental to the algebra of events and yields the complement of the union, or intersection, of two events, namely:

$$(A \cup B)^c = A^c \cap B^c$$
$$(A \cap B)^c = A^c \cup B^c$$

(2.3)

Finally, we define the notion of **partition**. A collection of n events $E_1, \cdots, E_n$ is called a **partition** of the sample space S, if

(i) $\quad \bigcup_{i=1}^{n} E_i = S$

(ii) $\quad E_i \cap E_j = \emptyset$ for all $i \neq j$ $(i, j = 1, \cdots, n)$

That is, the events in any partition are mutually **disjoint**, and their union exhaust all the sample space.

Example 2.5 The experiment is to generate on the computer a random number, U, in the interval $(0, 1)$. A random number in $(0, 1)$ can be obtained as

$$U = \sum_{j=1}^{\infty} I_j 2^{-j},$$

where I_j is the random result of tossing a coin, i.e.,

$$I_j = \begin{cases} 1, & \text{if Head} \\ 0, & \text{if Tail.} \end{cases}$$

For generating random numbers from a set of integers, the summation index j is bounded by a finite number N. This method is, however, not practical for generating random numbers on a continuous interval. Computer programs generate "pseudo-random" numbers. Methods for generating random numbers are described in various books on simulation (see Bratley et al. 1983). The most cosmmonly applied is the **linear congruential generator**. This method is based on the recursive equation

$$U_i = (aU_{i-1} + c) \mod m, \quad i = 1, 2, \cdots.$$

The parameters a, c, and m depend on the computer's architecture. In many programs, $a = 65539$, $c = 0$, and $m = 2^{31} - 1$. The first integer X_0 is called the "seed." Different choices of the parameters a, c, and m yield "pseudo-random" sequences with different statistical properties.

The sample space of this experiment is

$$S = \{u : 0 \le u \le 1\}$$

Let E_1 and E_2 be the events

$$E_1 = \{u : 0 \le u \le 0.5\},$$
$$E_2 = \{u : 0.35 \le u \le 1\}.$$

The union of these two events is

$$E_3 = E_1 \cup E_2 = \{u : 0 \le u \le 1\} = S.$$

The intersection of these events is

$$E_4 = E_1 \cap E_2 = \{u : 0.35 \le u < 0.5\}.$$

Thus, E_1 and E_2 are **not** disjoint.

The complementary events are

$$E_1^c = \{u : 0.5 \le u < 1\} \text{ and } E_2^c = \{u : u < 0.35\}$$

$E_1^c \cap E_2^c = \emptyset$, i.e., the complementary events are disjoint. By De Morgan's law

$$(E_1 \cap E_2)^c = E_1^c \cup E_2^c$$
$$= \{u : u < 0.35 \text{ or } u \ge 0.5\}.$$

However,

$$\emptyset = S^c = (E_1 \cup E_2)^c = E_1^c \cap E_2^c.$$

Finally, the following is a partition of S:

$$B_1 = \{u : u < 0.1\}, \qquad\qquad B_2 = \{u : 0.1 \le u < 0.2\},$$
$$B_3 = \{u : 0.2 \le u < 0.5\}, \qquad\qquad B_4 = \{u : 0.5 \le u < 1\}.$$

Notice that $B_4 = E_1^c$ ■

Different identities can be derived by the above rules of operations on events; a few will be given as exercises.

2.1.3 Probabilities of Events

A probability function $\Pr\{\cdot\}$ assigns to events of S real numbers, following the following basic axioms:

1. $\Pr\{E\} \ge 0$
2. $\Pr\{S\} = 1$.
3. **If $E_1, \cdots, E_n$ $(n \ge 1)$ are mutually disjoint events, then**

$$\Pr\left\{ \bigcup_{i=1}^n E_i \right\} = \sum_{i=1}^n \Pr\{E_i\}.$$

From these three basic axioms, we deduce the following results.

Result 1.   If $A \subset B$ then

$$\Pr\{A\} \le \Pr\{B\}.$$

Indeed, since $A \subset B$, $B = A \cup (A^c \cap B)$. Moreover, $A \cap A^c \cap B = \emptyset$. Hence, by axioms 1 and 3, $\Pr\{B\} = \Pr\{A\} + \Pr\{A^c \cap B\} \ge \Pr\{A\}$.

Thus, if E is any event, since $E \subset S$, $0 \le \Pr\{E\} \le 1$.

Result 2.   For any event E, $\Pr\{E^c\} = 1 - \Pr\{E\}$.

Indeed $S = E \cup E^c$. Since $E \cap E^c = \emptyset$,

$$1 = \Pr\{S\} = \Pr\{E\} + \Pr\{E^c\}. \tag{2.4}$$

This implies the result.

Result 3.   **For any events** A, B

$$\Pr\{A \cup B\} = \Pr\{A\} + \Pr\{B\} - \Pr\{A \cap B\}. \tag{2.5}$$

Indeed, we can write

$$A \cup B = A \cup A^c \cap B,$$

where $A \cap (A^c \cap B) = \emptyset$. Thus, by the third axiom,

$$\Pr\{A \cup B\} = \Pr\{A\} + \Pr\{A^c \cap B\}.$$

Moreover, $B = A^c \cap B \cup A \cap B$, where again $A^c \cap B$ and $A \cap B$ are disjoint. Thus, $\Pr\{B\} = \Pr\{A^c \cap B\} + \Pr\{A \cap B\}$, or $\Pr\{A^c \cap B\} = \Pr\{B\} - \Pr\{A \cap B\}$. Substituting this above we obtain the result.

Result 4.   **If** $B_1, \cdots, B_n$ $(n \geq 1)$ **is a partition of** S, **then for any event** E,

$$\Pr\{E\} = \sum_{i=1}^{n} \Pr\{E \cap B_i\}.$$

Indeed, by the distributive law,

$$E = E \cap S = E \cap \left(\bigcup_{i=1}^{n} B_i \right)$$

$$= \bigcup_{i=1}^{n} E B_i.$$

Finally, since $B_1, \cdots, B_n$ are mutually disjoint, $(E B_i) \cap (E B_j) = E \cap B_i \cap B_j = \emptyset$ for all $i \neq j$. Therefore, by the third axiom

$$\Pr\{E\} = \Pr \left\{ \bigcup_{i=1}^{n} E B_i \right\} = \sum_{i=1}^{n} \Pr\{E B_i\} \tag{2.6}$$

Example 2.6 Fuses are used to protect electronic devices from unexpected power surges. Modern fuses are produced on glass plates through processes of metal deposition and photographic lythography. On each plate several hundred fuses are simultaneously produced. At the end of the process, the plates undergo precise cutting with special saws. A certain fuse is handled on one of three alternative cutting machines. Machine M_1 yields 200 fuses per hour, machine M_2 yields 250 fuses per hour, and machine M_3 yields 350 fuses per hour. The fuses are then mixed together. The proportions of defective parts that are typically produced on these machines are 0.01, 0.02, and 0.005, respectively. A fuse is chosen at random from the production of a given hour. What is the probability that it is compliant with the amperage requirements (nondefective)?

Let E_i be the event that the chosen fuse is from machine M_i ($i = 1, 2, 3$). Since the choice of the fuse is random, each fuse has the same probability $\frac{1}{800}$ to be chosen. Hence, $\Pr\{E_1\} = \frac{1}{4}$, $\Pr\{E_2\} = \frac{5}{16}$ and $\Pr\{E_3\} = \frac{7}{16}$.

Let G denote the event that the selected fuse is non-defective. For example, for machine M_1, $\Pr\{G\} = 1 - 0.01 = 0.99$. We can assign $\Pr\{G \cap M_1\} = 0.99 \times 0.25 = 0.2475$, $\Pr\{G \cap M_2\} = 0.98 \times \frac{5}{16} = 0.3062$ and $\Pr\{G \cap M_3\} = 0.995 \times \frac{7}{16} = 0.4353$. Hence, the probability of selecting a non-defective fuse is, according to Result 4,

$$\Pr\{G\} = \Pr\{G \cap M_1\} + \Pr\{G \cap M_2\} + \Pr\{G \cap M_3\} = 0.989.$$

∎

Example 2.7 Consider the problem of generating random numbers, discussed in Example 2.5. Suppose that the probability function assigns any interval $I(a, b) = \{u : a < u < b\}, 0 \leq a < b \leq 1$, the probability

$$\Pr\{I(a, b)\} = b - a.$$

Let $E_3 = I(0.1, 0.4)$ and $E_4 = I(0.2, 0.5)$. $C = E_3 \cup E_4 = I(0.1, 0.5)$. Hence,

$$\Pr\{C\} = 0.5 - 0.1 = 0.4.$$

On the other hand, $\Pr\{E_3 \cap E_4\} = 0.4 - 0.2 = 0.2$.

$$\Pr\{E_3 \cup E_4\} = \Pr\{E_3\} + \Pr\{E_4\} - \Pr\{E_3 \cap E_4\}$$
$$= (0.4 - 0.1) + (0.5 - 0.2) - 0.2 = 0.4.$$

This illustrates Result 3.   ∎

2.1.4 Probability Functions for Random Sampling

Consider a finite population P, and suppose that the random experiment is to select a random sample from P, with or without replacement. More specifically let $L_N = \{w_1, w_2, \cdots, w_N\}$ be a **list** of the elements of P, where N is its size. w_j ($j = 1, \cdots, N$) is an identification number of the j-th element.

Suppose that a sample of size n is drawn from L_N [respectively, P] **with** replacement. Let W_1 denote the first element selected from L_N. If j_1 is the index of this element, then $W_1 = w_{j_1}$. Similarly, let W_i ($i = 1, \ldots, n$) denote the i-th element of the sample. The corresponding sample space is the collection

$$S = \{(W_1, \cdots, W_n) : W_i \in L_N, \quad i = 1, 2, \cdots, n\}$$

of all samples, with replacement from L_N. The total number of possible samples is N^n. Indeed, w_{j_1} could be any one of the elements of L_N and so are $w_{j_2}, \cdots, w_{j_n}$. With each one of the N possible choices of w_{j_1}, we should combine the N possible choices of w_{j_2} and so on. Thus, there are N^n possible ways of selecting a sample of size n, with replacement. The sample points are the elements of S (possible samples). The sample is called **random with replacement** (RSWR) if each one of these N^n possible samples is assigned the same probability, $1/N^n$, for being selected.

Let $M(i)$ $(i = 1, \cdots, N)$ be the number of samples in S, which contain the i-th element of L_N (at least once). Since sampling is **with** replacement

$$M(i) = N^n - (N - 1)^n.$$

Indeed, $(N - 1)^n$ is the number of samples with replacement, which do not include w_i. Since all samples are equally probable, the probability that a RSWR $\mathbf{S}_n$ includes w_i $(i = 1, \cdots, N)$ is

$$\Pr\{w_i \in \mathbf{S}_n\} = \frac{N^n - (N - 1)^n}{N^n}$$

$$= 1 - \left(1 - \frac{1}{N}\right)^n.$$

If $n > 1$, then the above probability is larger than $1/N$ which is the probability of selecting the element W_i in any given trial but smaller than n/N. Notice also that this probability does not depend on i, i.e., all elements of L_N have the same probability to be included in a RSWR. It can be shown that the probability that w_i is included in the sample exactly once is $\frac{n}{N}\left(1 - \frac{1}{N}\right)^{n-1}$.

If sampling is **without** replacement, the number of sample points in S is $N(N - 1) \cdots (N - n + 1)/n!$, since the order of selection is immaterial. The number of sample points which include w_i is $M(i) = (N-1)(N-2) \cdots (N-n+1)/(n-1)!$. A sample $\mathbf{S}_n$ is called **random without replacement** (RSWOR) if all possible samples are equally probable. Thus, under RSWOR,

$$\Pr\{w_i \in \mathbf{S}_n\} = \frac{n!M(i)}{N(N - 1) \cdots (N - n + 1)} = \frac{n}{N},$$

for all $i = 1, \cdots, N$.

We consider now events, which depend on the attributes of the elements of a population. Suppose that we sample to obtain information on the number of defective (nonstandard) elements in a population. The attribute in this case is "the element complies to the requirements of the standard." Suppose that M out of N

elements in L_N is non-defective (have the attribute). Let E_j be the event that j out of the n elements in the sample is non-defective. Notice that $E_0, \cdots, E_n$ is a partition of the sample space. What is the probability, under RSWR, of E_j? Let K_j^n denote the number of sample points in which j out of n are G elements (non-defective) and $(n - j)$ elements are D (defective). To determine K_j^n we can proceed as follows.

Choose first j G's and $(n - j)$ D's from the population. This can be done in $M^j (N - M)^{n-j}$ different ways. We have now to assign the j G's into j out of n components of the vector $(w_1, \cdots, w_n)$. This can be done in $n(n - 1) \cdots (n - j + 1)/j!$ possible ways. This is known as the number of combinations of j out of n, i.e.,

$$\binom{n}{j} = \frac{n!}{j!(n - j)!}, \quad j = 0, 1, \cdots, n \tag{2.7}$$

where $k! = 1 \cdot 2 \cdot \cdots \cdot k$ is the product of the first k positive integers, $0! = 1$. Hence, $K_j^n = \binom{n}{j} M^j (N - M)^{n-j}$. Since every sample is equally probable, under RSWR,

$$\Pr\{E_{j:n}\} = K_j^n / N^n = \binom{n}{j} P^j (1 - P)^{n-j}, \quad j = 0, \cdots, n \tag{2.8}$$

where $P = M/N$. If sampling is without replacement, then

$$K_j^n = \binom{M}{j}\binom{N - M}{n - j}$$

and

$$\Pr\{E_j\} = \frac{\binom{M}{j}\binom{N-M}{n-j}}{\binom{N}{n}}. \tag{2.9}$$

These results are valid since the order of selection is immaterial for the event E_j.

These probabilities of E_j under RSWR and RSWOR are called, respectively, the **binomial** and **hypergeometric** probabilities.

Example 2.8 The experiment consists of randomly transmitting a sequence of binary signals, 0 or 1. What is the probability that three out of six signals are 1s? Let E_3 denote this event.

The sample space of six signals consists of 2^6 points. Each point is equally probable. The probability of E_3 is

$$\Pr\{E_3\} = \binom{6}{3}\frac{1}{2^6} = \frac{6 \cdot 5 \cdot 4}{1 \cdot 2 \cdot 3 \cdot 64}$$

$$= \frac{20}{64} = \frac{5}{16} = 0.3125.$$

∎

Example 2.9 Two out of ten television sets are defective. A RSWOR of $n = 2$ sets is chosen. What is the probability that the two sets in the sample are good (non-defective)? This is the hypergeometric probability of E_0 when $M = 2$, $N = 10$, $n = 2$, i.e.,

$$\Pr\{E_0\} = \frac{\binom{8}{2}}{\binom{10}{2}} = \frac{8 \cdot 7}{10 \cdot 9} = 0.622.$$

∎

2.1.5 Conditional Probabilities and Independence of Events

In this section, we discuss the notion of conditional probabilities. When different events are related, the realization of one event may provide us relevant information to improve our probability assessment of the other event(s). In Sect. 2.1.3 we gave an example with three machines which manufacture the same part but with different production rates and different proportions of defective parts in the output of those machines. The random experiment was to choose at random a part from the mixed yield of the three machines.

We saw earlier that the probability that the chosen part is non-defective is 0.989. If we can identify, before the quality test, from which machine the part came, the probabilities of non-defective would be conditional on this information.

The probability of choosing at random a non-defective part from machine M_1 is 0.99. If we are given the information that the machine is M_2, the probability is 0.98 and given machine M_3 the probability is 0.995. These probabilities are called **conditional probabilities**. The information given changes our probabilities.

We define now formally the concept of conditional probability.

Let A and B be two events such that $\Pr\{B\} > 0$. The conditional probability of A, given B, is

$$\Pr\{A \mid B\} = \frac{\Pr\{A \cap B\}}{\Pr\{B\}}. \tag{2.10}$$

Example 2.10 The random experiment is to measure the length of a steel bar.

The sample space is $S = (19.5, 20.5)$ [cm]. The probability function assigns any subinterval a probability equal to its length. Let $A = (19.5, 20.1)$ and $B = (19.8, 20.5)$. $\Pr\{B\} = 0.7$. Suppose that we are told that the length belongs to the interval B, and we have to guess whether it belongs to A. We compute the conditional probability

$$\Pr\{A \mid B\} = \frac{\Pr\{A \cap B\}}{\Pr\{B\}} = \frac{0.3}{0.7} = 0.4286.$$

On the other hand, if the information that the length belongs to B is not given, then $\Pr\{A\} = 0.6$. Thus, there is a difference between the conditional and nonconditional probabilities. This indicates that the two events A and B are dependent. ∎

Definition Two events A, B are called **independent** if

$$\Pr\{A \mid B\} = \Pr\{A\}.$$

If A and B are independent events, then

$$\Pr\{A\} = \Pr\{A \mid B\} = \frac{\Pr\{A \cap B\}}{\Pr\{B\}}$$

or, equivalently,

$$\Pr\{A \cap B\} = \Pr\{A\} \Pr\{B\}.$$

If there are more than two events, $A_1, A_2, \cdots, A_n$, we say that the events are **pairwise independent** if

$$\Pr\{A_i \cap A_j\} = \Pr\{A_i\} \Pr\{A_j\} \text{ for all } i \neq j, \ i, j = 1, \cdots, n.$$

The n events are said to be **mutually independent** if, for any subset of k events, $k = 2, \ldots, n$, indexed by $A_{i_1}, \ldots, A_{i_k}$,

$$\Pr\{A_{i_1} \cap A_{i_2} \cdots \cap A_{i_k}\} = \Pr\{A_{i_1}\} \cdots \Pr\{A_{i_n}\}.$$

In particular, if n events are mutually independent, then

$$\Pr\left\{ \bigcap_{i=1}^{n} A_i \right\} = \prod_{i=1}^{n} \Pr\{A_i\}. \tag{2.11}$$

One can show examples of events which are pairwise independent but **not** mutually independent.

We can further show (see exercises) that if two events are independent then the corresponding complementary events are independent. Furthermore, if n events are mutually independent, then any pair of events is pairwise independent, every three events are triplewise independent, etc.

Example 2.11 Five identical parts are manufactured in a given production process. Let $E_1, \cdots, E_5$ be the events that these five parts comply with the quality specifications (non-defective). Under the model of mutual independence, the probability that **all** the five parts are indeed non-defective is

$$\Pr\{E_1 \cap E_2 \cap \cdots \cap E_5\} = \Pr\{E_1\}\Pr\{E_2\}\cdots\Pr\{E_5\}.$$

Since these parts come from the same production process, we can assume that $\Pr\{E_i\} = p$, all $i = 1, \cdots, 5$. Thus, the probability that **all** the five parts are non-defective is p^5.

What is the probability that one part is defective and all the other four are non-defective? Let A_1 be the event that one out of five parts is defective. In order to simplify the notation, we write the intersection of events as their product. Thus,

$$A_1 = E_1^c E_2 E_3 E_4 E_5 \ \cup \ E_1 E_2^c E_3 E_4 E_5 \ \cup \ E_1 E_2 E_3^c E_4 E_5 \ \cup$$

$$E_1 E_2 E_3 E_4^c E_5 \ \cup \ E_1 E_2 E_3 E_4 E_5^c.$$

A_1 is the union of five **disjoint** events. Therefore,

$$\Pr\{A_1\} = \Pr\{E_1^c E_2 \cdots E_5\} + \cdots + \Pr\{E_1 E_2 \cdots E_5^c\}$$

$$= 5p^4(1 - p).$$

Indeed, since $E_1, \cdots, E_5$ are **mutually** independent events ,

$$\Pr\{E_1^c E_2 \cdots E_5\} = \Pr\{E_1^c\}\Pr\{E_2\}\cdots\Pr\{E_5\} = (1 - p)p^4.$$

Also,

$$\Pr\{E_1 E_2^c E_3 E_4 E_5\} = (1 - p)p^4,$$

etc. Generally, if J_5 denotes the number of defective parts among the five ones,

$$\Pr\{J_5 = i\} = \binom{5}{i} p^{(5-i)}(1 - p)^i, \quad i = 0, 1, 2, \cdots, 5.$$

■

2.1.6 Bayes' Theorem and Its Application

Bayes' theorem, which is derived in the present section, provides us with a fundamental formula for weighing the evidence in the data concerning unknown parameters, or some unobservable events.

Suppose that the results of a random experiment depend on some event(s) which is (are) not directly observable. The observable event is related to the unobservable one(s) via the conditional probabilities. More specifically, suppose that $\{B_1, \cdots, B_m\}$ ($m \geq 2$) is a partition of the sample space. The events

$B_1, \cdots, B_m$ are not directly observable, or verifiable. The random experiment results in an event A (or its complement). The conditional probabilities $\Pr\{A \mid B_i\}$, $i = 1, \cdots, m$ are known. The question is whether, after observing the event A, we can assign probabilities to the events $B_1, \cdots, B_m$. In order to weigh the evidence that A has on $B_1, \cdots, B_m$, we first assume some probabilities $\Pr\{B_i\}, i = 1, \cdots, m$, which are called **prior probabilities**. The prior probabilities express our degree of belief in the occurrence of the events B_i $(i = 1, \cdots, m)$. After observing the event A, we convert the prior probabilities of B_i $(i = 1, \cdots, m)$ to **posterior probabilities** $\Pr\{B_i \mid A\}, i = 1, \cdots, m$ by using **Bayes' theorem**

$$\Pr\{B_i \mid A\} = \frac{\Pr\{B_i\}\Pr\{A \mid B_i\}}{\sum_{j=1}^{m} \Pr\{B_j\}\Pr\{A \mid B_j\}}, \quad i = 1, \cdots, m. \tag{2.12}$$

These posterior probabilities reflect the weight of evidence that the event A has concerning $B_1, \cdots, B_m$.

Bayes' theorem can be obtained from the basic rules of probability. Indeed, assuming that $\Pr\{A\} > 0$,

$$\begin{aligned} \Pr\{B_i \mid A\} &= \frac{\Pr\{A \cap B_i\}}{\Pr\{A\}} \\ &= \frac{\Pr\{B_i\}\Pr\{A \mid B_i\}}{\Pr\{A\}}. \end{aligned}$$

Furthermore, since $\{B_1, \cdots, B_m\}$ is a partition of the sample space,

$$\Pr\{A\} = \sum_{j=1}^{m} \Pr\{B_j\}\Pr\{A \mid B_j\}.$$

Substituting this expression above, we obtain Bayes' theorem.

The following example illustrates the applicability of Bayes' theorem to a problem of decision-making.

Example 2.12 Two vendors B_1, B_2 produce ceramic plates for a given production process of hybrid micro-circuits. The parts of vendor B_1 have probability $p_1 = 0.10$ of being defective. The parts of vendor B_2 have probability $p_2 = 0.05$ of being defective. A delivery of $n = 20$ parts arrive, but the label which identifies the vendor is missing. We wish to apply Bayes' theorem to assign a probability that the package came from vendor B_1.

Suppose that it is a priori, equally likely that the package was mailed by vendor B_1 or vendor B_2. Thus, the prior probabilities are $\Pr\{B_1\} = \Pr\{B_2\} = 0.5$. We inspect the 20 parts in the package and find $J_{20} = 3$ defective items. A is the event $\{J_{20} = 3\}$. The conditional probabilities of A, given B_i $(i = 1, 2)$, are

$$\Pr\{A \mid B_1\} = \binom{20}{3} p_1^3 (1 - p_1)^{17}$$

$$= 0.1901.$$

Similarly

$$\Pr\{A \mid B_2\} = \binom{20}{3} p_2^3 (1 - p_2)^{17}$$

$$= 0.0596.$$

According to Bayes' theorem,

$$\Pr\{B_1 \mid A\} = \frac{0.5 \times 0.1901}{0.5 \times 0.1901 + 0.5 \times 0.0596} = 0.7613$$

$$\Pr\{B_2 \mid A\} = 1 - \Pr\{B_1 \mid A\} = 0.2387.$$

Thus, after observing three defective parts in a sample of $n = 20$ ones, we believe that the delivery came from vendor B_1. The posterior probability of B_1, given A, is more than three times higher than that of B_2 given A. The a priori odds of B_1 against B_2 were 1:1. The a posteriori odds are 19:6. ∎

In the context of a graph representing directed links between variables, a directed acyclic graph (DAG) represents a qualitative causality model. The model parameters are derived by applying the Markov property, where the conditional probability distribution at each node depends only on its parents. For discrete random variables, this conditional probability is often represented by a table, listing the local probability that a child node takes on each of the feasible values—for each combination of values of its parents. The joint distribution of a collection of variables can be determined uniquely by these local conditional probability tables. A Bayesian network (BN) is represented by a DAG. A BN reflects a simple **conditional independence** statement, namely, that each variable is independent of its non-descendants in the graph given the state of its parents. This property is used to reduce, sometimes significantly, the number of parameters that are required to characterize the joint probability distribution of the variables. This reduction provides an efficient way to compute the posterior probabilities given the evidence present in the data. Moreover, conditioning on target variables, at the end of the DAG, and applying Bayes' theorem, provides us a diagnostic representation of the variable profiles leading to the conditioned value. Overall, BNs provide both predictive and diagnostic capabilities in analyzing multivariate datasets (see Sect. 8.3 for more details on Bayesian networks).

2.2   Random Variables and Their Distributions

Random variables are formally defined as **real-valued functions**, $X(w)$, **over the sample space**, S, **such that events** $\{w : X(w) \leq x\}$ **can be assigned probabilities, for all** $-\infty < x < \infty$, where w are the elements of S.

Example 2.13 Suppose that S is the sample space of all RSWOR of size n, from a finite population, P, of size N. $1 \leq n < N$. The elements w of S are subsets of distinct elements of the population P. A random variable $X(w)$ is some function which assigns w a finite real number, e.g., the number of "defective" elements of w. In the present example, $X(w) = 0, 1, \cdots, n$ and

$$\Pr\{X(w) = j\} = \frac{\binom{M}{j}\binom{N-M}{n-j}}{\binom{N}{n}}, \quad j = 0, \cdots, n,$$

where M is the number of "defective" elements of P.   ■

Example 2.14 Another example of random variable is the compressive strength of a concrete cube of a certain dimension. In this example, the random experiment is to manufacture a concrete cube according to a specified process. The sample space S is the space of all cubes, w, that can be manufactured by this process. $X(w)$ is the compressive strength of w. The probability function assigns each event $\{w : X(w) \leq \xi\}$ a probability, according to some mathematical model which satisfies the laws of probability. Any continuous nondecreasing function $F(x)$, such that $\lim_{x \to -\infty} F(x) = 0$ and $\lim_{x \to \infty} F(x) = 1$, will do the job. For example, for compressive strength of concrete cubes, the following model has been shown to fit experimental results

$$\Pr\{X(w) \leq x\} = \begin{cases} 0, & x \leq 0 \\ \frac{1}{\sqrt{2\pi}\sigma} \int_0^x \frac{1}{y} \exp\left\{-\frac{(\ln y - \mu)^2}{2\sigma^2}\right\} \mathrm{d}y, & 0 < x < \infty. \end{cases}$$

The constants μ and σ, $-\infty < \mu < \infty$, and $0 < \sigma < \infty$ are called **parameters** of the model. Such parameters characterize the manufacturing process.   ■

We distinguish between **two types** of random variables: **discrete** and **continuous**. **Discrete random variables**, $X(w)$, are random variables having a finite or countable range. For example, the number of "defective" elements in a random sample is a discrete random variable. The number of blemishes on a ceramic plate is a discrete random variable. A **continuous random variable** is one whose range consists of whole intervals of possible values. The weight, length, compressive strength, tensile strength, cycle time, output voltage, etc. are continuous random variables.

2.2.1 Discrete and Continuous Distributions

2.2.1.1 Discrete Random Variables

Suppose that a discrete random variable can assume the distinct values $x_0, \cdots, x_k$ (k is finite or infinite). The function

$$p(x) = \Pr\{X(w) = x\}, \quad -\infty < x < \infty \tag{2.13}$$

is called the **probability distribution function** (p.d.f.) of X.

Notice that if x is not one of the values in the specified range $S_X = \{x_j; j = 0, 1, \cdots, k\}$, then $\{X(w) = x\} = \phi$ and $p(x) = 0$. Thus, $p(x)$ assumes positive values only on the specified sequence S_X (S_X is also called the sample space of X), such that

$$p(x_j) \geq 0, j = 0, \cdots, k$$

$$\sum_{j=0}^{k} p(x_j) = 1. \tag{2.14}$$

Example 2.15 Suppose that the random experiment is to cast a die once. The sample points are six possible faces of the die, $\{w_1, \cdots, w_6\}$. Let $X(w_j) = j$, $j = 1, \cdots, 6$, be the random variable, representing the face number. The probability model yields

$$p(x) = \begin{cases} \frac{1}{6}, & \text{if } x = 1, 2, \cdots, 6 \\ 0, & \text{otherwise.} \end{cases}$$

∎

Example 2.16 Consider the example of Sect. 2.1.5, of drawing independently $n = 5$ parts from a production process and counting the number of "defective" parts in this sample. The random variable is $X(w) = J_5$. $S_X = \{0, 1, \cdots, 5\}$ and the p.d.f. is

$$p(x) = \begin{cases} \binom{5}{x} p^{5-x} (1 - p)^x, & x = 0, 1, \cdots, 5 \\ 0, & \text{otherwise.} \end{cases}$$

∎

The probability of the event $\{X(w) \leq x\}$, for any $-\infty < x < \infty$, can be computed by summing the probabilities of the values in S_X, which belong to the interval $(-\infty, x]$. This sum is called the **cumulative distribution function** (c.d.f.) of X and denoted by

Fig. 2.1 The graph of the p.d.f. $P(x) = \sum_{j=0}^{[x]} \frac{\binom{5}{j}}{2^5}$ random variable

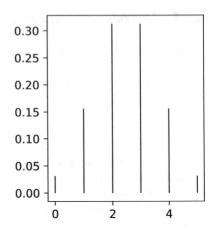

$$P(x) = \Pr\{X(w) \le x\} \tag{2.15}$$

$$= \sum_{\{x_j \le x\}} p(x_j), \tag{2.16}$$

where $x_j \in S_X$.

The c.d.f. corresponding to Example 2.16 is

$$P(x) = \begin{cases} 0, & x < 0 \\ \sum_{j=0}^{[x]} \binom{5}{j} p^{5-j}(1-p)^j, & 0 \le x < 5 \\ 1, & 5 \le x \end{cases}$$

where $[x]$ denotes the **integer part of** x., i.e., the **largest integer** smaller or equal to x.

Generally the graph of the p.d.f. of a discrete variable is a bar chart (see Fig. 2.1). The corresponding c.d.f. is a step function, as shown in Fig. 2.2.

2.2.1.2 Continuous Random Variables

In the case of continuous random variables, the model assigns the variable under consideration a function $F(x)$ which is:

 (i) continuous;
 (ii) Nondecreasing, i.e., if $x_1 < x_2$ then $F(x_1) \le F(x_2)$ and
(iii) $\lim_{x \to -\infty} F(x) = 0$ and $\lim_{x \to \infty} F(x) = 1$.

Such a function can serve as a **cumulative distribution function** (c.d.f.), for X.

An example of a c.d.f. for a continuous random variable which assumes non-negative values, e.g., the operation total time until a part fails, is

Fig. 2.2 The graph of the c.d.f of $P(x)$

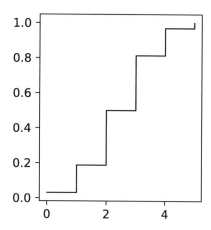

$$F(x) = \begin{cases} 0, & \text{if } x \le 0 \\ 1 - e^{-x}, & \text{if } x > 0. \end{cases}$$

This function is continuous, monotonically increasing, and $\lim_{x \to \infty} F(x) = 1 - \lim_{x \to \infty} e^{-x} = 1$ (see Fig. 2.3). If the c.d.f. of a continuous random variable can be represented as

$$F(x) = \int_{-\infty}^{x} f(y)\, dy, \tag{2.17}$$

for some $f(y) \ge 0$, then we say that $F(x)$ is **absolutely continuous** and $f(x) = \frac{d}{dx} F(x)$. (The derivative $f(x)$ may not exist on a finite number of x values, in any finite interval.) The function $f(x)$ is called the **probability density function** (p.d.f.) of X.

In the above example of total operational time, the p.d.f. is

$$f(x) = \begin{cases} 0, & \text{if } x < 0 \\ e^{-x}, & \text{if } x \ge 0. \end{cases}$$

Thus, as in the discrete case, we have $F(x) = \Pr\{X \le x\}$. It is now possible to write

$$\Pr\{a \le X < b\} = \int_{a}^{b} f(t)\, dt = F(b) - F(a) \tag{2.18}$$

or

$$\Pr\{X \ge b\} = \int_{b}^{\infty} f(t)\, dt = 1 - F(b). \tag{2.19}$$

Thus, if X has the exponential c.d.f.

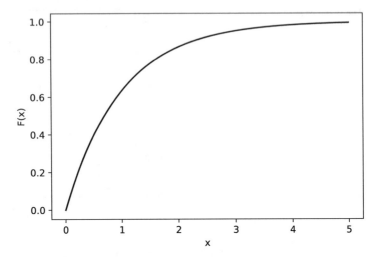

Fig. 2.3 c.d.f. of $F(x) = 1 - e^{-x}$

$$\Pr\{1 \leq X \leq 2\} = F(2) - F(1) = e^{-1} - e^{-2} = 0.2325.$$

There are certain phenomena which require more complicated modeling. The random variables under consideration may not have purely discrete or purely absolutely continuous distribution. There are many random variables with c.d.f.s which are absolutely continuous within certain intervals and have jump points (points of discontinuity) at the end points of the intervals. Distributions of such random variables can be expressed as mixtures of purely discrete c.d.f., $F_d(x)$, and of absolutely continuous c.d.f., $F_{ac}(x)$, i.e.,

$$F(x) = pF_d(x) + (1 - p)F_{ac}(x), \quad -\infty < x < \infty, \tag{2.20}$$

where $0 \leq p \leq 1$ (Fig. 2.4).

Example 2.17 A distribution which is a mixture of discrete and continuous distributions is obtained, for example, when a measuring instrument is not sensitive enough to measure small quantities or large quantities which are outside its range. This could be the case for a weighing instrument which assigns the value 0 [mg] to any weight smaller than 1 [mg], the value 1 [g] to any weight greater than 1 gram, and the correct weight to values in between.

Another example is the total number of minutes, within a given working hour, that a service station is busy serving customers. In this case the c.d.f. has a jump at 0, of height p, which is the probability that the service station is idle at the beginning of the hour and no customer arrives during that hour. In this case,

$$F(x) = p + (1 - p)G(x), \quad 0 \leq x < \infty,$$

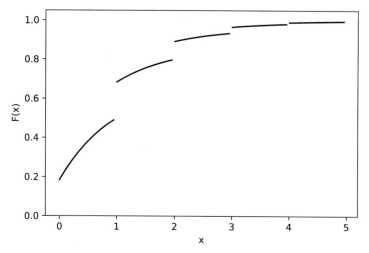

Fig. 2.4 c.d.f. of the mixture distribution $F(x) = 0.5(1 - e^{-x}) + 0.5e^{-1}\sum_{j=0}^{[x]}\frac{1}{j!}$

where $G(x)$ is the c.d.f. of the total service time, $G(0) = 0$.   ■

2.2.2   Expected Values and Moments of Distributions

The **expected value** of a function $g(X)$, under the distribution $F(x)$, is

$$E_F\{g(X)\} = \begin{cases} \int_{-\infty}^{\infty} g(x)f(x)\,dx, & \text{if } X \text{ is continuous} \\ \sum_{j=0}^{k} g(x_j)p(x_j), & \text{if } X \text{ is discrete.} \end{cases}$$

In particular,

$$\mu_l(F) = E_F\{X^l\}, \quad l = 1, 2, \cdots \tag{2.21}$$

is called the l-th **moment** of $F(x)$. $\mu_1(F) = E_F\{X\}$ is the expected value of X, or the **population mean**, according to the model $F(x)$.

Moments around $\mu_1(F)$ are called **central moments**, which are

$$\mu_l^*(F) = E\{(X - \mu_1(F))^l\}, \quad l = 1, 2, 3, \cdots. \tag{2.22}$$

Obviously, $\mu_1^*(F) = 0$. The **second central moment** is called the **variance** of $F(x)$, $V_F\{X\}$.

In the following, the notation $\mu_l(F)$ will be simplified to μ_l, if there is no room for confusion.

Expected values of a function $g(X)$, and in particular the moments, may not exist, since an integral $\int_{-\infty}^{\infty} x^l f(x)dx$ may not be well defined. Example of such a case is the distribution, called the **Cauchy distribution**, with p.d.f.

$$f(x) = \frac{1}{\pi} \cdot \frac{1}{1+x^2}, \quad -\infty < x < \infty.$$

Notice that under this model, moments do not exist for any $l = 1, 2, \cdots$. Indeed, the integral

$$\frac{1}{\pi} \int_{-\infty}^{\infty} \frac{x}{1+x^2} \, dx$$

does not exist. If the second moment exists, then

$$V\{X\} = \mu_2 - \mu_1^2.$$

Example 2.18 Consider the random experiment of casting a die once. The random variable, X, is the face number. Thus, $p(x) = \frac{1}{6}$, $x = 1, \cdots, 6$ and

$$\mu_1 = E\{X\} = \frac{1}{6} \sum_{j=1}^{6} j = \frac{6(6+1)}{2 \times 6} = \frac{7}{2} = 3.5$$

$$\mu_2 = \frac{1}{6} \sum_{j=1}^{6} j^2 = \frac{6(6+1)(2 \times 6 + 1)}{6 \times 6} = \frac{7 \times 13}{6} = \frac{91}{6} = 15.167.$$

The variance is

$$V\{X\} = \frac{91}{6} - \left(\frac{7}{2}\right)^2 = \frac{182 - 147}{12} = \frac{35}{12}.$$

■

Example 2.19 X has a continuous distribution with p.d.f.

$$f(x) = \begin{cases} 0, & \text{otherwise} \\ 1, & \text{if } 1 \leq x \leq 2. \end{cases}$$

Thus,

$$\mu_1 = \int_1^2 x \, dx = \frac{1}{2} \left(x^2 \Big|_1^2 \right) = \frac{1}{2}(4-1) = 1.5$$

$$\mu_2 = \int_1^2 x^2 \, dx = \frac{1}{3} \left(x^3 \Big|_1^2 \right) = \frac{7}{3}$$

$$V\{X\} = \mu_2 - \mu_1^2 = \frac{7}{3} - \frac{9}{4} = \frac{28-27}{12} = \frac{1}{12}.$$

∎

The following is a useful formula when X assumes only positive values, i.e., $F(x) = 0$ for all $x \leq 0$,

$$\mu_1 = \int_0^\infty (1 - F(x)) \, dx, \tag{2.23}$$

for continuous c.d.f. $F(x)$. Indeed,

$$\mu_1 = \int_0^\infty x f(x) \, dx$$

$$= \int_0^\infty \left(\int_0^x dy \right) f(x) \, dx$$

$$= \int_0^\infty \left(\int_y^\infty f(x) \, dx \right) dy$$

$$= \int_0^\infty (1 - F(y)) \, dy.$$

For example, suppose that $f(x) = \mu e^{-\mu x}$, for $x \geq 0$. Then $F(x) = 1 - e^{-\mu x}$ and

$$\int_0^\infty (1 - F(x)) \, dx = \int_0^\infty e^{-\mu x} \, dx = \frac{1}{\mu}.$$

When X is discrete, assuming the values $\{1, 2, 3, \ldots\}$, then we have a similar formula

$$E\{X\} = 1 + \sum_{i=1}^\infty (1 - F(i)).$$

2.2.3 The Standard Deviation, Quantiles, Measures of Skewness, and Kurtosis

The **standard deviation** of a distribution $F(x)$ is $\sigma = (V\{X\})^{1/2}$. The standard deviation is used as a measure of dispersion of a distribution. An important theorem in probability theory, called the **Chebychev Theorem**, relates the standard deviation to the probability of deviation from the mean. More formally, the theorem states that if σ exists, then

$$\Pr\{|X - \mu_1| > \lambda\sigma\} \le \frac{1}{\lambda^2}. \tag{2.24}$$

Thus, by this theorem, the probability that a random variable will deviate from its expected value by more than three standard deviations is less than 1/9, whatever the distribution is. This theorem has important implications, which will be highlighted later.

The **p-th quantile of a distribution** $F(x)$ **is the smallest value of** x, ξ_p **such that** $F(x) \ge p$. We also write $\xi_p = F^{-1}(p)$.

For example, if $F(x) = 1 - e^{-\lambda x}$, $0 \le x < \infty$, where $0 < \lambda < \infty$, then ξ_p is such that

$$F(\xi_p) = 1 - e^{-\lambda \xi_p} = p.$$

Solving for ξ_p we get

$$\xi_p = -\frac{1}{\lambda} \cdot \ln(1 - p).$$

The **median** of $F(x)$ is $Ff^{-1}(.5) = \xi_{.5}$. Similarly $\xi_{.25}$ and $\xi_{.75}$ are the first and third quartiles of F.

A distribution $F(x)$ is **symmetric** about the mean $\mu_1(F)$ if

$$F(\mu_1 + \delta) = 1 - F(\mu_1 - \delta)$$

for all $\delta \ge 0$.

In particular, if F is symmetric, then $F(\mu_1) = 1 - F(\mu_1)$ or $\mu_1 = F^{-1}(.5) = \xi_{.5}$. Accordingly, the mean and median of a symmetric distribution coincide. In terms of the p.d.f., a distribution is symmetric about its mean if

$$f(\mu_1 + \delta) = f(\mu_1 - \delta), \quad \text{for all } \delta \ge 0.$$

A commonly used index of **skewness** (asymmetry) is

$$\beta_3 = \frac{\mu_3^*}{\sigma^3}, \tag{2.25}$$

where μ_3^* is the third central moment of F. One can prove that **if $F(x)$ is symmetric, then $\beta_3 = 0$.** If $\beta_3 > 0$ we say that $F(x)$ is positively skewed; otherwise, it is negatively skewed.

Example 2.20   Consider the **binomial** distribution, with p.d.f.

$$p(x) = \binom{n}{x} p^x (1-p)^{n-x}, \quad x = 0, 1, \cdots, n.$$

In this case

$$
\begin{aligned}
\mu_1 &= \sum_{x=0}^{n} x \binom{n}{x} p^x (1-p)^{n-x} \\
&= np \sum_{x=1}^{n} \binom{n-1}{x-1} p^{x-1} (1-p)^{n-1-(x-1)} \\
&= np \sum_{j=0}^{n-1} \binom{n-1}{j} p^j (1-p)^{n-1-j} \\
&= np.
\end{aligned}
$$

Indeed,

$$
\begin{aligned}
x \binom{n}{x} &= x \frac{n!}{x!(n-x)!} = \frac{n!}{(x-1)!((n-1)-(x-1))!} \\
&= n \binom{n-1}{x-1}.
\end{aligned}
$$

Similarly, we can show that

$$\mu_2 = n^2 p^2 + np(1-p),$$

and

$$\mu_3 = np[n(n-3)p^2 + 3(n-1)p + 1 + 2p^2].$$

The third central moment is

$$
\begin{aligned}
\mu_3^* &= \mu_3 - 3\mu_2\mu_1 + 2\mu_1^3 \\
&= np(1-p)(1-2p).
\end{aligned}
$$

Furthermore,

$$V\{X\} = \mu_2 - \mu_1^2$$
$$= np(1 - p).$$

Hence,

$$\sigma = \sqrt{np(1 - p)}$$

and the index of asymmetry is

$$\beta_3 = \frac{\mu_3^*}{\sigma_3} = \frac{np(1 - p)(1 - 2p)}{(np(1 - p))^{3/2}}$$
$$= \frac{1 - 2p}{\sqrt{np(1 - p)}}.$$

Thus, if $p = \frac{1}{2}$ then $\beta_3 = 0$ and the distribution is symmetric. If $p < \frac{1}{2}$ the distribution is positively skewed, and it is negatively skewed if $p > \frac{1}{2}$. ∎

In Chap. 1 we mentioned also the index of **kurtosis** (steepness). This is given by

$$\beta_4 = \frac{\mu_4^*}{\sigma^4}. \tag{2.26}$$

Example 2.21 Consider the exponential c.d.f.

$$F(x) = \begin{cases} 0, & \text{if } x < 0 \\ 1 - e^{-x}, & \text{if } x \geq 0. \end{cases}$$

The p.d.f. is $f(x) = e^{-x}, x \geq 0$. Thus, for this distribution

$$\mu_1 = \int_0^\infty x e^{-x} \, dx = 1$$

$$\mu_2 = \int_0^\infty x^2 e^{-x} \, dx = 2$$

$$\mu_3 = \int_0^\infty x^3 e^{-x} \, dx = 6$$

$$\mu_4 = \int_0^\infty x^4 e^{-x} \, dx = 24.$$

Therefore,

$$V\{X\} = \mu_2 - \mu_1^2 = 1,$$

$$\sigma = 1$$

$$\mu_4^* = \mu_4 - 4\mu_3\mu_1 + 6\mu_2\mu_1^2 - 3\mu_1^4$$

$$= 24 - 4 \times 6 \times 1 + 6 \times 2 \times 1 - 3 = 9.$$

Finally, the index of kurtosis is

$$\beta_4 = 9.$$

∎

2.2.4   Moment Generating Functions

The **moment generating function** (m.g.f.) of a distribution of X is defined as a function of a real variable t,

$$M(t) = E\{e^{tX}\}. \tag{2.27}$$

$M(0) = 1$ for all distributions. $M(t)$, however, may not exist for some $t \neq 0$. To be useful, it is sufficient that $M(t)$ will exist in some interval containing $t = 0$.

For example, if X has a continuous distribution with p.d.f.

$$f(x) = \begin{cases} \dfrac{1}{b-a}, & \text{if } a \leq x \leq b, \ a < b \\ 0, & \text{otherwise} \end{cases}$$

then

$$M(t) = \frac{1}{b-a} \int_a^b e^{tx} \, dx = \frac{1}{t(b-a)} \left(e^{tb} - e^{ta} \right).$$

This is a differentiable function of t, for all t, $-\infty < t < \infty$.

On the other hand, if for $0 < \lambda < \infty$,

$$f(x) = \begin{cases} \lambda e^{-\lambda x}, & 0 \leq x < \infty \\ 0, & x < 0 \end{cases}$$

then

$$M(t) = \lambda \int_0^\infty e^{tx - \lambda x} \, dx$$

$$= \frac{\lambda}{\lambda - t}, \quad t < \lambda.$$

This m.g.f. exists only for $t < \lambda$. The m.g.f. $M(t)$ is a transform of the distribution $F(x)$, and the correspondence between $M(t)$ and $F(x)$ is one-to-one. In the above example, $M(t)$ is the Laplace transform of the p.d.f. $\lambda e^{-\lambda x}$. This correspondence is often useful in identifying the distributions of some statistics, as will be shown later.

Another useful property of the m.g.f. $M(t)$ is that often we can obtain the moments of $F(x)$ by differentiating $M(t)$. More specifically, consider the r-th order derivative of $M(t)$. Assuming that this derivative exists, and differentiation can be interchanged with integration (or summation), then

$$M^{(r)}(t) = \frac{d^r}{dt^r} \int e^{tx} f(x) \, dx = \int \left(\frac{d^r}{dt^r} e^{tx} \right) f(x) \, dx$$

$$= \int x^r e^{tx} f(x) \, dx.$$

Thus, if these operations are justified, then

$$M^{(r)}(t) \bigg|_{t=0} = \int x^r f(x) \, dx = \mu_r. \tag{2.28}$$

In the following sections, we will illustrate the usefulness of the m.g.f.

2.3 Families of Discrete Distribution

In the present section, we discuss several families of discrete distributions and illustrate possible application in modeling industrial phenomena.

2.3.1 The Binomial Distribution

Consider n **identical** independent trials. In each trial the probability of "success" is fixed at some value p, and successive events of "success" or "failure" are **independent**. Such trials are called **Bernoulli trials**. The distribution of the number of "successes," J_n, is binomial with p.d.f.

$$b(j; n, p) = \binom{n}{j} p^j (1-p)^{n-j}, \quad j = 0, 1, \cdots, n. \tag{2.29}$$

This p.d.f. was derived in Example 2.11 as a special case.

A binomial random variable, with parameters (n, p), will be designated as $B(n, p)$. n is a given integer and p belongs to the interval $(0, 1)$. The collection of all such binomial distributions is called the **binomial family**.

The binomial distribution is a proper model whenever we have a sequence of independent binary events (0 and 1, or "success" and "failure") with the same probability of "success."

Example 2.22 We draw a random sample of $n = 10$ items from a mass production line of light bulbs. Each light bulb undergoes an inspection, and if it complies with the production specifications, we say that the bulb is compliant (successful event). Let $X_i = 1$ if the i-th bulb is compliant and $X_i = 0$ otherwise. If we can assume that the probability of $\{X_i = 1\}$ is the same, p, for all bulbs and if the n events are mutually independent, then the number of bulbs in the sample which complies with the specifications, i.e., $J_n = \sum_{i=1}^{n} X_i$, has the binomial p.d.f. $b(i; n, p)$. Notice that if we draw a sample at random **with** replacement, RSWR, from a lot of size N, which contains M compliant units, then the distribution of J_n is $B\left(n, \frac{M}{N}\right)$.

Indeed, if sampling is with replacement, the probability that the i-th item selected is compliant is $p = \frac{M}{N}$ for all $i = 1, \cdots, n$. Furthermore, selections are **independent** of each other. ∎

The binomial c.d.f. will be denoted by $B(i; n, p)$. Recall that

$$B(i; n, p) = \sum_{j=0}^{i} b(j; n, p), \tag{2.30}$$

$i = 0, 1, \cdots, n$. The m.g.f. of $B(n, p)$ is

$$M(t) = E\{e^{tX}\}$$

$$= \sum_{j=0}^{n} \binom{n}{j} (pe^t)^j (1-p)^{n-j} \tag{2.31}$$

$$= (pe^t + (1-p))^n, \quad -\infty < t < \infty.$$

Notice that

$$M'(t) = n(pe^t + (1-p))^{n-1} pe^t$$

and

$$M''(t) = n(n-1)p^2 e^{2t} (pe^t + (1-p))^{n-2} + npe^t (pe^t + (1-p))^{n-1}.$$

Table 2.1 Values of the
p.d.f. and c.d.f. of $B(30, 0.6)$

i	$b(i; 30, 0.6)$	$B(i; 30, 0.6)$
8	0.0002	0.0002
9	0.0006	0.0009
10	0.0020	0.0029
11	0.0054	0.0083
12	0.0129	0.0212
13	0.0269	0.0481
14	0.0489	0.0971
15	0.0783	0.1754
16	0.1101	0.2855
17	0.1360	0.4215
18	0.1474	0.5689
19	0.1396	0.7085
20	0.1152	0.8237
21	0.0823	0.9060
22	0.0505	0.9565
23	0.0263	0.9828
24	0.0115	0.9943
25	0.0041	0.9985
26	0.0012	0.9997
27	0.0003	1.0000

The expected value and variance of $B(n, p)$ are

$$E\{J_n\} = np, \tag{2.32}$$

and

$$V\{J_n\} = np(1 - p). \tag{2.33}$$

This was shown in Example 2.20 and can be verified directly by the above formulae of $M'(t)$ and $M''(t)$. To obtain the values of $b(i; n, p)$, we can use Python. For example, suppose we wish to tabulate the values of the p.d.f. $b(i; n, p)$ and those of the c.d.f. $B(i; n, p)$ for $n = 30$ and $p = 0.60$. Below commands generate a data frame with values as illustrated in Table 2.1.

```
x = list(range(0, 31))
rv = stats.binom(30, 0.6)
df = pd.DataFrame({
    'i': x,
    'b': rv.pmf(x),
    'B': rv.cdf(x),
})
```

After tabulating the values of the c.d.f., we can obtain the quantiles (or fractiles) of the distribution. Recall that in the discrete case, the p-th quantile of a random

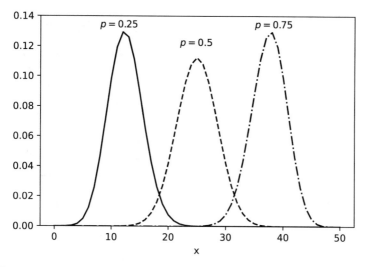

Fig. 2.5 p.d.f. of $B(50, p)$, $p = 0.25, 0.50, 0.75$

variable X is

$$x_p = \text{smallest } x \text{ such that } F(x) \geq p.$$

Thus, from Table 2.1 we find that the lower, median, and upper quartile of $B(30, .6)$ are $Q_1 = 16$, $M_e = 18$, and $Q_3 = 20$. These values can also be obtained directly with the scipy package.

```
stats.binom(30, 0.6).ppf(0.5)
```

In Fig. 2.5 we present the p.d.f. of three binomial distributions, with $n = 50$ and $p = 0.25, 0.50$, and 0.75. We see that if $p = 0.25$, the p.d.f. is positively skewed. When $p = 0.5$ it is symmetric, and when $p = 0.75$ it is negatively skewed. This is in accordance with the index of skewness β_3, which was presented in Example 2.20.

2.3.2 The Hypergeometric Distribution

Let J_n denote the number of units, in a RSWOR of size n, from a population of size N, having a certain property. The number of population units before sampling, having this property, is M. The distribution of J_n is called the hypergeometric distribution. We denote a random variable having such a distribution by $H(N, M, n)$. The p.d.f. of J_n is

Table 2.2 The p.d.f. and c.d.f. of $H(75, 15, 10)$

j	$h(j; 75, 15, 10)$	$H(j; 75, 15, 10)$
0	0.0910	0.0910
1	0.2675	0.3585
2	0.3241	0.6826
3	0.2120	0.8946
4	0.0824	0.9770
5	0.0198	0.9968
6	0.0029	0.9997
7	0.0003	1.0000

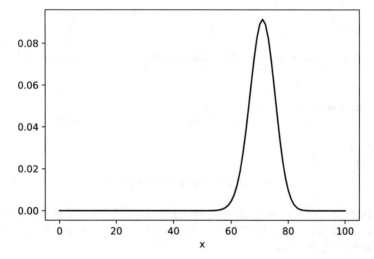

Fig. 2.6 The p.d.f. $h(i; 500, 350, 100)$

$$h(j; N, M, n) = \frac{\binom{M}{j}\binom{N-M}{n-j}}{\binom{N}{n}}, \quad j = 0, \cdots, n. \tag{2.34}$$

This formula was shown already in Sect. 2.1.4.

The c.d.f. of $H(N, M, n)$ will be designated by $H(j; N, M, n)$. In Table 2.2 we present the p.d.f. and c.d.f. of $H(75, 15, 10)$.

```
x = list(range(0, 8))
rv = stats.hypergeom(M=75, n=15, N=10)
df = pd.DataFrame({
    'j': x,
    'h': rv.pmf(x),
    'H': rv.cdf(x),
})
```

In Fig. 2.6 we show the p.d.f. of $H(500, 350, 100)$.

The expected value and variance of $H(N, M, n)$ are

Table 2.3 The p.d.f. of $H(500, 350, 20)$ and $B(20, 0.7)$

j	$h(i; 500, 350, 20)$	$b(i; 20, 0.7)$
5	0.00003	0.00004
6	0.00016	0.00022
7	0.00082	0.00102
8	0.00333	0.00386
9	0.01093	0.01201
10	0.02928	0.03082
11	0.06418	0.06537
12	0.11491	0.11440
13	0.16715	0.16426
14	0.19559	0.19164
15	0.18129	0.17886
16	0.12999	0.13042
17	0.06949	0.07160
18	0.02606	0.02785
19	0.00611	0.00684
20	0.00067	0.00080

$$E\{J_n\} = n\frac{M}{N} \tag{2.35}$$

and

$$V\{J_n\} = n\frac{M}{N}\left(1 - \frac{M}{N}\right)\left(1 - \frac{n-1}{N-1}\right). \tag{2.36}$$

Notice that when $n = N$, the variance of J_n is $V\{J_N\} = 0$. Indeed, if $n = N$, $J_N = M$, which is not a random quantity. Derivation of these formulae is given in Sect. 5.2.2. There is no simple expression for the m.g.f.

If the sample size n is small relative to N, i.e., $n/N << 0.1$, the hypergeometric p.d.f. can be approximated by that of the binomial $B\left(n, \frac{M}{N}\right)$. In Table 2.3 we compare the p.d.f. of $H(500, 350, 20)$ to that of $B(20, 0.7)$.

The expected value and variance of the binomial and the hypergeometric distributions are compared in Table 2.4. We see that the expected values have the same formula but that the variance formulae differ by the correction factor $(N - n)/(N - 1)$ which becomes 1 when $n = 1$ and 0 when $n = N$.

Example 2.23 At the end of a production day, printed circuit boards (PCBs) soldered by wave soldering process are subjected to sampling audit. A RSWOR of size n is drawn from the lot, which consists of all the PCBs produced on that day. If the sample has any defective PCB, another RSWOR of size $2n$ is drawn from the lot. If there are more than three defective boards in the combined sample, the lot is sent for rectification, in which every PCB is inspected. If the lot consists of $N = 100$ PCBs, and the number of defective ones is $M = 5$, what is the probability that the lot will be rectified, when $n = 10$?

Table 2.4 The expected value and variance of the hypergeometric and binomial distribution

	Hypergeometric $H(a; N, M, n)$	Binomial $B\left(n, \dfrac{M}{N}\right)$
Expected value	$n\dfrac{M}{N}$	$n\dfrac{M}{N}$
Variance	$n\dfrac{M}{N}\left(1 - \dfrac{M}{N}\right)\left(1 - \dfrac{n-1}{N-1}\right)$	$n\dfrac{M}{N}\left(1 - \dfrac{M}{N}\right)$

Let J_1 be the number of defective items in the first sample. If $J_1 > 3$, then the lot is rectified without taking a second sample. If $J_1 = 1, 2,$ or 3, a second sample is drawn. Thus, if R denotes the event "the lot is sent for rectification,"

$$\Pr\{R\} = 1 - H(3; 100, 5, 10)$$

$$+ \sum_{i=1}^{3} h(i; 100, 5, 10) \times [1 - H(3 - i; 90, 5 - i, 20)$$

$$= 0.00025 + 0.33939 \times 0.03313$$

$$+ 0.07022 \times 0.12291$$

$$+ 0.00638 \times 0.397 = 0.0227.$$

∎

2.3.3 The Poisson Distribution

A third discrete distribution that plays an important role in quality control is the Poisson distribution, denoted by $P(\lambda)$. It is sometimes called the distribution of rare events, since it is used as an approximation to the binomial distribution when the sample size, n, is large and the proportion of defectives, p, is small. The parameter λ represents the "rate" at which defectives occur, i.e., the expected number of defectives per time interval or per sample. The Poisson probability distribution function is given by the formula

$$p(j; \lambda) = \frac{e^{-\lambda}\lambda^j}{j!}, \qquad j = 0, 1, 2, \cdots \tag{2.37}$$

and the corresponding c.d.f. is

$$P(j; \lambda) = \sum_{i=0}^{j} p(i; \lambda), \qquad j = 0, 1, 2, \cdots. \tag{2.38}$$

Table 2.5 Binomial distributions for $np = 2$ and the Poisson distribution with $\lambda = 2$

	Binomial				Poisson
	$n = 20$	$n = 40$	$n = 100$	$n = 1000$	
k	$p = 0.1$	$p = 0.05$	$p = 0.02$	$p = 0.002$	$\lambda = 2$
0	0.121577	0.128512	0.132620	0.135065	0.135335
1	0.270170	0.270552	0.270652	0.270670	0.270671
2	0.285180	0.277672	0.273414	0.270942	0.270671
3	0.190120	0.185114	0.182276	0.180628	0.180447
4	0.089779	0.090122	0.090208	0.090223	0.090224
5	0.031921	0.034151	0.035347	0.036017	0.036089
6	0.008867	0.010485	0.011422	0.011970	0.012030
7	0.001970	0.002680	0.003130	0.003406	0.003437
8	0.000356	0.000582	0.000743	0.000847	0.000859
9	0.000053	0.000109	0.000155	0.000187	0.000191

Example 2.24 Suppose that a machine produces aluminum pins for airplanes. The probability p that a single pin emerges defective is small, say $p = 0.002$. In 1 h, the machine makes $n = 1000$ pins (considered here to be a random sample of pins). The number of defective pins produced by the machine in 1 h has a binomial distribution with a mean of $\mu = np = 1000(0.002) = 2$, so the rate of defective pins for the machine is $\lambda = 2$ pins per hour. In this case, the binomial probabilities are very close to the Poisson probabilities. This approximation is illustrated below in Table 2.5, by considering processes which produce defective items at a rate of $\lambda = 2$ parts per hour, based on various sample sizes. In Exercise 2.46 the student is asked to prove that the binomial p.d.f. converges to that of the Poisson with mean λ when $n \to \infty$, $p \to 0$ but $np \to \lambda$. ∎

The m.g.f. of the Poisson distribution is

$$M(t) = e^{-\lambda} \sum_{j=0}^{\infty} e^{tj} \frac{\lambda^j}{j!}$$

$$= e^{-\lambda} \cdot e^{\lambda e^t} = e^{-\lambda(1-e^t)}, \quad -\infty < t < \infty. \tag{2.39}$$

Thus,

$$M'(t) = \lambda M(t) e^t$$

$$M''(t) = \lambda^2 M(t) e^{2t} + \lambda M(t) e^t \tag{2.40}$$

$$= (\lambda^2 e^{2t} + \lambda e^t) M(t).$$

Hence, the mean and variance of the Poisson distribution are

$$\mu = E\{X\} = \lambda$$

and   (2.41)

$$\sigma^2 = V\{X\} = \lambda.$$

The Poisson distribution is used not only as an approximation to the binomial. It is a useful model for describing the number of "events" occurring in a unit of time (or area, volume, etc.) when those events occur "at random." The rate at which these events occur is denoted by λ. An example of a Poisson random variable is the number of decaying atoms, from a radioactive substance, detected by a Geiger counter in a fixed period of time. If the rate of detection is 5 per second, then the number of atoms detected in a second has a Poisson distribution with mean $\lambda = 5$. The number detected in 5 s, however, will have a Poisson distribution with $\lambda = 25$. A rate of 5 per second equals a rate of 25 per 5 s. Other examples of Poisson random variables include:

1. The number of blemishes found in a unit area of a finished surface (ceramic plate).
2. The number of customers arriving at a store in 1 h.
3. The number of defective soldering points found on a circuit board.

The p.d.f., c.d.f., and quantiles of the Poisson distribution can be computed using Python. In Fig. 2.7 we illustrate the p.d.f. for three values of λ.

```
x = np.linspace(0, 50, 51)
distributions = pd.DataFrame({
    'x': x,
    'density5': stats.poisson(mu=5).pmf(x),
    'density10': stats.poisson(mu=10).pmf(x),
    'density15': stats.poisson(mu=15).pmf(x),
})
ax = distributions.plot(x='x', y='density5', color='black')
distributions.plot(x='x', y='density10', color='black', ls='--', ax=ax)
distributions.plot(x='x', y='density15', color='black', ls='-.', ax=ax)
ax.text(8, 0.17, '$\lambda=5$')
ax.text(14, 0.12, '$\lambda=10$')
ax.text(19, 0.10, '$\lambda=15$')
ax.get_legend().remove()
plt.show()
```

2.3.4 The Geometric and Negative Binomial Distributions

Consider a sequence of **independent** trials, each one having the same probability for "success," say p. Let N be a random variable which counts the number of trials until the first "success" is realized, including the successful trial. N may assume positive integer values with probabilities

$$\Pr\{N = n\} = p(1 - p)^{n-1}, \quad n = 1, 2, \cdots. \tag{2.42}$$

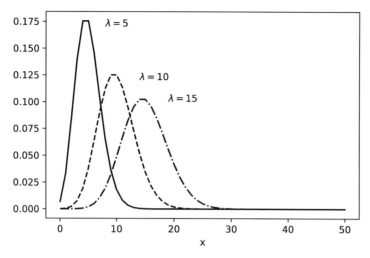

Fig. 2.7 Poisson p.d.f. $\lambda = 5, 10, 15$

This probability function is the p.d.f. of the **geometric** distribution.
Let $g(n; p)$ designate the p.d.f. The corresponding c.d.f. is

$$G(n; p) = 1 - (1 - p)^n, \quad n = 1, 2, \cdots .$$

From this we obtain that the α-quantile $(0 < \alpha < 1)$ is given by

$$N_\alpha = \left[\frac{\log(1 - \alpha)}{\log(1 - p)} \right] + 1,$$

where $[x]$ designates the integer part of x.
The expected value and variance of the geometric distribution are

$$E\{N\} = \frac{1}{p},$$

and

$$V\{N\} = \frac{1 - p}{p^2}.$$

(2.43)

Indeed, the m.g.f. of the geometric distribution is

$$M(t) = pe^t \sum_{j=0}^{\infty} (e^t(1-p))^j$$

$$= \frac{pe^t}{1 - e^t(1-p)}, \quad \text{if } t < -\log(1-p).$$
(2.44)

Thus, for $t < -\log(1-p)$,

$$M'(t) = \frac{pe^t}{(1 - e^t(1-p))^2}$$

and

$$M''(t) = \frac{pe^t}{(1 - e^t(1-p))^2} + \frac{2p(1-p)e^{2t}}{(1 - e^t(1-p))^3}.$$

Hence,

$$\mu_1 = M'(0) = \frac{1}{p}$$

$$\mu_2 = M''(0) = \frac{2-p}{p^2},$$
(2.45)

and the above formulae of $E\{X\}$ and $V\{X\}$ are obtained.

The geometric distribution is applicable in many problems. We illustrate one such application in the following example.

Example 2.25 An insertion machine stops automatically if there is failure in handling a component during an insertion cycle. A cycle starts immediately after the insertion of a component and ends at the insertion of the next component. Suppose that the probability of stopping is $p = 10^{-3}$ per cycle. Let N be the number of cycles until the machine stops. It is assumed that events at different cycles are mutually independent. Thus, N has a geometric distribution and $E\{N\} = 1000$. We expect a run of 1000 cycles between consecutive stopping. The number of cycles, N, however is a random variable with standard deviation of $\sigma = \left(\frac{1-p}{p^2}\right)^{1/2} = 999.5$. This high value of σ indicates that we may see very short runs and also long ones. Indeed, for $\alpha = 0.5$ the quantiles of N are $N_{0.05} = 52$ and $N_{0.95} = 2995$. ∎

The number of failures until the first success, $N - 1$, has a shifted geometric distribution, which is a special case of the family of **negative binomial** distribution.

We say that a non-negative integer valued random variable X has a negative binomial distribution, with parameters (p, k), where $0 < p < 1$ and $k = 1, 2, \cdots$, if its p.d.f. is

$$g(j; p, k) = \binom{j + k - 1}{k - 1} p^k (1 - p)^j, \tag{2.46}$$

$j = 0, 1, \cdots$. The shifted geometric distribution is the special case of $k = 1$.

A more general version of the negative binomial distribution can be formulated, in which $k - 1$ is replaced by a positive real parameter. A random variable having the above negative binomial will be designated by $NB(p, k)$. The $NB(p, k)$ represents the number of failures observed until the k-th success. The expected value and variance of $NB(p, k)$ are

$$E\{X\} = k\frac{1 - p}{p},$$

and $\tag{2.47}$

$$V\{X\} = k\frac{1 - p}{p^2}.$$

In Fig. 2.8 we present the p.d.f. of $NB(p, k)$. The negative binomial distributions have been applied as a model of the distribution for the periodic demand of parts in inventory theory.

```
x = np.linspace(0, 100, 101)
distributions = pd.DataFrame({
    'x': x,
    'density_2': stats.nbinom(n=5, p=0.2).pmf(x),
    'density_1': stats.nbinom(n=5, p=0.1).pmf(x),
})
ax = distributions.plot(x='x', y='density_2', color='black')
distributions.plot(x='x', y='density_1', color='black', ls='--', ax=ax)
ax.text(35, 0.04, '$p=0.2$')
ax.text(70, 0.015, '$p=0.1$')
ax.set_xlabel('$i$')
ax.set_ylabel('$nb(i,5,p)$')
ax.get_legend().remove()
plt.show()
```

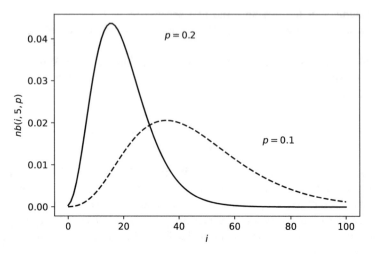

Fig. 2.8 p.d.f. of $NB(p, 5)$ with $p = 0.10, 0.20$

2.4   Continuous Distributions

2.4.1   *The Uniform Distribution on the Interval* (a, b), $a < b$

We denote a random variable having this distribution by $U(a, b)$. The p.d.f. is given by

$$f(x; a, b) = \begin{cases} 1/(b - a), & a \leq x \leq b \\ 0, & \text{elsewhere,} \end{cases} \tag{2.48}$$

and the c.d.f. is

$$F(x; a, b) = \begin{cases} 0, & \text{if } x < a \\ (x - a)/(b - a), & \text{if } a \leq x < b \\ 1, & \text{if } b \leq x \end{cases} \tag{2.49}$$

The expected value and variance of $U(a, b)$ are

$$\mu = (a + b)/2,$$

and   (2.50)

$$\sigma^2 = (b - a)^2/12.$$

The p-th quantile is $x_p = a + p(b - a)$.

To verify the formula for μ, we set

$$\mu = \frac{1}{b-a} \int_a^b x \, dx = \frac{1}{b-a} \left| \frac{1}{2} x^2 \right|_a^b = \frac{1}{2(b-a)}(b^2 - a^2)$$
$$= \frac{a+b}{2}.$$

Similarly,

$$\mu_2 = \frac{1}{b-a} \int_a^b x^2 \, dx = \frac{1}{b-a} \left| \frac{1}{3} x^3 \right|_a^b$$
$$= \frac{1}{3(b-a)}(b^3 - a^3) = \frac{1}{3}(a^2 + ab + b^2).$$

Thus,

$$\sigma^2 = \mu_2 - \mu_1^2 = \frac{1}{3}(a^2 + ab + b^2) - \frac{1}{4}(a^2 + 2ab + b^2)$$
$$= \frac{1}{12}(4a^2 + 4ab + 4b^2 - 3a^2 - 6ab - 3b^2)$$
$$= \frac{1}{12}(b-a)^2.$$

We can get these moments also from the m.g.f., which is

$$M(t) = \frac{1}{t(b-a)}(e^{tb} - e^{ta}), \quad -\infty < t < \infty.$$

Moreover, for values of t close to 0

$$M(t) = 1 + \frac{1}{2}t(b+a) + \frac{1}{6}t^2(b^2 + ab + a^2) + \cdots.$$

2.4.2 The Normal and Log-Normal Distributions

2.4.2.1 The Normal Distribution

The normal or Gaussian distribution denoted by $N(\mu, \sigma)$ occupies a central role in statistical theory. Its density function (p.d.f.) is given by the formula

$$n(x; \mu, \sigma) = \frac{1}{\sigma\sqrt{2\pi}} \exp\left\{ -\frac{1}{2\sigma^2}(x - \mu)^2 \right\}. \qquad (2.51)$$

This p.d.f. is symmetric around the location parameter, μ. σ is a scale parameter. The m.g.f. of $N(0, 1)$ is

$$
\begin{aligned}
M(t) &= \frac{1}{\sqrt{2\pi}} e^{tx - \frac{1}{2}x^2} \, dx \\
&= \frac{e^{t^2/2}}{\sqrt{2\pi}} \int_{-\infty}^{\infty} e^{-\frac{1}{2}(x^2 - 2tx + t^2)} \, dx \qquad (2.52) \\
&= e^{t^2/2}.
\end{aligned}
$$

Indeed, $\frac{1}{\sqrt{2\pi}} \exp\left\{-\frac{1}{2}(x - t)^2\right\}$ is the p.d.f. of $N(t, 1)$. Furthermore,

$$
\begin{aligned}
M'(t) &= t M(t) \\
M''(t) &= t^2 M(t) + M(t) = (1 + t^2) M(t) \\
M'''(t) &= (t + t^3) M(t) + 2t M(t) \\
&= (3t + t^3) M(t) \\
M^{(4)}(t) &= (3 + 6t^2 + t^4) M(t).
\end{aligned}
$$

Thus, by substituting $t = 0$ we obtain that

$$
\begin{aligned}
E\{N(0, 1)\} &= 0, \\
V\{N(0, 1)\} &= 1, \\
\mu_3^* &= 0, \qquad (2.53) \\
\mu_4^* &= 3.
\end{aligned}
$$

To obtain the moments in the general case of $N(\mu, \sigma^2)$, we write $X = \mu + \sigma N(0, 1)$. Then

$$
\begin{aligned}
E\{X\} &= E\{\mu + \sigma N(0, 1)\} \\
&= \mu + \sigma E\{N(0, 1)\} = \mu \\
V\{X\} &= E\{(X - \mu)^2\} = \sigma^2 E\{N^2(0, 1)\} = \sigma^2 \\
\mu_3^* &= E\{(X - \mu)^3\} = \sigma^3 E\{N^3(0, 1)\} = 0 \\
\mu_4^* &= E\{(X - \mu)^4\} = \sigma^4 E\{N^4(0, 1)\} = 3\sigma^4.
\end{aligned}
$$

Thus, the index of kurtosis in the normal case is $\beta_4 = 3$.

The graph of the p.d.f. $n(x; \mu, \sigma)$ is a symmetric bell-shaped curve that is centered at μ (shown in Fig. 2.9). The spread of the density is determined by the

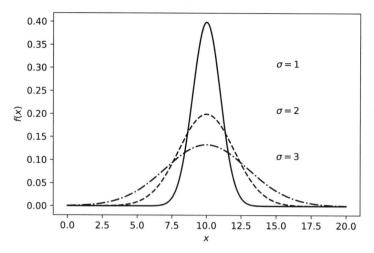

Fig. 2.9 The p.d.f. of $N(\mu, \sigma)$, $\mu = 10$, $\sigma = 1, 2, 3$

variance σ^2 in the sense that most of the area under the curve (in fact, 99.7% of the area) lies between $\mu - 3\sigma$ and $\mu + 3\sigma$. Thus, if X has a normal distribution with mean $\mu = 25$ and standard deviation $\sigma = 2$, the probability is 0.997 that the observed value of X will fall between 19 and 31.

Areas (i.e., probabilities) under the normal p.d.f. are found in practice using a table or appropriate statistical software like `scipy`. Since it is not practical to have a table for each pair of parameters μ and σ, we use the standardized form of the normal random variable. A random variable Z is said to have a **standard normal distribution** if it has a normal distribution with mean zero and variance one. The standard normal density function is $\phi(x) = n(x; 0, 1)$ and the standard cumulative distribution function is denoted by $\Phi(x)$. This function is also called the **standard normal** integral, i.e.,

$$\Phi(x) = \int_{-\infty}^{x} \phi(t) \, dt = \int_{-\infty}^{x} \frac{1}{\sqrt{2\pi}} e^{-\frac{1}{2}t^2} \, dt. \tag{2.54}$$

The c.d.f., $\Phi(x)$, represents the area over the x-axis under the standard normal p.d.f. to the left of the value x (see Fig. 2.10).

If we wish to determine the probability that a standard normal random variable is less than 1.5, for example, we use the following Python code

```
stats.norm(loc=0, scale=1).cdf(1.5)
```

We find that $\Pr\{Z \le 1.5\} = \Phi(1.5) = 0.9332$. To obtain the probability that Z lies between 0.5 and 1.5, we first find the probability that Z is less than 1.5 and then subtract from this number the probability that Z is less than 0.5. This yields

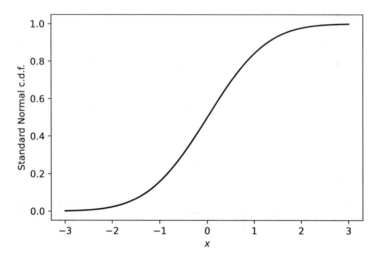

Fig. 2.10 Standard normal c.d.f.

$$\Pr\{0.5 < Z < 1.5\} = \Pr\{Z < 1.5\} - \Pr\{Z < 0.5\}$$
$$= \Phi(1.5) - \Phi(0.5) = 0.9332 - 0.6915 = 0.2417.$$

Many tables of the normal distribution do not list values of $\Phi(x)$ for $x < 0$. This is because the normal density is symmetric about $x = 0$, and we have the relation (Fig. 2.11)

$$\Phi(-x) = 1 - \Phi(x), \quad \text{for all } x. \tag{2.55}$$

Thus, to compute the probability that Z is less than -1, for example, we write

$$\Pr\{Z < -1\} = \Phi(-1) = 1 - \Phi(1) = 1 - 0.8413 = 0.1587.$$

The **p-th quantile (percentile of quantile)** of the standard normal distribution is the number z_p that satisfies the statement

$$\Phi(z_p) = \Pr\{Z \le z_p\} = p. \tag{2.56}$$

If X has a normal distribution with mean μ and standard deviation σ we denote the p-th quantile of the distribution by x_p. We can show that x_p is related to the standard normal quantile by

$$x_p = \mu + z_p \sigma.$$

The p-th quantile of the normal distribution can be obtained by using `scipy.stats`.

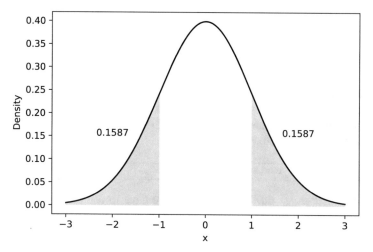

Fig. 2.11 The symmetry of the normal distribution

```
stats.norm(loc=0, scale=1).ppf(0.95)
```

In this command we used $p = 0.95$. The printed result is $z_{.95} = 1.6449$. We can use any value of μ (loc) and σ (scale) in the command. Thus, for $\mu = 10$ and $\sigma = 1.5$

```
stats.norm(loc=10, scale=1.5).ppf(0.95)
```

$$x_{.95} = 10 + z_{.95} \times \sigma = 12.4673.$$

Now suppose that X is a random variable having a normal distribution with mean μ and variance σ^2, that is, X has a $N(\mu, \sigma)$ distribution. We define the **standardized form** of X as

$$Z = \frac{X - \mu}{\sigma}.$$

By subtracting the mean from X and then dividing by the standard deviation, we transform X to a standard normal random variable (i.e., Z has expected value zero and standard deviation one). This will allow us to use the standard normal table to compute probabilities involving X. Thus, to compute the probability that X is less than a, we write

$$\Pr\{X \leq a\} = \Pr\left\{\frac{X - \mu}{\sigma} < \frac{a - \mu}{\sigma}\right\}$$

$$= \Pr \left\{ Z < \frac{a - \mu}{\sigma} \right\} = \Phi \left(\frac{a - \mu}{\sigma} \right).$$

Example 2.26 Let X represent the length (with cap) of a randomly selected aluminum pin. Suppose we know that X has a normal distribution with mean $\mu = 60.02$ and standard deviation $\sigma = 0.048$ [mm]. What is the probability that the length with cap of a randomly selected pin will be less than 60.1 [mm]? The corresponding scipy command is

```
stats.norm(loc=60.02, scale=0.048).cdf(60.1)
```

and we obtain $\Pr\{X \leq 60.1\} = 0.9522$. If we have to use the table of $\Phi(Z)$ we write

$$\Pr\{X \leq 60.1\} = \Phi \left(\frac{60.1 - 60.02}{0.048} \right)$$

$$= \Phi(1.667) = 0.9522.$$

Continuing with the example, consider the following question: If a pin is considered "acceptable" when its length is between 59.9 and 60.1 mm, what proportion of pins is expected to be rejected? To answer this question, we first compute the probability of accepting a single pin. This is the probability that X lies between 59.9 and 60.1, i.e.,

$$\Pr\{50.9 < X < 60.1\} = \Phi \left(\frac{60.1 - 60.02}{0.048} \right) - \Phi \left(\frac{59.9 - 60.02}{0.048} \right)$$

$$= \Phi(1.667) - \Phi(-2.5)$$

$$= 0.9522 - 0.0062 = 0.946.$$

Thus, we expect that 94.6% of the pins will be accepted and that 5.4% of them will be rejected. ∎

2.4.2.2　The Log-Normal Distribution

A random variable X is said to have a **log-normal distribution**, $LN(\mu, \sigma^2)$, if $Y = \log X$ has the normal distribution $N(\mu, \sigma^2)$.

The log-normal distribution has been applied for modeling distributions of strength variables, like the tensile strength of fibers (see Chap. 1), the compressive strength of concrete cubes, etc. It has also been used for random quantities of pollutants in water or air and other phenomena with skewed distributions.

The p.d.f. of $LN(\mu, \sigma)$ is given by the formula

$$f(x; \mu, \sigma^2) = \begin{cases} \dfrac{1}{\sqrt{2\pi}\sigma x} \exp\left\{-\dfrac{1}{2\sigma^2}(\log x - \mu)^2\right\}, & 0 < x < \infty \\ 0, & x \le 0. \end{cases} \tag{2.57}$$

The c.d.f. is expressed in terms of the standard normal integral as

$$F(x) = \begin{cases} 0, & x \le 0 \\ \Phi\left(\dfrac{\log x - \mu}{\sigma}\right), & 0 < x < \infty. \end{cases} \tag{2.58}$$

The expected value and variance of $LN(\mu, \sigma)$ are

$$E\{X\} = e^{\mu + \sigma^2/2}$$

and   (2.59)

$$V\{X\} = e^{2\mu + \sigma^2}(e^{\sigma^2} - 1).$$

One can show that the third central moment of $LN(\mu, \sigma^2)$ is

$$\mu_3^* = e^{3\mu + \frac{3}{2}\sigma^2}\left(e^{3\sigma^2} - 3e^{\sigma^2} + 2\right).$$

Hence, the **index of skewness** of this distribution is

$$\beta_3 = \frac{\mu_3^*}{\sigma^3} = \frac{e^{3\sigma^2} - 3e^{\sigma^2} + 2}{(e^{\sigma^2} - 1)^{3/2}}. \tag{2.60}$$

It is interesting that the index of skewness does not depend on μ and is positive for all $\sigma^2 > 0$. This index of skewness grows very fast as σ^2 increases. This is shown in Fig. 2.12.

2.4.3 The Exponential Distribution

We designate this distribution by $E(\beta)$. The p.d.f. of $E(\beta)$ is given by the formula

$$f(x; \beta) = \begin{cases} 0, & \text{if } x < 0 \\ (1/\beta)e^{-x/\beta}, & \text{if } x \ge 0, \end{cases} \tag{2.61}$$

where β is a positive parameter, i.e., $0 < \beta < \infty$. In Fig. 2.13 we present these p.d.f.s for various values of β.

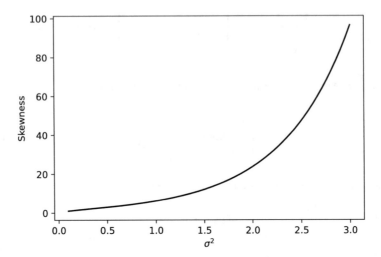

Fig. 2.12 The index of skewness of $LN(\mu, \sigma)$

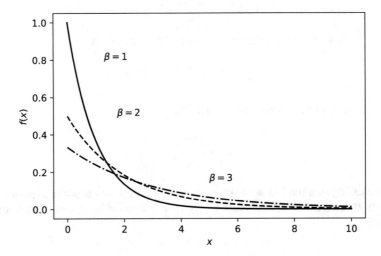

Fig. 2.13 The p.d.f. of $E(\beta)$, $\beta = 1, 2, 3$

The corresponding c.d.f. is

$$F(x; \beta) = \begin{cases} 0, & \text{if } x < 0 \\ 1 - e^{-x/\beta}, & \text{if } x \geq 0. \end{cases} \tag{2.62}$$

The expected value and the variance of $E(\beta)$ are

$$\mu = \beta,$$

and

$$\sigma^2 = \beta^2.$$

Indeed,

$$\mu = \frac{1}{\beta} \int_0^\infty x e^{-x/\beta} \, dx.$$

Making the change of variable to $y = x/\beta$, $dx = \beta \, dy$, we obtain

$$\mu = \beta \int_0^\infty y e^{-y} \, dy$$

$$= \beta.$$

Similarly

$$\mu_2 = \frac{1}{\beta} \int_0^\infty x^2 e^{-x/\beta} \, dx = \beta^2 \int_0^\infty y^2 e^{-y} \, dy$$

$$= 2\beta^2.$$

Hence,

$$\sigma^2 = \beta^2.$$

The p-th quantile is $x_p = -\beta \ln(1 - p)$.

The exponential distribution is related to the Poisson model in the following way: If the number of events occurring in a period of time follows a Poisson distribution with rate λ, then the time between occurrences of events has an exponential distribution with parameter $\beta = 1/\lambda$. The exponential model can also be used to describe the lifetime (i.e., time to failure) of certain electronic systems. For example, if the mean life of a system is 200 h, then the probability that it will work at least 300 h without failure is

$$\Pr\{X \geq 300\} = 1 - \Pr\{X < 300\}$$

$$= 1 - F(300) = 1 - (1 - e^{-300/200}) = 0.223.$$

The exponential distribution is positively skewed, and its index of skewness is

$$\beta_3 = \frac{\mu_3^*}{\sigma^3} = 2,$$

irrespective of the value of β. We have seen before that the kurtosis index is $\beta_4 = 9$.

2.4.4   The Gamma and Weibull Distributions

Two important distributions for studying the reliability and failure rates of systems are the gamma and the Weibull distributions. We will need these distributions in our study of reliability methods (Chapter 9 in the Industrial Statistics book). These distributions are discussed here as further examples of continuous distributions.

Suppose we use in a manufacturing process a machine which mass produces a particular part. In a random manner, it produces defective parts at a rate of λ per hour. The number of defective parts produced by this machine in a time period $[0, t]$ is a random variable $X(t)$ having a Poisson distribution with mean λt, i.e.,

$$\Pr\{X(t) = j\} = (\lambda t)^j e^{-(\lambda t)}/j!, \quad j = 0, 1, 2, \cdots . \tag{2.63}$$

Suppose we wish to study the distribution of the time until the k-th defective part is produced. Call this continuous random variable Y_k. We use the fact that the k-th defect will occur before time t (i.e., $Y_k \leq t$) if and only if at least k defects occur up to time t (i.e., $X(t) \geq k$). Thus the c.d.f. for Y_k is

$$\begin{aligned} G(t; k, \lambda) &= \Pr\{Y_k \leq t\} \\ &= \Pr\{X(t) \geq k\} \\ &= 1 - \sum_{j=0}^{k-1} (\lambda t)^j e^{-\lambda t}/j!. \end{aligned} \tag{2.64}$$

The corresponding p.d.f. for Y_k is

$$g(t; k, \lambda) = \frac{\lambda^k}{(k-1)!} t^{k-1} e^{-\lambda t}, \quad \text{for } t \geq 0. \tag{2.65}$$

This p.d.f. is a member of a general family of distributions which depend on two parameters, ν and β, and are called the **gamma** distributions $G(\nu, \beta)$. The p.d.f. of a gamma distribution $G(\nu, \beta)$ is

$$g(x; \nu, \beta) = \begin{cases} \dfrac{1}{\beta^\nu \Gamma(\nu)} x^{\nu-1} e^{-x/\beta}, & x \geq 0, \\ 0, & x < 0 \end{cases} \tag{2.66}$$

The `scipy` function gamma computes c.d.f of a gamma distribution having $\nu =$ shape (a) and $\beta =$ scale

```
stats.gamma(a=1, scale=1).cdf(1)
```

where $0 < \nu, \beta < \infty$, $\Gamma(\nu)$ is called the **gamma function** of ν and is defined as the integral

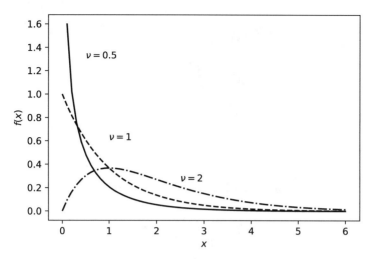

Fig. 2.14 The gamma densities, with $\beta = 1$ and $\nu = 0.5, 1, 2$

$$\Gamma(\nu) = \int_0^\infty x^{\nu-1} e^{-x} \, dx, \quad \nu > 0. \tag{2.67}$$

The gamma function satisfies the relationship

$$\Gamma(\nu) = (\nu - 1)\Gamma(\nu - 1), \quad \text{for all } \nu > 1. \tag{2.68}$$

Hence, for every positive integer k, $\Gamma(k) = (k - 1)!$. Also, $\Gamma\left(\frac{1}{2}\right) = \sqrt{\pi}$. We note also that the exponential distribution, $E(\beta)$, is a special case of the gamma distribution with $\nu = 1$. Some gamma p.d.f.s are presented in Fig. 2.14. The value of $\Gamma(\nu)$ can be computed in Python by the following commands which compute $\Gamma(5)$. Generally, replace 5 in line 2 by ν.

```
from scipy.special import gamma
gamma(5)
```

The expected value and variance of the gamma distribution $G(\nu, \beta)$ are, respectively,

$$\mu = \nu\beta,$$

and $\tag{2.69}$

$$\sigma^2 = \nu\beta^2.$$

To verify these formulae, we write

$$\mu = \frac{1}{\beta^{\nu} \Gamma(\nu)} \int_0^{\infty} x \, x^{\nu-1} e^{-x/\beta} \, dx$$

$$= \frac{\beta^{\nu+1}}{\beta^{\nu} \Gamma(\nu)} \int_0^{\infty} y^{\nu} e^{-y} \, dy$$

$$= \beta \frac{\Gamma(\nu + 1)}{\Gamma(\nu)} = \nu\beta.$$

Similarly,

$$\mu_2 = \frac{1}{\beta^{\nu} \Gamma(\nu)} \int_0^{\infty} x^2 \, x^{\nu-1} e^{-x/\beta} \, dx$$

$$= \frac{\beta^{\nu+2}}{\beta^{\nu} \Gamma(\nu)} \int_0^{\infty} y^{\nu+1} e^{-y} \, dy$$

$$= \beta^2 \frac{\Gamma(\nu + 2)}{\Gamma(\nu)} = (\nu + 1)\nu\beta^2.$$

Hence,

$$\sigma^2 = \mu_2 - \mu_1^2 = \nu\beta^2.$$

An alternative way is to differentiate the m.g.f.

$$M(t) = (1 - t\beta)^{-\nu}, \quad t < \frac{1}{\beta}. \tag{2.70}$$

Weibull distributions are often used in reliability models in which the system either "ages" with time or becomes "younger" (see Chapter 9, Industrial Statistics book). The Weibull family of distributions will be denoted by $W(\alpha, \beta)$. The parameters α and β, $\alpha, \beta > 0$ are called the shape and the scale parameters, respectively. The p.d.f. of $W(\alpha, \beta)$ is given by

$$w(t; \alpha, \beta) = \begin{cases} \dfrac{\alpha t^{\alpha-1}}{\beta^{\alpha}} e^{-(t/\beta)^{\alpha}}, & t \geq 0, \\ 0, & t < 0. \end{cases} \tag{2.71}$$

The corresponding c.d.f. is

$$W(t; \alpha, \beta) = \begin{cases} 1 - e^{-(t/\beta)^{\alpha}}, & t \geq 0 \\ 0, & t < 0. \end{cases} \tag{2.72}$$

Notice that $W(1, \beta) = E(\beta)$. The mean and variance of this distribution are

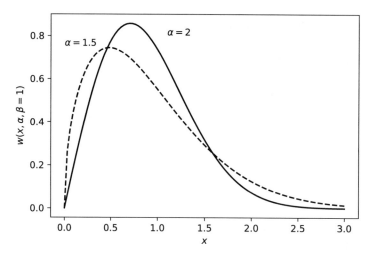

Fig. 2.15 Weibull density functions, $\alpha = 1.5, 2$

$$\mu = \beta \cdot \Gamma \left(1 + \frac{1}{\alpha} \right) \tag{2.73}$$

and

$$\sigma^2 = \beta^2 \left\{ \Gamma \left(1 + \frac{2}{\alpha} \right) - \Gamma^2 \left(1 + \frac{1}{\alpha} \right) \right\} \tag{2.74}$$

respectively. The values of $\Gamma(1 + (1/\alpha))$ and $\Gamma(1 + (2/\alpha))$ can be computed in Python. If, for example, $\alpha = 2$, then

$$\mu = \beta \sqrt{\pi}/2 = 0.8862\beta$$
$$\sigma^2 = \beta^2 (1 - \pi/4) = 0.2145\beta^2,$$

since

$$\Gamma \left(1 + \frac{1}{2} \right) = \frac{1}{2} \cdot \Gamma \left(\frac{1}{2} \right) = \frac{1}{2} \sqrt{\pi},$$

and

$$\Gamma \left(1 + \frac{2}{2} \right) = \Gamma(2) = 1.$$

In Fig. 2.15 we present three p.d.f. of $W(\alpha, \beta)$ for $\alpha = 1.5, 2.0$ and $\beta = 1$.

2.4.5 The Beta Distributions

Distributions having p.d.f. of the form

$$
f(x; \nu_1, \nu_2) = \begin{cases} \frac{1}{B(\nu_1,\nu_2)} x^{\nu_1-1}(1-x)^{\nu_2-1}, & 0 < x < 1, \\ 0, & \text{otherwise} \end{cases} \tag{2.75}
$$

where, for ν_1, ν_2 positive,

$$
B(\nu_1, \nu_2) = \int_0^1 x^{\nu_1-1}(1-x)^{\nu_2-1} \, dx \tag{2.76}
$$

are called beta distributions. The function $B(\nu_1, \nu_2)$ is called the beta integral. One can prove that

$$
B(\nu_1, \nu_2) = \frac{\Gamma(\nu_1)\Gamma(\nu_2)}{\Gamma(\nu_1 + \nu_2)}. \tag{2.77}
$$

The parameters ν_1 and ν_2 are shape parameters. Notice that when $\nu_1 = 1$ and $\nu_2 = 1$, the beta distribution reduces to $U(0, 1)$. We designate distributions of this family by Beta(ν_1, ν_2). The c.d.f. of Beta(ν_1, ν_2) is denoted also by $I_x(\nu_1, \nu_2)$, which is known as the **incomplete beta function ratio**, i.e.,

$$
I_x(\nu_1, \nu_2) = \frac{1}{B(\nu_1, \nu_2)} \int_0^x u^{\nu_1-1}(1-u)^{\nu_2-1} \, du, \tag{2.78}
$$

for $0 \leq x \leq 1$. Notice that $I_x(\nu_1, \nu_2) = 1 - I_{1-x}(\nu_2, \nu_1)$. The density functions of the p.d.f. Beta$(2.5, 5.0)$ and Beta$(2.5, 2.5)$ are plotted in Fig. 2.16. Notice that if $\nu_1 = \nu_2$ then the p.d.f. is symmetric around $\mu = \frac{1}{2}$. There is no simple formula for the m.g.f. of Beta(ν_1, ν_2). However, the m-th moment is equal to

$$
\begin{aligned}
\mu_m &= \frac{1}{B(\nu_1, \nu_2)} \int_0^1 u^{m+\nu_1-1}(1-u)^{\nu_2-1} \, du \\
&= \frac{B(\nu_1 + m, \nu_2)}{B(\nu_1, \nu_2)} \\
&= \frac{\nu_1(\nu_1 + 1) \cdots (\nu_1 + m - 1)}{(\nu_1 + \nu_2)(\nu_1 + \nu_2 + 1) \cdots (\nu_1 + \nu_2 + m - 1)}.
\end{aligned} \tag{2.79}
$$

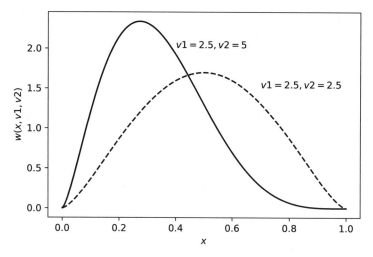

Fig. 2.16 Beta densities, $\nu_1 = 2.5, \nu_2 = 2.5$; $\nu_1 = 2.5, \nu_2 = 5.0$

Hence,

$$E\{\text{Beta}(\nu_1, \nu_2)\} = \frac{\nu_1}{\nu_1 + \nu_2}$$

$$V\{\text{Beta}(\nu_1, \nu_2)\} = \frac{\nu_1 \nu_2}{(\nu_1 + \nu_2)^2(\nu_1 + \nu_2 + 1)}.$$

(2.80)

The beta distribution has an important role in the theory of statistics. As will be seen later, many methods of statistical inference are based on the order statistics (see Sect. 2.7). The distribution of the order statistics is related to the beta distribution. Moreover, since the beta distribution can get a variety of shapes, it has been applied in many cases in which the variable has a distribution on a finite domain. By introducing a location and a scale parameter, one can fit a shifted-scaled beta distribution to various frequency distributions.

2.5 Joint, Marginal, and Conditional Distributions

2.5.1 Joint and Marginal Distributions

Let $X_1, \ldots, X_k$ be random variables which are jointly observed at the same experiments. In Sect. 2.6 we present various examples of bivariate and multivariate frequency distributions. In the present section, we present only the fundamentals of the theory, mainly for future reference. We make the presentation here, focusing on continuous random variables. The theory holds generally for discrete or for mixture of continuous and discrete random variables.

A function $F(x_1, \ldots, x_k)$ is called the joint c.d.f. of $X_1, \ldots, X_k$ if

$$F(x_1, \ldots, x_k) = \Pr\{X_1 \le x_1, \ldots, X_k \le x_k\} \tag{2.81}$$

for all $(x_1, \ldots, x_k) \in \mathbb{R}^k$ (the Euclidean k-space). By letting one or more variables tend to infinity, we obtain the joint c.d.f. of the remaining variables. For example,

$$\begin{aligned} F(x_1, \infty) &= \Pr\{X_1 \le x_1, X_2 \le \infty\} \\ &= \Pr\{X_1 \le x_1\} = F_1(x_1). \end{aligned} \tag{2.82}$$

The c.d.f.s of the individual variables are called the **marginal** distributions. $F_1(x_1)$ is the marginal c.d.f. of X_1.

A non-negative function $f(x_1, \cdots, x_k)$ is called the **joint p.d.f.** of $X_1, \cdots, X_k$, if

(i)

$$f(x_1, \cdots, x_k) \ge 0 \text{ for all } (x_1, \cdots, x_k), \text{ where } -\infty < x_i < \infty \ (i = 1, \cdots, k)$$

(ii)

$$\int_{-\infty}^{\infty} \cdots \int_{-\infty}^{\infty} f(x_1, \ldots, x_k) \, dx_1 \cdots dx_k = 1.$$

and

(iii)

$$F(x_1, \ldots, x_k) = \int_{-\infty}^{x_1} \cdots \int_{-\infty}^{x_k} f(y_1, \ldots, y_k) \, dy_1 \cdots dy_k.$$

The **marginal p.d.f.** of X_i $(i = 1, \cdots, k)$ can be obtained from the joint p.d.f. $f(x_1, \ldots, x_k)$, by integrating the joint p.d.f. with respect to all x_j, $j \ne i$. For example, if $k = 2$, $f(x_1, x_2)$ is the joint p.d.f. of X_1, X_2. The marginal p.d.f. of X_1 is

$$f_1(x_1) = \int_{-\infty}^{\infty} f(x_1, x_2) \, dx_2.$$

Similarly, the marginal p.d.f. of X_2 is

$$f_2(x_2) = \int_{-\infty}^{\infty} f(x_1, x_2) \, dx_1.$$

Indeed, the marginal c.d.f. of X_i is

$$F(x_1) = \int_{-\infty}^{x_1} \int_{-\infty}^{\infty} f(y_1, y_2)\, dy_1\, dy_2.$$

Differentiating $F(x_1)$ with respect to x_1, we obtain the marginal p.d.f. of X_1, i.e.,

$$f(x_1) = \frac{d}{dx_1} \int_{-\infty}^{x_1} \int_{-\infty}^{\infty} f(y_1, y_2)\, dy_1\, dy_2$$

$$= \int_{-\infty}^{\infty} f(x_1, y_2)\, dy_2.$$

If $k = 3$, we can obtain the marginal joint p.d.f. of a pair of random variables by integrating with respect to the third variable. For example, the joint marginal p.d.f. of (X_1, X_2) can be obtained from that of (X_1, X_2, X_3) as

$$f_{1,2}(x_1, x_2) = \int_{-\infty}^{\infty} f(x_1, x_2, x_3)\, dx_3.$$

Similarly,

$$f_{1,3}(x_1, x_3) = \int_{-\infty}^{\infty} f(x_1, x_2, x_3)\, dx_2,$$

and

$$f_{2,3}(x_2, x_3) = \int_{-\infty}^{\infty} f(x_1, x_2, x_3)\, dx_1.$$

Example 2.27 The present example is theoretical and is designed to illustrate the above concepts.

Let (X, Y) be a pair of random variables having a joint uniform distribution on the region

$$T = \{(x, y) : 0 \le x, y,\ x + y \le 1\}.$$

T is a triangle in the (x, y)-plane with vertices at $(0, 0)$, $(1, 0)$, and $(0, 1)$. According to the assumption of uniform distribution, the joint p.d.f. of (X, Y) is

$$f(x, y) = \begin{cases} 2, & \text{if } (x, y) \in T \\ 0, & \text{otherwise.} \end{cases}$$

The marginal p.d.f. of X is obtained as

$$f_1(x) = 2 \int_0^{1-x} dy = 2(1-x), \quad 0 \le x \le 1.$$

Obviously, $f_1(x) = 0$ for x outside the interval $[0, 1]$. Similarly, the marginal p.d.f. of Y is

$$f_2(y) = \begin{cases} 2(1-y), & 0 \le y \le 1 \\ 0, & \text{otherwise.} \end{cases}$$

Both X and Y have the same marginal Beta$(1, 2)$ distribution. Thus,

$$E\{X\} = E\{Y\} = \frac{1}{3}$$

and

$$V\{X\} = V\{Y\} = \frac{1}{18}.$$

■

2.5.2 Covariance and Correlation

Given any two random variables (X_1, X_2) having a joint distribution with p.d.f. $f(x_1, x_2)$, the **covariance** of X_1 and X_2 is defined as

$$\text{Cov}(X_1, X_2) = \int_{-\infty}^{\infty} \int_{-\infty}^{\infty} (x_1 - \mu_1)(x_2 - \mu_2) f(x_1, x_2) \, dx_1 \, dx_2, \tag{2.83}$$

where

$$\mu_i = \int_{-\infty}^{\infty} x f_i(x) \, dx, \quad i = 1, 2,$$

is the expected value of X_i. Notice that

$$\begin{aligned} \text{Cov}(X_1, X_2) &= E\{(X_1 - \mu_1)(X_2 - \mu_2)\} \\ &= E\{X_1 X_2\} - \mu_1 \mu_2. \end{aligned}$$

The **correlation** between X_1 and X_2 is defined as

$$\rho_{12} = \frac{\text{Cov}(X_1, X_2)}{\sigma_1 \sigma_2}, \tag{2.84}$$

where σ_i $(i = 1, 2)$ is the standard deviation of X_i.

Example 2.28 In continuation of the previous example, we compute $\text{Cov}(X, Y)$.

We have seen that $E\{X\} = E\{Y\} = \frac{1}{3}$. We compute now the expected value of their product

$$E\{XY\} = 2 \int_0^1 x \int_0^{1-x} y \, dy$$

$$= 2 \int_0^1 x \cdot \frac{1}{2}(1 - x)^2 \, dx$$

$$= B(2, 3) = \frac{\Gamma(2)\Gamma(3)}{\Gamma(5)} = \frac{1}{12}.$$

Hence,

$$\text{Cov}(X, Y) = E\{XY\} - \mu_1 \mu_2 = \frac{1}{12} - \frac{1}{9}$$

$$= -\frac{1}{36}.$$

Finally, the correlation between X, Y is

$$\rho_{XY} = -\frac{1/36}{1/18} = -\frac{1}{2}.$$

∎

The following are some properties of the covariance

(i)

$$|\text{Cov}(X_1, X_2)| \leq \sigma_1 \sigma_2,$$

where σ_1 and σ_2 are the standard deviations of X_1 and X_2, respectively.

(ii) If c is any constant, then,

$$\text{Cov}(X, c) = 0. \tag{2.85}$$

(iii) For any constants a_1 and a_2,

$$\text{Cov}(a_1 X_1, a_2 X_2) = a_1 a_2 \text{Cov}(X_1, X_2). \tag{2.86}$$

(iv) For any constants $a, b, c,$ and d,

$$\text{Cov}(a X_1 + b X_2, c X_3 + d X_4) = ac \, \text{Cov}(X_1, X_3) + ad \, \text{Cov}(X_1, X_4)$$

$$+ \, bc \, \text{Cov}(X_2, X_3) + bd \, \text{Cov}(X_2, X_4).$$

Property (iv) can be generalized to be

$$\text{Cov}\left(\sum_{i=1}^{m} a_i X_i, \sum_{j=1}^{n} b_j Y_j\right) = \sum_{i=1}^{m} \sum_{j=1}^{n} a_i b_j \, \text{Cov}(X_i, Y_j). \tag{2.87}$$

From property (i) above, we deduce that $-1 \leq \rho_{12} \leq 1$. The correlation obtains the values ± 1 only if the two variables are linearly dependent.

Definition of Independence

Random variables $X_1, \cdots, X_k$ are said to be **mutually independent** if, for every $(x_1, \cdots, x_k)$,

$$f(x_1, \cdots, x_k) = \prod_{i=1}^{k} f_i(x_i), \tag{2.88}$$

where $f_i(x_i)$ is the marginal p.d.f. of X_i. The variables X, Y of Example 2.28 are dependent, since $f(x, y) \neq f_1(x) f_2(y)$.

If two random variables are independent, then their correlation (or covariance) is zero. The converse is generally not true. Zero correlation **does not** imply independence.

We illustrate this in the following example.

Example 2.29 Let (X, Y) be discrete random variables having the following joint p.d.f.

$$p(x, y) = \begin{cases} \dfrac{1}{3}, & \text{if } X = -1, Y = 0 \text{ or } X = 0, Y = 0 \text{ or } X = 1, Y = 1 \\ 0, & \text{elsewhere.} \end{cases}$$

In this case the marginal p.d.f. are

$$p_1(x) = \begin{cases} \dfrac{1}{3}, & x = -1, 0, 1 \\ 0, & \text{otherwise} \end{cases}$$

$$p_2(y) = \begin{cases} \dfrac{1}{3}, & y = 0 \\ \dfrac{2}{3}, & y = 1. \end{cases}$$

$p(x, y) \neq p_1(x) p_2(y)$ if $X = 1, Y = 1$, for example. Thus, X and Y are dependent. On the other hand, $E\{X\} = 0$ and $E\{XY\} = 0$. Hence, $\text{cov}(X, Y) = 0$. ∎

The following result is very important for independent random variables.

If $X_1, X_2, \ldots, X_k$ **are mutually independent, then, for any integrable functions** $g_1(X_1), \ldots, g_k(X_k)$,

$$E\left\{\prod_{i=1}^{k} g_i(X_i)\right\} = \prod_{i=1}^{k} E\{g_i(X_i)\}. \tag{2.89}$$

Indeed,

$$E\left\{\prod_{i=1}^{k} g_i(X_i)\right\} = \int \cdots \int g_1(x_1) \cdots g_k(x_k) \cdot f(x_1, \ldots, x_k)\, dx_1 \cdots dx_k$$

$$= \int \cdots \int g_1(x_1) \cdots g_k(x_k) f_1(x_1) \ldots f_k(x_k)\, dx_1 \cdots dx_k$$

$$= \int g_1(x_1) f_1(x_1)\, dx_1 \cdot \int g_2(x_2) f_2(x_2)\, dx_2 \cdots \int g_k(x_k) f_k(x_k)\, dx_k$$

$$= \prod_{i=1}^{k} E\{g_i(X_i)\}.$$

2.5.3   Conditional Distributions

If (X_1, X_2) are two random variables having a joint p.d.f. $f(x_1, x_2)$ and marginals ones, $f_1(\cdot)$, and $f_2(\cdot)$, respectively, then the **conditional** p.d.f. of X_2, given $\{X_1 = x_1\}$, where $f_1(x_1) > 0$, is defined to be

$$f_{2\cdot 1}(x_2 \mid x_1) = \frac{f(x_1, x_2)}{f_1(x_1)}. \tag{2.90}$$

Notice that $f_{2\cdot 1}(x_2 \mid x_1)$ is a p.d.f. Indeed $f_{2\cdot 1}(x_2 \mid x_1) \geq 0$ for all x_2, and

$$\int_{-\infty}^{\infty} f_{2\cdot 1}(x_2 \mid x_1)\, dx_2 = \frac{\int_{-\infty}^{\infty} f(x_1, x_2)\, dx_2}{f_1(x_1)}$$

$$= \frac{f_1(x_1)}{f_1(x_1)} = 1.$$

The **conditional expectation** of X_2, given $\{X_1 = x_1\}$ such that $f_1(x_1) > 0$, is the expected value of X_2 with respect to the conditional p.d.f. $f_{2\cdot 1}(x_2 \mid x_1)$, i.e.,

$$E\{X_2 \mid X_1 = x_1\} = \int_{-\infty}^{\infty} x f_{2\cdot 1}(x \mid x_1)\, dx.$$

Similarly, we can define the **conditional variance** of X_2, given $\{X_1 = x_1\}$, as the variance of X_2, with respect to the conditional p.d.f. $f_{2 \cdot 1}(x_2 \mid x_1)$. If X_1 and X_2 are independent, then, by substituting $f(x_1, x_2) = f_1(x_1) f_2(x_s 2)$ we obtain

$$f_{2 \cdot 1}(x_2 \mid x_1) = f_2(x_2),$$

and

$$f_{1 \cdot 2}(x_1 \mid x_2) = f_1(x_1).$$

Example 2.30 Returning to Examples 2.27 and 2.28, we compute the conditional distribution of Y, given $\{X = x\}$, for $0 < x < 1$.

According to the above definition, the conditional p.d.f. of Y, given $\{X = x\}$, for $0 < x < 1$, is

$$f_{Y|X}(y \mid x) = \begin{cases} \dfrac{1}{1 - x}, & \text{if } 0 < y < (1 - x) \\ 0, & \text{otherwise.} \end{cases}$$

Notice that this is a uniform distribution over $(0, 1 - x)$, $0 < x < 1$. If $x \notin (0, 1)$, then the conditional p.d.f. does not exist. This is, however, an event of probability zero. From the above result, the conditional expectation of Y, given $\{X = x\}$, for $0 < x < 1$, is

$$E\{Y \mid X = x\} = \frac{1 - x}{2}.$$

The conditional variance is

$$V\{Y \mid X = x\} = \frac{(1 - x)^2}{12}.$$

In a similar fashion, we show that the conditional distribution of X, given $Y = y$, $0 < y < 1$, is uniform on $(0, 1 - y)$. ∎

One can immediately prove that if X_1 and X_2 are independent, then the conditional distribution of X_1 given $\{X_2 = x_2\}$, when $f_2(x_2) > 0$, is just the marginal distribution of X_1. Thus, X_1 and X_2 are independent if and only if

$$f_{2 \cdot 1}(x_2 \mid x_1) = f_2(x_2) \text{ for all } x_2$$

and

$$f_{1 \cdot 2}(x_1 \mid x_2) = f_1(x_1) \text{ for all } x_1,$$

provided that the conditional p.d.f. is well defined.

Notice that for a pair of random variables (X, Y), $E\{Y \mid X = x\}$ changes with x, as shown in Example 2.30, if X and Y are dependent. Thus, we can consider $E\{Y \mid X\}$ to be a random variable, which is a function of X. It is interesting to compute the expected value of this function of X, i.e.,

$$E\{E\{Y \mid X\}\} = \int E\{Y \mid X = x\} f_1(x) \, dx$$

$$= \int \left\{ \int y f_{Y \cdot X}(y \mid x) \, dy \right\} f_1(x) \, dx$$

$$= \int \int y \frac{f(x, y)}{f_1(x)} f_1(x) \, dy \, dx.$$

If we can interchange the order of integration (whenever $\int |y| f_2(y) \, dy < \infty$), then

$$E\{E\{Y \mid X\}\} = \int y \left\{ \int f(x, y) \, dx \right\} dy$$

$$= \int y f_2(y) \, dy \qquad (2.91)$$

$$= E\{Y\}.$$

This result, known as **the law of the iterated expectation**, is often very useful. An example of the use of the law of the iterated expectation is the following.

Example 2.31 Let (J, N) be a pair of random variables. The conditional distribution of J, given $\{N = n\}$, is the binomial $B(n, p)$. The marginal distribution of N is Poisson with mean λ. What is the expected value of J?

By the law of the iterated expectation,

$$E\{J\} = E\{E\{J \mid N\}\}$$

$$= E\{Np\} = p E\{N\} = p\lambda.$$

One can show that the marginal distribution of J is Poisson, with mean $p\lambda$.   ∎

Another important result relates variances and conditional variances, that is, if (X, Y) is a pair of random variables having finite variances, then

$$V\{Y\} = E\{V\{Y \mid X\}\} + V\{E\{Y \mid X\}\}. \qquad (2.92)$$

We call this relationship the **law of total variance**.

Example 2.32 Let (X, Y) be a pair of independent random variables having finite variances σ_X^2 and σ_Y^2 and expected values μ_X, μ_Y. Determine the variance of $W = XY$. By the law of total variance,

$$V\{W\} = E\{V\{W \mid X\}\} + V\{E\{W \mid X\}\}.$$

Since X and Y are independent

$$V\{W \mid X\} = V\{XY \mid X\} = X^2 V\{Y \mid X\}$$
$$= X^2 \sigma_Y^2.$$

Similarly,

$$E\{W \mid X\} = X \mu_Y.$$

Hence,

$$V\{W\} = \sigma_Y^2 E\{X^2\} + \mu_Y^2 \sigma_X^2$$
$$= \sigma_Y^2 (\sigma_X^2 + \mu_X^2) + \mu_Y^2 \sigma_X^2$$
$$= \sigma_X^2 \sigma_Y^2 + \mu_X^2 \sigma_Y^2 + \mu_Y^2 \sigma_X^2.$$

∎

2.6 Some Multivariate Distributions

2.6.1 The Multinomial Distribution

The multinomial distribution is a generalization of the binomial distribution to cases of n **independent** trials in which the results are classified to k possible categories (e.g., excellent, good, average, poor). The random variables $(J_1, J_2, \cdots, J_k)$ are the number of trials yielding results in each one of the k categories. These random variables are dependent, since $J_1 + J_2 + \cdots + J_k = n$. Furthermore, let $p_1, p_2, \cdots, p_k$; $p_i \geq 0$, $\sum_{i=1}^{k} p_i = 1$ be the probabilities of the k categories. The binomial distribution is the special case of $k = 2$. Since $J_k = n - (J_1 + \cdots + J_{k-1})$, the joint probability function is written as a function of $k - 1$ arguments, and its formula is

$$p(j_1, \cdots, j_{k-1}) = \binom{n}{j_1, \cdots, j_{k-1}} p_1^{j_1} \cdots p_{k-1}^{j_{k-1}} p_k^{j_k} \tag{2.93}$$

for $j_1, \cdots, j_{k-1} \geq 0$ such that $\sum_{i=1}^{k-1} j_i \leq n$. In this formula,

$$\binom{n}{j_1, \cdots, j_{k-1}} = \frac{n!}{j_1! j_2! \cdots j_k!}, \tag{2.94}$$

and $j_k = n - (j_1 + \cdots + j_{k-1})$. For example, if $n = 10$, $k = 3$, $p_1 = 0.3$, $p_2 = 0.4$, $p_3 = 0.3$,

$$p(5, 2) = \frac{10!}{5!2!3!}(0.3)^5(0.4)^2(0.3)^3$$

$$= 0.02645.$$

The marginal distribution of each one of the k variables is binomial, with parameters n and p_i ($i = 1, \cdots, k$). The joint marginal distribution of (J_1, J_2) is trinomial, with parameters n, p_1, p_2, and $(1 - p_1 - p_2)$. Finally, the conditional distribution of $(J_1, \cdots, J_r)$, $1 \leq r < k$, given $\{J_{r+1} = j_{r+1}, \cdots, J_k = j_k\}$, is $(r + 1)$-nomial, with parameters $n_r = n - (j_{r+1} + \cdots + j_k)$, and $p'_1, \cdots, p'_r, p'_{r+1}$, where

$$p'_i = \frac{p_i}{(1 - p_{r+1} - \cdots - p_k)}, \quad i = 1, \cdots, r$$

and

$$p'_{r+1} = 1 - \sum_{i=1}^{r} p'_i.$$

Finally, we can show that, for $i \neq j$,

$$\text{Cov}(J_i, J_j) = -np_i p_j. \tag{2.95}$$

Example 2.33 An insertion machine is designed to insert components into computer-printed circuit boards. Every component inserted on a board is scanned optically. An insertion is either error-free or its error is classified into the following two main categories: mis-insertion (broken lead, off pad, etc.) or wrong component. Thus, we have altogether three general categories. Let

$$J_1 = \text{\# of error free components};$$

$$J_2 = \text{\# of misinsertion};$$

$$J_3 = \text{\# of wrong components}.$$

The probabilities that an insertion belongs to one of these categories is $p_1 = 0.995$, $p_2 = 0.001$, and $p_2 = 0.004$.

The insertion rate of this machine is $n = 3500$ components per hour of operation. Thus, we expect during 1 hour of operation $n \times (p_2 + p_3) = 175$ insertion errors.

Given that there are 16 insertion errors during a particular hour of operation, the conditional distribution of the number of mis-insertions is binomial $B\left(16, \frac{0.01}{0.05}\right)$.

Thus,

$$E\{J_2 \mid J_2 + J_3 = 16\} = 16 \times 0.2 = 3.2.$$

On the other hand,

$$E\{J_2\} = 3500 \times 0.001 = 3.5.$$

We see that the information concerning the total number of insertion errors makes a difference.

Finally,

$$Cov(J_2, J_3) = -3500 \times 0.001 \times 0.004$$
$$= -0.014$$

$$V\{J_2\} = 3500 \times 0.001 \times 0.999 = 3.4965$$

and

$$V\{J_3\} = 3500 \times 0.004 \times 0.996 = 13.944.$$

Hence, the correlation between J_2 and J_3 is

$$\rho_{2,3} = \frac{-0.014}{\sqrt{3.4965 \times 13.944}} = -0.0020.$$

This correlation is quite small.   ∎

2.6.2   The Multi-Hypergeometric Distribution

Suppose that we draw from a population of size N a RSWOR of size n. Each one of the n units in the sample is classified to one of k categories. Let $J_1, J_2, \cdots, J_k$ be the number of sample units belonging to each one of these categories. $J_1 + \cdots + J_k = n$. The distribution of $J_1, \cdots, J_k$ is k-variate hypergeometric. If $M_1, \cdots, M_k$ are the number of units in the population in these categories, before the sample is drawn, then the joint p.d.f. of $J_1, \cdots, J_k$ is

$$p(j_1, \cdots, j_{k-1}) = \frac{\binom{M_1}{j_1}\binom{M_2}{j_2}\cdots\binom{M_k}{j_k}}{\binom{N}{n}}, \tag{2.96}$$

where $j_k = n - (j_1 + \cdots + j_{k-1})$. This distribution is a generalization of the hypergeometric distribution $H(N, M, n)$. The hypergeometric distribution $H(N, M_i, n)$

is the marginal distribution of J_i $(i = 1, \cdots, k)$. Thus,

$$E\{J_i\} = n\frac{M_i}{N}, \quad i = 1, \cdots, k$$

$$V\{J_i\} = n\frac{M_i}{N}\left(1 - \frac{M_i}{N}\right)\left(1 - \frac{n-1}{N-1}\right), \quad i = 1, \cdots, k \tag{2.97}$$

and for $i \neq j$

$$\text{Cov}(J_i, J_j) = -n\frac{M_i}{N} \cdot \frac{M_j}{N}\left(1 - \frac{n-1}{N-1}\right).$$

Example 2.34 A lot of 100 spark plugs contains 20 plugs from vendor V_1, 50 plugs from vendor V_2, and 30 plugs from vendor V_3.

A random sample of $n = 20$ plugs is drawn from the lot without replacement.

Let J_i be the number of plugs in the sample from the vendor V_i, $i = 1, 2, 3$. Accordingly,

$$\Pr\{J_1 = 5, J_2 = 10\} = \frac{\binom{20}{5}\binom{50}{10}\binom{30}{5}}{\binom{100}{20}}$$

$$= 0.00096.$$

If we are told that 5 out of the 20 plugs in the sample are from vendor V_3, then the conditional distribution of J_1 is

$$\Pr\{J_1 = j_1 \mid J_3 = 5\} = \frac{\binom{20}{j_1}\binom{50}{15-j_1}}{\binom{70}{15}}, \quad j_1 = 0, \cdots, 15.$$

Indeed, given $J_3 = 5$, then J_1 can assume only the values $0, 1, \cdots, 15$. The conditional probability that j_1 out of the 15 remaining plugs in the sample are from vendor V_1 is the same like that of choosing a RSWOR of size 15 from a lot of size $70 = 20 + 50$, with 20 plugs from vendor V_1. ∎

2.6.3 The Bivariate Normal Distribution

The bivariate normal distribution is the joint distribution of two continuous random variables (X, Y) having a joint p.d.f.

Fig. 2.17 Bivariate normal
p.d.f.

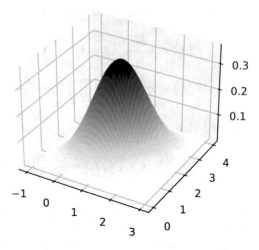

$$f(x, y; \mu, \eta, \sigma_X, \sigma_Y, \rho) = \frac{1}{2\pi \sigma_X \sigma_Y \sqrt{1 - \rho^2}} \cdot$$

$$\exp\left\{-\frac{1}{2(1 - \rho^2)}\left[\left(\frac{x - \mu}{\sigma_x}\right)^2 - 2\rho \frac{x - \mu}{\sigma_X} \cdot \frac{y - \eta}{\sigma_Y} + \left(\frac{y - \eta}{\sigma_Y}\right)^2\right]\right\}$$

$$-\infty < x, y < \infty.$$

$$(2.98)$$

$\mu, \eta, \sigma_X, \sigma_Y$, and ρ are parameters of this distribution.

Integration of y yields that the marginal distribution of X is $N(\mu, \sigma_x^2)$. Similarly, the marginal distribution of Y is $N(\eta, \sigma_Y^2)$. Furthermore, ρ is the correlation between X and Y. Notice that if $\rho = 0$, then the joint p.d.f. becomes the product of the two marginal ones, i.e.,

$$f(x, y); \mu, \eta, \sigma_X, \sigma_Y, 0) = \frac{1}{\sqrt{2\pi}\sigma_X} \exp\left\{-\frac{1}{2}\left(\frac{x - \mu}{\sigma_X}\right)^2\right\} \cdot$$

$$\frac{1}{\sqrt{2\pi}\sigma_Y} \exp\left\{-\frac{1}{2}\left(\frac{y - \eta}{\sigma_Y}\right)^2\right\}, \quad \text{for all} \ -\infty < x, y < \infty.$$

Hence, if $\rho = 0$, then X and Y are independent. On the other hand, if $\rho \neq 0$, then $f(x, y; \mu, \eta, \sigma_X, \sigma_Y, \rho) \neq f_1(x; \mu, \sigma_X) f_2(y; \eta, \sigma_Y)$, and the two random variables are dependent.

In Fig. 2.17 we present the bivariate p.d.f. for $\mu = \eta = 0$, $\sigma_X = \sigma_Y = 1$ and $\rho = 0.5$.

One can verify also that the conditional distribution of Y, given $\{X = x\}$, is normal with mean

$$\mu_{Y \cdot x} = \eta + \rho \frac{\sigma_Y}{\sigma_X}(x - \mu) \qquad (2.99)$$

and variance

$$\sigma_{Y \cdot x}^2 = \sigma_Y^2(1 - \rho^2). \qquad (2.100)$$

It is interesting to see that $\mu_{Y \cdot x}$ is a linear function of x. We can say that $\mu_{Y \cdot x} = E\{Y \mid X = x\}$ is, in the bivariate normal case, the theoretical (linear) regression of Y on X (see Chap. 4). Similarly,

$$\mu_{X \cdot y} = \mu + \rho \frac{\sigma_X}{\sigma_Y}(y - \eta),$$

and

$$\sigma_{X \cdot y}^2 = \sigma_X^2(1 - \rho^2).$$

If $\mu = \eta = 0$ and $\sigma_X = \sigma_Y = 1$, we have the **standard** bivariate normal distribution. The joint c.d.f. in the standard case is denoted by $\Phi_2(x, y; \rho)$ and its formula is

$$\Phi_2(x, y; \rho) = \frac{1}{2\pi\sqrt{1 - \rho^2}} \int_{-\infty}^{x} \int_{-\infty}^{y} \exp\left\{-\frac{1}{2(1 - \rho^2)}(z_1^2 - 2\rho z_1 z_2 + z^2)\right\} dz_1 \, dz_2$$

$$= \int_{-\infty}^{x} \phi(z_1) \Phi\left(\frac{y - \rho z_1}{\sqrt{1 - \rho^2}}\right) dz_1$$

$$(2.101)$$

Values of $\Phi_2(x, y; \rho)$ can be obtained by numerical integration. If one has to compute the bivariate c.d.f. in the general case, the following formula is useful:

$$F(x, y; \mu, \eta, \sigma_X, \sigma_Y, \rho) = \Phi_2\left(\frac{x - \mu}{\sigma_X}, \frac{y - \eta}{\sigma_Y}; \rho\right).$$

For computing $\Pr\{a \leq X \leq b, c \leq Y \leq d\}$ we use the formula

$$\Pr\{a \leq X \leq b, c \leq Y \leq d\} = F(b, d; -)$$
$$- F(a, d; -) - F(b, c; -) + F(a, c; -).$$

Example 2.35 Suppose that (X, Y) deviations in component placement on PCB by an automatic machine have a bivariate normal distribution with means $\mu = \eta = 0$, standard deviations $\sigma_X = 0.00075$ and $\sigma_Y = 0.00046$ [Inch], and $\rho = 0.160$. The placement errors are within the specifications if $|X| < 0.001$ [Inch] and $|Y| < 0.001$ [Inch]. What proportion of components are expected to have X, Y deviations compliant with the specifications? The standardized version of the spec limits is

$$Z_1 = \frac{0.001}{0.00075} = 1.33 \text{ and } Z_2 = \frac{0.001}{0.00046} = 2.174. \text{ We compute}$$

$$\Pr\{|X| < 0.001, |Y| < 0.001\} = \Phi_2(1.33, 2.174, .16) - \Phi_2(-1.33, 2.174, .16)$$
$$- \Phi_2(1.33, -2.174; .16) + \Phi_2(-1.33, -2.174; .16)$$
$$= 0.793.$$

This is the expected proportion of good placements.   ∎

2.7 Distribution of Order Statistics

As defined in Chap. 1, the order statistics of the sample are the sorted data. More specifically, let $X_1, \cdots, X_n$ be identically distributed independent (i.i.d.) random variables. The order statistics are $X_{(i)}, i = 1, \cdots, n$, where

$$X_{(1)} \leq X_{(2)} \leq \cdots \leq X_{(n)}.$$

In the present section, we discuss the distributions of these order statistics, when $F(x)$ is (absolutely) continuous, having a p.d.f. $f(x)$.

We start with the extremal statistics $X_{(1)}$ and $X_{(n)}$.

Since the random variables X_i ($i = 1, \cdots, n$) are i.i.d., the c.d.f. of $X_{(1)}$ is

$$F_{(1)}(x) = \Pr\{X_{(1)} \leq x\}$$

$$= 1 - \Pr\{X_{(1)} \geq x\} = 1 - \prod_{i=1}^{n} \Pr\{X_i \geq x\}$$

$$= 1 - (1 - F(x))^n.$$

By differentiation we obtain that the p.d.f. of $X_{(1)}$ is

$$f_{(1)}(x) = nf(x)[1 - F(x)]^{n-1}. \tag{2.102}$$

Similarly, the c.d.f. of the sample maximum $X_{(n)}$ is

$$F_{(n)}(x) = \prod_{i=1}^{n} \Pr\{X_i \leq x\}$$

$$= (F(x))^n.$$

The p.d.f. of $X_{(n)}$ is

$$f_{(n)}(x) = nf(x)(F(x))^{n-1}. \tag{2.103}$$

Fig. 2.18 Series and parallel
systems

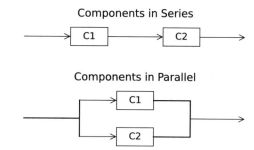

Example 2.36

(i) A switching circuit consists of n modules, which operate independently and
which are connected in **series** (see Fig. 2.18). Let X_i be the time till failure of
the i-th module. The system fails when any module fails. Thus, the time till
failure of the system is $X_{(1)}$. If all X_i are exponentially distributed with mean
life β, then the c.d.f. of $X_{(1)}$ is

$$F_{(1)}(x) = 1 - e^{-nx/\beta}, \quad x \ge 0.$$

Thus, $X_{(1)}$ is distributed like $E\left(\dfrac{\beta}{n}\right)$. It follows that the expected time till

failure of the circuit is $E\{X_{(1)}\} = \dfrac{\beta}{n}$.

(ii) If the modules are connected in parallel, then the circuit fails at the instant the
last of the n modules fail, which is $X_{(n)}$. Thus, if X_i is $E(\beta)$, the c.d.f. of $X_{(n)}$
is

$$F_{(n)}(x) = (1 - e^{-(x/\beta)})^n.$$

The expected value of $X_{(n)}$ is

$$E\{X_{(n)}\} = \frac{n}{\beta} \int_0^\infty xe^{-x/\beta}(1 - e^{-x/\beta})^{n-1}\, dx$$

$$= n\beta \int_0^\infty ye^{-y}(1 - e^{-y})^{n-1}\, dy$$

$$= n\beta \sum_{j=0}^{n-1}(-1)^j \binom{n-1}{j} \int_0^\infty ye^{-(1+j)y}\, dy$$

$$= n\beta \sum_{j=1}^{n}(-1)^{j-1} \binom{n-1}{j-1} \frac{1}{j^2}.$$

Furthermore, since $n\binom{n-1}{j-1} = j\binom{n}{j}$, we obtain that

$$E\{X_{(n)}\} = \beta \sum_{j=1}^{n} (-1)^{j-1} \binom{n}{j} \frac{1}{j}.$$

One can also show that this formula is equivalent to

$$E\{X_{(n)}\} = \beta \sum_{j=1}^{n} \frac{1}{j}.$$

Accordingly, if the parallel circuit consists of three modules, and the time till failure of each module is exponential with $\beta = 1000$ [hr], the expected time till failure of the system is 1833.3 [hr].   ∎

Generally, the distribution of $X_{(i)}$ ($i = 1, \cdots, n$) can be obtained by the following argument. The event $\{X_{(i)} \leq x\}$ is equivalent to the event that the number of X_i values in the random example which are smaller or equal to x is at least i.

Consider n independent and identical trials, in which "success" is that $\{X_i \leq x\}$ ($i = 1, \cdots, n$). The probability of "success" is $F(x)$. The distribution of the number of successes is $B(n, F(x))$. Thus, the c.d.f. of $X_{(i)}$ is

$$F_{(i)}(x) = \Pr\{X_{(i)} \leq x\} = 1 - B(i - 1; n, F(x))$$

$$= \sum_{j=i}^{n} \binom{n}{j} (F(x))^{j} (1 - F(x))^{n-j}.$$

Differentiating this c.d.f. with respect to x yields the p.d.f. of $X_{(i)}$, namely,

$$f_{(i)}(x) = \frac{n!}{(i-1)!(n-i)!} f(x)(F(x))^{i-1}(1 - F(x))^{n-i}. \tag{2.104}$$

Notice that if X has a uniform distribution on $(0, 1)$, then the distribution of $X_{(i)}$ is like that of Beta $(i, n - i + 1)$, $i = 1, \cdots, n$. In a similar manner, one can derive the joint p.d.f. of $(X_{(i)}, X_{(j)})$, $1 \leq i < j \leq n$, etc. This joint p.d.f. is given by

$$f_{(i),(j)}(x, y) = \frac{n!}{(i-1)!(j-1-i)!(n-j)!} f(x)f(y) \cdot$$
$$\cdot (F(x))^{i-1}[F(y) - F(x)]^{j-i-1}(1 - F(y))^{n-j}, \tag{2.105}$$

for $-\infty < x < y < \infty$.

2.8   Linear Combinations of Random Variables

Let $X_1, X_2, \cdots, X_n$ be random variables having a joint distribution, with joint p.d.f. $f(x_1, \cdots, x_n)$. Let $\alpha_1, \cdots, \alpha_n$ be given constants. Then

$$W = \sum_{i=1}^{n} \alpha_i X_i$$

is a linear combination of the Xs. The p.d.f. of W can generally be derived, using various methods. We discuss in the present section only the formulae of the expected value and variance of W.

It is straightforward to show that

$$E\{W\} = \sum_{i=1}^{n} \alpha_i E\{X_i\}. \tag{2.106}$$

That is, the expected value of a linear combination is the same linear combination of the expectations.

The formula for the variance is somewhat more complicated and is given by

$$V\{W\} = \sum_{i=1}^{n} \alpha_i^2 V\{X_i\} + \sum \sum_{i \neq j} \alpha_i \alpha_j \text{cov}(X_i, X_j). \tag{2.107}$$

Example 2.37 Let $X_1, X_2, \cdots, X_n$ be i.i.d. random variables, with common expectations μ and common finite variances σ^2. The sample mean $\bar{X}_n = \frac{1}{n} \sum_{i=1}^{n} X_i$ is a particular linear combination, with

$$\alpha_1 = \alpha_2 = \cdots = \alpha_n = \frac{1}{n}.$$

Hence,

$$E\{\bar{X}_n\} = \frac{1}{n} \sum_{i=1}^{n} E\{X_i\} = \mu$$

and, since $X_1, X_2, \cdots, X_n$ are mutually independent, $\text{cov}(X_i, X_j) = 0$, all $i \neq j$. Hence,

$$V\{\bar{X}_n\} = \frac{1}{n^2} \sum_{i=1}^{n} V\{X_i\} = \frac{\sigma^2}{n}.$$

Thus, we have shown that in a random sample of n i.i.d. random variables, the sample mean has the same expectation as that of the individual variables, but its sample variance is reduced by a factor of $1/n$.

Moreover, from Chebychev's inequality, for any $\epsilon > 0$

$$\Pr\{|\bar{X}_n - \mu| > \epsilon\} < \frac{\sigma^2}{n\epsilon^2}.$$

Therefore, since $\lim_{n \to \infty} \frac{\sigma^2}{n\epsilon^2} = 0$,

$$\lim_{n \to \infty} \Pr\{|\bar{X}_n - \mu| > \epsilon\} = 0.$$

This property is called the **convergence in probability** of $\bar{X}_n$ to μ. ∎

Example 2.38 Let U_1, U_2, U_3 be three i.i.d. random variables having uniform distributions on $(0, 1)$. We consider the statistic

$$W = \frac{1}{4}U_{(1)} + \frac{1}{2}U_{(2)} + \frac{1}{4}U_{(3)},$$

where $0 < U_{(1)} < U_{(2)} < U_{(3)} < 1$ are the order statistics. We have seen in Sect. 2.7 that the distribution of $U_{(i)}$ is like that of $\text{Beta}(i, n - i + 1)$. Hence,

$$E\{U_{(1)}\} = E\{\text{Beta}(1, 3)\} = \frac{1}{4}$$

$$E\{U_{(2)}\} = E\{\text{Beta}(2, 2)\} = \frac{1}{2}$$

$$E\{U_{(3)}\} = E\{\text{Beta}(3, 1)\} = \frac{3}{4}.$$

It follows that

$$E\{W\} = \frac{1}{4} \cdot \frac{1}{4} + \frac{1}{2} \cdot \frac{1}{2} + \frac{1}{4} \cdot \frac{3}{4} = \frac{1}{2}.$$

To find the variance of W, we need more derivations.
   First,

$$V\{U_{(1)}\} = V\{\text{Beta}(1, 3)\} = \frac{3}{4^2 \times 5} = \frac{3}{80}$$

$$V\{U_{(2)}\} = V\{\text{Beta}(2, 2)\} = \frac{4}{4^2 \times 5} = \frac{1}{20}$$

$$V\{U_{(3)}\} = V\{\text{Beta}(3, 1)\} = \frac{3}{4^2 \times 5} = \frac{3}{80}.$$

We need to find $\text{Cov}(U_{(1)}, U_{(2)})$, $\text{Cov}(U_{(1)}, U_{(3)})$, and $\text{Cov}(U_{(2)}, U_{(3)})$. From the joint p.d.f. formula of order statistics, the joint p.d.f. of $(U_{(1)}, U_{(2)})$ is

$$f_{(1),(2)}(x, y) = 6(1 - y), \quad 0 < x \le y < 1.$$

Hence,

$$E\{U_{(1)}U_{(2)}\} = 6 \int_0^1 x \left(\int_0^1 y(1 - y)\, dy \right) dx$$

$$= \frac{6}{40}.$$

Thus,

$$\mathrm{Cov}(U_{(1)}, U_{(2)}) = \frac{6}{40} - \frac{1}{4} \cdot \frac{1}{2}$$

$$= \frac{1}{40}.$$

Similarly, the p.d.f. of $(U_{(1)}, U_{(3)})$ is

$$f_{(1),(3)}(x, y) = 6(y - x), \quad 0 < x \le y < 1.$$

Thus,

$$E\{U_{(1)}U_{(3)}\} = 6 \int_0^1 x \left(\int_x^1 y(y - x)\, dy \right) dx$$

$$= 6 \int_0^1 x \left(\frac{1}{3}(1 - x^3) - \frac{x}{2}(1 - x^2) \right) dx$$

$$= \frac{1}{5},$$

$$E\{U_{(1)}U_{(3)}\} = 6 \int_0^1 x \left(\int_x^1 y(y - x)\, dy \right) dx$$

$$= 6 \int_0^1 x \left. \frac{2y^3 - 3xy^2}{6} \right|_x^1 dx$$

$$= 6 \int_0^1 x \left(\frac{2 - 3x}{6} - \frac{2x^3 - 3x^3}{6} \right) dx$$

$$= 6 \int_0^1 x \frac{2 - 3x + x^3}{6}\, dx$$

$$= \int_0^1 x(2 - 3x + x^3)\, dx$$

$$= \left. \frac{x^5}{5} - x^3 + x^2 \right|_0^1$$

$$= \frac{1}{5},$$

and

$$\text{Cov}(U_{(1)}, U_{(3)}) = \frac{1}{5} - \frac{1}{4} \times \frac{3}{4} = \frac{1}{80}.$$

The p.d.f. of $(U_{(2)}, U_{(3)})$ is

$$f_{(2),(3)}(x, y) = 6x, \quad 0 < x \le y \le 1,$$

and

$$\text{Cov}(U_{(2)}, U_{(3)}) = \frac{1}{40}.$$

Finally,

$$V\{W\} = \frac{1}{16} \cdot \frac{3}{80} + \frac{1}{4} \cdot \frac{1}{20} + \frac{1}{16} \cdot \frac{3}{80}$$

$$+ 2 \cdot \frac{1}{4} \cdot \frac{1}{2} \cdot \frac{1}{40} + 2 \cdot \frac{1}{4} \cdot \frac{1}{4} \cdot \frac{1}{80}$$

$$+ 2 \cdot \frac{1}{2} \cdot \frac{1}{4} \cdot \frac{1}{40}$$

$$= \frac{1}{32} = 0.03125.$$

∎

The following is a useful result:

If $X_1, X_2, \cdots, X_n$ are mutually independent, then the m.g.f. of $T_n = \sum_{i=1}^{n} X_i$ is

$$M_{T_n}(t) = \prod_{i=1}^{n} M_{X_i}(t). \tag{2.108}$$

Indeed, as shown in Sect. 2.5.2, when $X_1, \ldots, X_n$ are independent, the expected value of the product of functions is the product of their expectations. Therefore,

$$M_{T_n}(t) = E\left\{ e^{t \sum_{i=1}^{n} X_i} \right\}$$

$$= E\left\{\prod_{i=1}^{n} e^{tX_i}\right\}$$

$$= \prod_{i=1}^{n} E\{e^{tX_i}\}$$

$$= \prod_{i=1}^{n} M_{X_i}(t).$$

The expected value of the product is equal to the product of the expectations, since $X_1, \cdots, X_n$ are mutually independent.

Example 2.39   In the present example, we illustrate some applications of the last result.

(i) Let $X_1, X_2, \cdots, X_k$ be independent random variables having binomial distributions like $B(n_i, p)$, $i = 1, \cdots, k$; then their sum T_k has the binomial distribution. To show this,

$$M_{T_k}(t) = \prod_{i=1}^{k} M_{X_i}(t)$$

$$= \left[e^t p + (1 - p)\right]^{\sum_{i=1}^{k} n_i}.$$

That is, T_k is distributed like $B\left(\sum_{i=1}^{k} n_i, p\right)$. This result is intuitively clear.

(ii) If $X_1, \cdots, X_n$ are independent random variables, having Poisson distributions with parameters λ_i $(i = 1, \cdots, n)$, then the distribution of $T_n = \sum_{i=1}^{n} X_i$ is Poisson with parameter $\mu_n = \sum_{i=1}^{n} \lambda_i$. Indeed,

$$M_{T_n}(t) = \prod_{j=1}^{n} \exp\{-\lambda_j(1 - e^t)\}$$

$$= \exp\left\{-\sum_{j=1}^{n} \lambda_j(1 - e^t)\right\}$$

$$= \exp\{-\mu_n(1 - e^t)\}.$$

(iii) Suppose $X_1, \cdots, X_n$ are independent random variables, and the distribution of X_i is normal $N(\mu_i, \sigma_i^2)$, then the distribution of $W = \sum_{i=1}^{n} \alpha_i X_i$ is normal like that of

$$N\left(\sum_{i=1}^{n} \alpha_i \mu_i, \sum_{i=1}^{n} \alpha_i^2 \sigma_i^2\right).$$

To verify this we recall that $X_i = \mu_i + \sigma_i Z_i$, where Z_i is $N(0, 1)$ ($i = 1, \cdots, n$). Thus,

$$M_{\alpha_i X_i}(t) = E\{e^{t(\alpha_i \mu_i + \alpha_i \sigma_i Z_i)}\}$$
$$= e^{t\alpha_i \mu_i} M_{Z_i}(\alpha_i \sigma_i t).$$

We derived before that $M_{Z_i}(u) = e^{u^2/2}$. Hence,

$$M_{\alpha_i X_i}(t) = \exp\left\{\alpha_i \mu_i t + \frac{\alpha_i^2 \sigma_i^2}{2}t^2\right\}.$$

Finally,

$$M_W(t) = \prod_{i=1}^{n} M_{\alpha_i X_i}(t)$$
$$= \exp\left\{\left(\sum_{i=1}^{n}\alpha_i \mu_i\right)t + \frac{\sum_{i=1}^{n}\alpha_i^2\sigma_i^2}{2}t^2\right\}.$$

This implies that the distribution of W is normal, with

$$E\{W\} = \sum_{i=1}^{n}\alpha_i \mu_i$$

and

$$V\{W\} = \sum_{i=1}^{n}\alpha_i^2\sigma_i^2.$$

(iv) If $X_1, X_2, \cdots, X_n$ are independent random variables, having gamma distribution like $G(\nu_i, \beta)$, respectively, $i = 1, \cdots, n$, then the distribution of $T_n = \sum_{i=1}^{n} X_i$ is gamma, like that of $G\left(\sum_{i=1}^{n} \nu_i, \beta\right)$. Indeed,

$$M_{T_n}(t) = \prod_{i=1}^{n}(1 - t\beta)^{-\nu_i}$$
$$= (1 - t\beta)^{-\sum_{i=1}^{n}\nu_i}.$$

■

2.9 Large Sample Approximations

2.9.1 The Law of Large Numbers

We have shown in Example 2.37 that the mean of a random sample, $\bar{X}_n$, converges in probability to the expected value of X, μ (the population mean). This is the **law of large numbers** (L.L.N.) which states that, if $X_1, X_2, \cdots$ are i.i.d. random variables and $E\{|X_1|\} < \infty$, then, for any $\epsilon > 0$,

$$\lim_{n \to \infty} \Pr\{|\bar{X}_n - \mu| > \epsilon\} = 0.$$

We also write,

$$\lim_{n \to \infty} \bar{X}_n = \mu, \quad \text{in probability.}$$

This is known as the **weak** L.L.N. There is a stronger law, which states that, under the above conditions,

$$\Pr\{\lim_{n \to \infty} \bar{X}_n = \mu\} = 1.$$

It is beyond the scope of the book to discuss the meaning of the strong L.L.N.

2.9.2 The Central Limit Theorem

The **central limit theorem** (CLT) is one of the most important theorems in probability theory. We formulate here the simplest version of this theorem, which is often sufficient for applications. The theorem states that if $\bar{X}_n$ is the sample mean of n i.i.d. random variables, then, if the population variance σ^2 is positive and finite, the sampling distribution of $\bar{X}_n$ is approximately normal, as $n \to \infty$. More precisely,

If $X_1, X_2, \cdots$ is a sequence of i.i.d. random variables, with $E\{X_1\} = \mu$ and $V\{X_1\} = \sigma^2, 0 < \sigma^2 < \infty$, then

$$\lim_{n \to \infty} \Pr\left\{\frac{(\bar{X}_n - \mu)\sqrt{n}}{\sigma} \leq z\right\} = \Phi(z), \qquad (2.109)$$

where $\Phi(z)$ is the c.d.f. of $N(0, 1)$.

The proof of this basic version of the CLT is based on a result in probability theory, stating that if $X_1, X_2, \cdots$ is a sequence of random variables having m.g.f.s, $M_n(T), n = 1, 2, \cdots$ and if $\lim_{n\to\infty} M_n(t) = M(t)$ is the m.g.f. of a random variable X^*, having a c.d.f. $F^*(x)$, then $\lim_{n\to\infty} F_n(x) = F^*(x)$, where $F_n(x)$ is the c.d.f. of X_n.

The m.g.f. of

$$Z_n = \frac{\sqrt{n}(\bar{X}_n - \mu)}{\sigma},$$

can be written as

$$M_{Z_n}(t) = E\left\{\exp\left\{\frac{t}{\sqrt{n}\sigma}\sum_{i=1}^{n}(X_i - \mu)\right\}\right\}$$

$$= \left(E\left\{\exp\left\{\frac{t}{\sqrt{n}\sigma}(X_1 - \mu)\right\}\right\}\right)^n,$$

since the random variables are independent. Furthermore, Taylor expansion of $\exp\left\{\frac{t}{\sqrt{n}\sigma}(X_1 - \mu)\right\}$ is

$$1 + \frac{t}{\sqrt{n}\sigma}(X_1 - \mu) + \frac{t^2}{2n\sigma^2}(X_1 - \mu)^2 + o\left(\frac{1}{n}\right),$$

for n large. Hence, as $n \to \infty$

$$E\left\{\exp\left\{\frac{t}{\sqrt{n}\sigma}(X_1 - \mu)\right\}\right\} = 1 + \frac{t^2}{2n} + o\left(\frac{1}{n}\right).$$

Hence,

$$\lim_{n\to\infty} M_{Z_n}(t) = \lim_{n\to\infty}\left(1 + \frac{t^2}{2n} + o\left(\frac{1}{n}\right)\right)^n$$

$$= e^{t^2/2},$$

which is the m.g.f. of $N(0, 1)$. This is a sketch of the proof. For rigorous proofs and extensions, see textbooks on probability theory.

2.9.3 *Some Normal Approximations*

The CLT can be applied to provide an approximation to the distribution of the sum of n i.i.d. random variables, by a standard normal distribution, when n is large. We list below a few such useful approximations.

(i) **Binomial Distribution**

   When n is large, then the c.d.f. of $B(n, p)$ can be approximated by

$$B(k; n, p) \cong \Phi\left(\frac{k + \frac{1}{2} - np}{\sqrt{np(1 - p)}}\right). \qquad (2.110)$$

   We add $\frac{1}{2}$ to k, in the argument of $\Phi(\cdot)$ to obtain a better approximation when n is not too large. This modification is called "correction for discontinuity." How large should n be to get a "good" approximation? A general rule is

$$n > \frac{9}{p(1 - p)}. \qquad (2.111)$$

(ii) **Poisson Distribution**

   The c.d.f. of Poisson with parameter λ can be approximated by

$$P(k; \lambda) \cong \Phi\left(\frac{k + \frac{1}{2} - \lambda}{\sqrt{\lambda}}\right), \qquad (2.112)$$

   if λ is large (greater than 30).

(iii) **Gamma Distribution**

   The c.d.f. of $G(\nu, \beta)$ can be approximated by

$$G(x; \nu, \beta) \cong \Phi\left(\frac{x - \nu\beta}{\beta\sqrt{\nu}}\right), \qquad (2.113)$$

   for large values of ν.

Example 2.40

(i) A lot consists of $n = 10,000$ screws. The probability that a screw is defective is $p = 0.01$. What is the probability that there are more than 120 defective screws in the lot?

The number of defective screws in the lot, J_n, has a distribution like $B(10000, 0.01)$. Hence,

$$\Pr\{J_{10000} > 120\} = 1 - B(120; 10000, 0.01)$$

$$\cong 1 - \Phi\left(\frac{120.5 - 100}{\sqrt{99}}\right)$$

$$= 1 - \Phi(2.06) = 0.0197.$$

(ii) In the production of industrial film, we find on the average 1 defect per 100 [ft]^2 of film. What is the probability that fewer than 100 defects will be found on 12,000 [ft]^2 of film?

   We assume that the number of defects per unit area of film is a Poisson random variable. Thus, our model is that the number of defects, X, per 12,000 [ft]^2 has a Poisson distribution with parameter $\lambda = 120$. Thus,

$$\Pr\{X < 100\} \cong \Phi\left(\frac{99.5 - 120}{\sqrt{120}}\right)$$

$$= 0.0306.$$

(iii) The time till failure, T, of radar equipment is exponentially distributed with mean time till failure (MTTF) of $\beta = 100$ [hr].

   A sample of $n = 50$ units is put on test. Let $\bar{T}_{50}$ be the sample mean. What is the probability that $\bar{T}_{50}$ will fall in the interval (95, 105) [hr]?

   We have seen that $\sum_{i=1}^{50} T_i$ is distributed like $G(50, 100)$, since $E(\beta)$ is distributed like $G(1, \beta)$. Hence, $\bar{T}_{50}$ is distributed like $\frac{1}{50} G(50, 100)$ which is $G(50, 2)$. By the normal approximation

$$\Pr\{95 < \bar{T}_{50} < 105\} \cong \Phi\left(\frac{105 - 100}{2\sqrt{50}}\right)$$

$$- \Phi\left(\frac{95 - 100}{2\sqrt{50}}\right) = 2\Phi(0.3536) - 1 = 0.2763.$$

■

2.10   Additional Distributions of Statistics of Normal Samples

In the present section, we assume that $X_1, X_2, \cdots, X_n$ are i.i.d. $N(\mu, \sigma^2)$ random variables. In the Sects. 2.10.1–2.10.3, we present the chi-squared, t-, and F-distributions which play an important role in the theory of statistical inference (Chap. 3).

2.10.1   Distribution of the Sample Variance

Writing $X_i = \mu + \sigma Z_i$, where $Z_1, \cdots, Z_n$ are i.i.d. $N(0, 1)$, we obtain that the sample variance S^2 is distributed like

$$S^2 = \frac{1}{n-1} \sum_{i=1}^{n} (X_i - \bar{X}_n)^2$$

$$= \frac{1}{n-1} \sum_{i=1}^{n} \left(\mu + \sigma Z_i - (\mu + \sigma \bar{Z}_n) \right)^2$$

$$= \frac{\sigma^2}{n-1} \sum_{i=1}^{n} (Z_i - \bar{Z}_n)^2.$$

One can show that $\sum_{i=1}^{n} (Z_i - \bar{Z}_n)^2$ is distributed like $\chi^2[n-1]$, where $\chi^2[v]$ is called a **chi-squared random variable with v degrees of freedom**. Moreover, $\chi^2[v]$ is distributed like $G\left(\frac{v}{2}, 2\right)$.

The α-th quantile of $\chi^2[v]$ is denoted by $\chi^2_\alpha[v]$. Accordingly, the c.d.f. of the sample variance is

$$\begin{aligned} H_{S^2}(x; \sigma^2) &= \Pr\left\{ \frac{\sigma^2}{n-1} \chi^2[n-1] \leq x \right\} \\ &= \Pr\left\{ \chi^2[n-1] \leq \frac{(n-1)x}{\sigma^2} \right\} \\ &= \Pr\left\{ G\left(\frac{n-1}{2}, 2\right) \leq \frac{(n-1)x}{\sigma^2} \right\} \\ &= G\left(\frac{(n-1)x}{2\sigma^2}; \frac{n-1}{2}, 1 \right). \end{aligned} \qquad (2.114)$$

The probability values of the distribution of $\chi^2[v]$, as well as the α-quantiles, can be computed using Python or read from appropriate tables.

The expected value and variance of the sample variance are

$$\begin{aligned} E\{S^2\} &= \frac{\sigma^2}{n-1} E\{\chi^2[n-1]\} \\ &= \frac{\sigma^2}{n-1} E\left\{ G\left(\frac{n-1}{2}, 2\right) \right\} \\ &= \frac{\sigma^2}{n-1} (n-1) = \sigma^2. \end{aligned}$$

Similarly

$$V\{S^2\} = \frac{\sigma^4}{(n-1)^2} V\{\chi^2[n-1]\}$$

$$= \frac{\sigma^4}{(n-1)^2} V\left\{G\left(\frac{n-1}{2}, 2\right)\right\}$$

$$= \frac{\sigma^4}{(n-1)^2} \cdot 2(n-1) \tag{2.115}$$

$$= \frac{2\sigma^4}{n-1}.$$

Thus applying Chebyshev's inequality, for any given $\epsilon > 0$,

$$\Pr\{|S^2 - \sigma^2| > \epsilon\} < \frac{2\sigma^4}{(n-1)\epsilon^2}.$$

Hence, S^2 converges in probability to σ^2. Moreover,

$$\lim_{n\to\infty} \Pr\left\{\frac{(S^2 - \sigma^2)}{\sigma^2\sqrt{2}}\sqrt{n-1} \le z\right\} = \Phi(z). \tag{2.116}$$

That is, the distribution of S^2 can be approximated by the normal distributions in large samples.

2.10.2 The "Student" t-Statistic

We have seen that

$$Z_n = \frac{\sqrt{n}(\bar{X}_n - \mu)}{\sigma}$$

has a $N(0, 1)$ distribution. As we will see in Chap. 3, when σ is unknown, we test hypotheses concerning μ by the statistic

$$t = \frac{\sqrt{n}(\bar{X}_n - \mu_0)}{S},$$

where S is the sample standard deviation. If $X_1, \cdots, X_n$ are i.i.d. like $N(\mu_0, \sigma^2)$, then the distribution of t is called the **student t-distribution with $\nu = n-1$ degrees of freedom**. The corresponding random variable is denoted by $t[\nu]$.

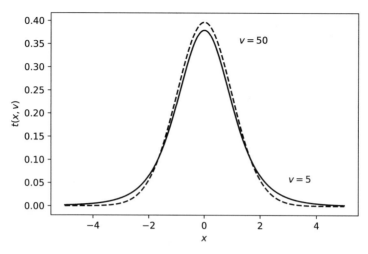

Fig. 2.19 Density functions of $t[\nu]$, $\nu = 5, 50$

The p.d.f. of $t[\nu]$ is symmetric about 0 (see Fig. 2.19). Thus,

$$E\{t[\nu]\} = 0, \quad \text{for } \nu \geq 2 \tag{2.117}$$

and

$$V\{t[\nu]\} = \frac{\nu}{\nu - 2}, \quad \nu \geq 3. \tag{2.118}$$

The α-quantile of $t[\nu]$ is denoted by $t_\alpha[\nu]$. It can be read from a table or determined using `scipy.stats.t`.

2.10.3   Distribution of the Variance Ratio

$$F = \frac{S_1^2 \, \sigma_2^2}{S_2^2 \, \sigma_1^2}.$$

Consider now two independent samples of size n_1 and n_2, respectively, which have been taken from normal populations having variances σ_1^2 and σ_2^2. Let

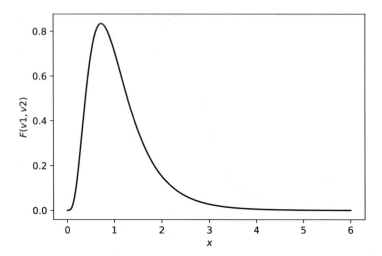

Fig. 2.20 Density function of $F(\nu_1, \nu_2)$

$$S_1^2 = \frac{1}{n_1 - 1} \sum_{i=1}^{n_1} (X_{1i} - \bar{X}_1)^2$$

and

$$S_2^2 = \frac{1}{n_2 - 1} \sum_{i=1}^{n_2} (X_{2i} - \bar{X}_2)^2$$

be the variances of the two samples where $\bar{X}_1$ and $\bar{X}_2$ are the corresponding sample means. The F-ratio has a distribution denoted by $F[\nu_1, \nu_2]$, with $\nu_1 = n_1 - 1$ and $\nu_2 = n_2 - 1$. This distribution is called the F-**distribution with ν_1 and ν_2 degrees of freedom**. A graph of the densities of $F[\nu_1, \nu_2]$ is given in Fig. 2.20.

The expected value and the variance of $F[\nu_1, \nu_2]$ are

$$E\{F[\nu_1, \nu_2]\} = \nu_2/(\nu_2 - 2), \quad \nu_2 > 2, \tag{2.119}$$

and

$$V\{F[\nu_1, \nu_2]\} = \frac{2\nu_2^2(\nu_1 + \nu_2 - 2)}{\nu_1(\nu_2 - 2)^2(\nu_2 - 4)}, \quad \nu_2 > 4. \tag{2.120}$$

The $(1 - \alpha)$-th quantile of $F[\nu_1, \nu_2]$, i.e., $F_{1-\alpha}[\nu_1, \nu_2]$, can be computed with `scipy` in Python. If we wish to obtain the α-th quantile $F_\alpha[\nu_1, \nu_2]$ for values of $\alpha < 0.5$, we can apply the relationship

$$F_{1-\alpha}[\nu_1, \nu_2] = \frac{1}{F_\alpha[\nu_2, \nu_1]}. \tag{2.121}$$

Thus, for example, to compute $F_{.05}[15, 10]$, we write

$$F_{0.05}[15, 10] = 1/F_{0.95}[10, 15] = 1/2.54 = 0.3937.$$

2.11 Chapter Highlights

The concepts and definitions introduced are:

- Sample space
- Elementary events
- Operations with events
- Disjoint events
- Probability of events
- Random sampling with replacement (RSWR)
- Random sampling without replacement (RSWOR)
- Conditional probabilities
- Independent events
- Bayes' theorem
- Prior probability
- Posterior probability
- Probability distribution function (p.d.f.)
- Discrete random variable
- Continuous random variable
- Cumulative distribution function
- Central moments
- Expected value
- Standard deviation
- Chebyshev's inequality
- Moment generating function
- Skewness
- Kurtosis
- Independent trials
- P-th quantile
- Joint distribution
- Marginal distribution
- Conditional distribution
- Mutual independence
- Conditional independence
- Law of total variance
- Law of iterated expectation

- Order statistics
- Convergence in probability
- Central limit theorem
- Law of large numbers

2.12   Exercises

Exercise 2.1 An experiment consists of making 20 observations on the quality of chips. Each observation is recorded as G or D.

  (i) What is the sample space, S, corresponding to this experiment?
 (ii) How many elementary events in S?
(iiii) Let A_n, $n = 0, \cdots, 20$, be the event that exactly n G observations are made. Write the events A_n formally. How many elementary events belong to A_n?

Exercise 2.2 An experiment consists of ten measurements $w_1, \cdots, w_{10}$ of the weights of packages. All packages under consideration have weights between 10 and 20 pounds. What is the sample space S? Let $A = \{(w_1, w_2, \cdots, w_{10}) : w_1 + w_2 = 25\}$. Let $B = \{(w_1, \cdots, w_{10}) : w_1 + w_2 \leq 25\}$. Describe the events A and B graphically. Show that $A \subset B$.

Exercise 2.3 Strings of 30 binary (0, 1) signals are transmitted.

  (i) Describe the sample space, S.
 (ii) Let A_{10} be the event that the first ten signals transmitted are all 1s. How many elementary events belong to A_{10}?
(iii) Let B_{10} be the event that exactly 10 signals, out of 30 transmitted, are 1s. How many elementary events belong to B_{10}? Does $A_{10} \subset B_{10}$?

Exercise 2.4 Prove De Morgan's laws

 (i) $(A \cup B)^c = A^c \cap B^c$.
(ii) $(A \cap B)^c = A^c \cup B^c$.

Exercise 2.5 Consider Exercise [3.1] Show that the events $A_0, A_1, \cdots, A_{20}$ are a partition of the sample space S.

Exercise 2.6 Let $A_1, \cdots, A_n$ be a partition of S. Let B be an event. Show that $B = \bigcup_{i=1}^{n} A_i B$, where $A_i B = A_i \cap B$, is a union of disjoint events.

Exercise 2.7 Develop a formula for the probability $\Pr\{A \cup B \cup C\}$, where A, B, and C are arbitrary events.

Exercise 2.8 Show that if $A_1, \cdots, A_n$ is a partition, then for any event B, $P\{B\} = \sum_{i=1}^{n} P\{A_i B\}$. [Use the result of 2.6.]

Exercise 2.9 An unbiased die has the numbers $1, 2, \cdots, 6$ written on its faces. The die is thrown twice. What is the probability that the two numbers shown on its upper face sum up to 10?

Exercise 2.10 The time till failure, T, of electronic equipment is a random quantity. The event $A_t = \{T > t\}$ is assigned the probability $\Pr\{A_t\} = \exp\{-t/200\}$, $t \geq 0$. What is the probability of the event $B = \{150 < T < 280\}$?

Exercise 2.11 A box contains 40 parts, 10 of type A, 10 of type B, 15 of type C, and 5 of type D. A random sample of eight parts is drawn without replacement. What is the probability of finding two parts of each type in the sample?

Exercise 2.12 How many samples of size $n = 5$ can be drawn from a population of size $N = 100$,

(i) with replacement?
(ii) without replacement?

Exercise 2.13 A lot of 1000 items contain $M = 900$ "good" ones and 100 "defective" ones. A random sample of size $n = 10$ is drawn from the lot. What is the probability of observing in the sample of at least eight good items,

(i) when sampling is with replacement?
(ii) when sampling is without replacement?

Exercise 2.14 In continuation of the previous exercise, what is the probability of observing in an RSWR at least one defective item?

Exercise 2.15 Consider the problem of Exercise 2.10. What is the conditional probability $\Pr\{T > 300 \mid T > 200\}$?

Exercise 2.16 A point (X, Y) is chosen at random within the unit square, i.e.,

$$S = \{(x, y) : 0 \leq x, y \leq 1\}.$$

Any set A contained in S having area given by

$$\text{Area}\{A\} = \iint_A dx \, dy$$

is an event, whose probability is the area of A. Define the events

$$B = \left\{ (x, y) : x > \frac{1}{2} \right\}$$

$$C = \{(x, y) : x^2 + y^2 \le 1\}$$

$$D = \{(x, y) : (x + y) \le 1\}.$$

(i) Compute the conditional probability $\Pr\{D \mid B\}$.
(ii) Compute the conditional probability $\Pr\{C \mid D\}$.

Exercise 2.17 Show that if A and B are independent events, then A^c and B^c are also independent events.

Exercise 2.18 Show that if A and B are disjoint events, then A and B are dependent events.

Exercise 2.19 Show that if A and B are independent events, then

$$\Pr\{A \cup B\} = \Pr\{A\}(1 - \Pr\{B\}) + \Pr\{B\}$$
$$= \Pr\{A\} + \Pr\{B\}(1 - \Pr\{A\}).$$

Exercise 2.20 A machine which tests whether a part is defective, D, or good, G, may err. The probabilities of errors are given by

$$\Pr\{A \mid G\} = 0.95,$$
$$\Pr\{A \mid D\} = 0.10,$$

where A is the event "the part is considered G after testing." If $\Pr\{G\} = 0.99$, what is the probability of D given A?

Additional Problems in Combinatorial and Geometric Probabilities

Exercise 2.21 Assuming 365 days in a year, if there are 10 people in a party, what is the probability that their birthdays fall on different days? Show that if there are more than 22 people in the party, the probability is greater than 1/2 that at least 2 will have birthdays on the same day.

Exercise 2.22 A number is constructed at random by choosing 10 digits from $\{0, \ldots, 9\}$ with replacement. We allow the digit 0 at any position. What is the probability that the number does not contain three specific digits?

Exercise 2.23 A caller remembers all the seven digits of a telephone number but is uncertain about the order of the last four. He keeps dialing the last four digits at random, without repeating the same number, until he reaches the right number. What is the probability that he will dial at least ten wrong numbers?

Exercise 2.24 One hundred lottery tickets are sold. There are four prizes and ten consolation prizes. If you buy five tickets, what is the probability that you win:

 (i) one prize?
 (ii) a prize and a consolation prize?
(iii) Something?

Exercise 2.25 Ten PCBs are in a bin, of which two of these are defective. The boards are chosen at random, one by one, without replacement. What is the probability that exactly five good boards will be found between the drawing of the first and second defective PCB?

Exercise 2.26 A random sample of 11 integers is drawn without replacement from the set $\{1, 2, \ldots, 20\}$. What is the probability that the sample median, Me, is equal to the integer k? $6 \leq k \leq 15$.

Exercise 2.27 A stick is broken at random into three pieces. What is the probability that these pieces can be the sides of a triangle?

Exercise 2.28 A particle is moving at a uniform speed on a circle of unit radius and is released at a random point on the circumference. Draw a line segment of length $2h$ ($h < 1$) centered at a point A of distance $a > 1$ from the center of the circle, O. Moreover, the line segment is perpendicular to the line connecting O with A. What is the probability that the particle will hit the line segment? [The particle flies along a straight line tangential to the circle.]

Exercise 2.29 A block of 100 bits is transmitted over a binary channel, with probability $p = 10^{-3}$ of bit error. Errors occur independently. Find the probability that the block contains at least three errors.

Exercise 2.30 A coin is tossed repeatedly until two "heads" occur. What is the probability that four tosses are required?

Exercise 2.31 Consider the sample space S of all sequences of ten binary numbers (0-1 signals). Define on this sample space two random variables and derive their probability distribution function, assuming the model that all sequences are equally probable.

Exercise 2.32 The number of blemishes on a ceramic plate is a discrete random variable. Assume the probability model, with p.d.f.

$$p(x) = e^{-5}\frac{5^x}{x!}, \quad x = 0, 1, \cdots$$

 (i) Show that $\sum_{x=0}^{\infty} p(x) = 1$
 (ii) What is the probability of at most one blemish on a plate?

(iii) What is the probability of no more than seven blemishes on a plate?

Exercise 2.33 Consider a distribution function of a mixed type with c.d.f.

$$
F_X(x) = \begin{cases}
0, & \text{if } x < -1 \\
0.3 + 0.2(x + 1), & \text{if } -1 \leq x < 0 \\
0.7 + 0.3x, & \text{if } 0 \leq x < 1 \\
1, & \text{if } 1 \leq x.
\end{cases}
$$

  (i) What is $\Pr\{X = -1\}$?
 (ii) What is $\Pr\{-0.5 < X < 0\}$?
(iii) What is $\Pr\{0 \leq X < 0.75\}$?
 (iv) What is $\Pr\{X = 1\}$?
  (v) Compute the expected value, $E\{X\}$, and variance, $V\{X\}$.

Exercise 2.34 A random variable has the Rayleigh distribution, with c.d.f.

$$
F(x) = \begin{cases}
0, & x < 0 \\
1 - e^{-x^2/2\sigma^2}, & x \geq 0
\end{cases}
$$

where σ^2 is a positive parameter. Find the expected value $E\{X\}$.

Exercise 2.35 A random variable X has a discrete distribution over the integers $\{1, 2, \ldots, N\}$ with equal probabilities. Find $E\{X\}$ and $V\{X\}$.

Exercise 2.36 A random variable has expectation $\mu = 10$ and standard deviation $\sigma = 0.5$. Use Chebyshev's inequality to find a lower bound to the probability

$$
\Pr\{8 < X < 12\}.
$$

Exercise 2.37 Consider the random variable X with c.d.f.

$$
F(x) = \frac{1}{2} + \frac{1}{\pi} \tan^{-1}(x), \quad -\infty < x < \infty.
$$

Find the 0.25-th, 0.50-th, and 0.75-th quantiles of this distribution.

Exercise 2.38 Show that the central moments μ_i^* relate to the moments μ_l around the origin, by the formula

$$\mu_l^* = \sum_{j=0}^{l-2} (-1)^j \binom{l}{j} \mu_{l-j} \mu_1^j + (-1)^{l-1}(l-1)\mu_1^l.$$

Exercise 2.39 Find the expected value μ_1 and the second moment μ_2 of the random variable whose c.d.f. is given in Exercise 2.33.

Exercise 2.40 A random variable X has a continuous uniform distribution over the interval (a, b), i.e.,

$$f(x) = \begin{cases} \dfrac{1}{b-a}, & \text{if } a \le x \le b \\ 0, & \text{otherwise.} \end{cases}$$

Find the moment generating function of X. Find the mean and variance by differentiating the m.g.f.

Exercise 2.41 Consider the moment generating function (m.g.f.) of the exponential distribution, i.e.,

$$M(t) = \frac{\lambda}{\lambda - t}, \quad t < \lambda.$$

(i) Find the first four moments of the distribution, by differentiating $M(t)$.
(ii) Convert the moments to central moments.
(iii) What is the index of kurtosis β_4?

Exercise 2.42 Using Python, prepare a table of the p.d.f. and c.d.f. of the binomial distribution $B(20, 0.17)$.

Exercise 2.43 What are the first quantile (Q_1), median (Me), and third quantile (Q_3) of $B(20, 0.17)$?

Exercise 2.44 Compute the mean $E\{X\}$ and standard deviation, σ, of $B(45, 0.35)$.

Exercise 2.45 A PCB is populated by 50 chips which are randomly chosen from a lot. The probability that an individual chip is non-defective is p. What should be the value of p so that no defective chip is installed on the board is $\gamma = 0.99$? [The answer to this question shows why the industry standards are so stringent.]

Exercise 2.46 Let $b(j; n, p)$ be the p.d.f. of the binomial distribution. Show that as $n \to \infty$, $p \to 0$ so that $np \to \lambda$, $0 < \lambda < \infty$, then

$$\lim_{\substack{n \to \infty \\ p \to 0 \\ np \to \lambda}} b(j; n, p) = e^{-\lambda} \frac{\lambda^j}{j!}, \quad j = 0, 1, \dots.$$

Exercise 2.47 Use the result of the previous exercise to find the probability that a block of 1000 bits, in a binary communication channel, will have less than four errors, when the probability of a bit error is $p = 10^{-3}$.

Exercise 2.48 Compute $E\{X\}$ and $V\{X\}$ of the hypergeometric distribution $H(500, 350, 20)$.

Exercise 2.49 A lot of size $N = 500$ items contains $M = 5$ defective ones. A random sample of size $n = 50$ is drawn from the lot without replacement (RSWOR). What is the probability of observing more than one defective item in the sample?

Exercise 2.50 Consider Example 2.23. What is the probability that the lot will be rectified if $M = 10$ and $n = 20$?

Exercise 2.51 Use the m.g.f. to compute the third and fourth central moments of the Poisson distribution $P(10)$. What is the index of skewness and kurtosis of this distribution?

Exercise 2.52 The number of blemishes on ceramic plates has a Poisson distribution with mean $\lambda = 1.5$. What is the probability of observing more than two blemishes on a plate?

Exercise 2.53 The error rate of an insertion machine is 380 PPM (per 10^6 parts inserted). What is the probability of observing more than six insertion errors in 2 h of operation, when the insertion rate is 4000 parts per hour?

Exercise 2.54 In continuation of the previous exercise, let N be the number of parts inserted until an error occurs. What is the distribution of N? Compute the expected value and the standard deviation of N.

Exercise 2.55 What are Q_1, Me, and Q_3 of the negative binomial N.B. (p, k) with $p = 0.01$ and $k = 3$?

Exercise 2.56 Derive the m.g.f. of $NB(p, k)$.

Exercise 2.57 Differentiate the m.g.f. of the geometric distribution, i.e.,

$$M(t) = \frac{pe^t}{(1 - e^t(1 - p))}, \quad t < -\log(1 - p),$$

to obtain its first four moments, and derive then the indices of skewness and kurtosis.

Exercise 2.58 The proportion of defective RAM chips is $p = 0.002$. You have to install 50 chips on a board. Each chip is tested before its installation. How many chips should you order so that, with probability greater than $\gamma = 0.95$, you will have at least 50 good chips to install?

Exercise 2.59 The random variable X assumes the values $\{1, 2, \ldots\}$ with probabilities of a geometric distribution, with parameter p, $0 < p < 1$. Prove the "memoryless" property of the geometric distribution, namely,

$$P[X > n + m \mid X > m] = P[X > n],$$

for all $n, m = 1, 2, \ldots$.

Exercise 2.60 Let X be a random variable having a continuous c.d.f. $F(x)$. Let $Y = F(X)$. Show that Y has a uniform distribution on $(0, 1)$. Conversely, if U has a uniform distribution on $(0, 1)$, then $X = F^{-1}(U)$ has the c.d.f. $F(x)$.

Exercise 2.61 Compute the expected value and the standard deviation of a uniform distribution $U(10, 50)$.

Exercise 2.62 Show that if U is uniform on $(0, 1)$, then $X = -\log(U)$ has an exponential distribution $E(1)$.

Exercise 2.63 Use Python to compute the probabilities, for $N(100, 15)$, of

(i) $92 < X < 108$;
(ii) $X > 105$;
(iii) $2X + 5 < 200$.

Exercise 2.64 The 0.9-quantile of $N(\mu, \sigma)$ is 15 and its 0.99-quantile is 20. Find the mean μ and standard deviation σ.

Exercise 2.65 A communication channel accepts an arbitrary voltage input v and outputs a voltage $v + E$, where $E \sim N(0, 1)$. The channel is used to transmit binary information as follows:

(i) to transmit 0, input $-v$
(ii) to transmit 1, input v
(iii) The receiver decides a 0 if the voltage Y is negative and 1 otherwise.

What should be the value of v so that the receiver's probability of bit error is $\alpha = 0.01$?

Exercise 2.66 Aluminum pins manufactured for an aviation industry have a random diameter, whose distribution is (approximately) normal with mean of $\mu = 10$ [mm] and standard deviation $\sigma = 0.02$ [mm]. Holes are automatically drilled on aluminum plates, with diameters having a normal distribution with mean μ_d [mm] and $\sigma = 0.02$ [mm]. What should be the value of μ_d so that the probability that a pin will not enter a hole (too wide) is $\alpha = 0.01$?

Exercise 2.67 Let $X_1, \ldots, X_n$ be a random sample (i.i.d.) from a normal distribution $N(\mu, \sigma^2)$. Find the expected value and variance of $Y = \sum_{i=1}^{n} i X_i$.

Exercise 2.68 Concrete cubes have compressive strength with log-normal distribution $LN(5, 1)$. Find the probability that the compressive strength X of a random concrete cube will be greater than 300 [kg/cm^2].

Exercise 2.69 Using the m.g.f. of $N(\mu, \sigma)$, derive the expected value and variance of $LN(\mu, \sigma)$. [Recall that $X \sim e^{N(\mu,\sigma)}$.]

Exercise 2.70 What are Q_1, Me, and Q_3 of $E(\beta)$?

Exercise 2.71 Show that if the life length of a chip is exponential $E(\beta)$, then only 36.7% of the chips will function longer than the mean time till failure β.

Exercise 2.72 Show that the m.g.f. of $E(\beta)$ is $M(t) = (1 - \beta t)^{-1}$, for $t < \dfrac{1}{\beta}$.

Exercise 2.73 Let X_1, X_2, X_3 be independent random variables having an identical exponential distribution $E(\beta)$. Compute $\Pr\{X_1 + X_2 + X_3 \geq 3\beta\}$.

Exercise 2.74 Establish the formula

$$G\left(t; k, \frac{1}{\lambda}\right) = 1 - e^{-\lambda t} \sum_{j=0}^{k-1} \frac{(\lambda t)^j}{j!},$$

by integrating in parts the p.d.f. of

$$G\left(k; \frac{1}{\lambda}\right).$$

Exercise 2.75 Use Python to compute $\Gamma(1.17)$, $\Gamma\left(\dfrac{1}{2}\right)$, $\Gamma\left(\dfrac{3}{2}\right)$.

Exercise 2.76 Using m.g.f., show that the sum of k independent exponential random variables, $E(\beta)$, has the gamma distribution $G(k, \beta)$.

Exercise 2.77 What is the expected value and variance of the Weibull distribution $W(2, 3.5)$?

Exercise 2.78 The time till failure (days) of an electronic equipment has the Weibull distribution $W(1.5, 500)$. What is the probability that the failure time will not be before 600 days?

Exercise 2.79 Compute the expected value and standard deviation of a random variable having the Beta distribution $\text{Beta}\left(\dfrac{1}{2}, \dfrac{3}{2}\right)$.

Exercise 2.80 Show that the index of kurtosis of $\text{Beta}(v, v)$ is $\beta_2 = \dfrac{3(1 + 2v)}{3 + 2v}$.

Exercise 2.81 The joint p.d.f. of two random variables (X, Y) is

$$f(x, y) = \begin{cases} \dfrac{1}{2}, & \text{if } (x, y) \in S \\ 0, & \text{otherwise} \end{cases}$$

where S is a square of area 2, whose vertices are $(1, 0)$, $(0, 1)$, $(-1, 0)$, $(0, -1)$.

(i) Find the marginal p.d.f. of X and Y.
(ii) Find $E\{X\}$, $E\{Y\}$, $V\{X\}$, $V\{Y\}$.

Exercise 2.82 Let (X, Y) have a joint p.d.f.

$$f(x, y) = \begin{cases} \dfrac{1}{y} \exp\left\{-y - \dfrac{x}{y}\right\}, & \text{if } 0 < x, y < \infty \\ 0, & \text{otherwise.} \end{cases}$$

Find $\text{COV}(X, Y)$ and the coefficient of correlation ρ_{XY}.

Exercise 2.83 Show that the random variables (X, Y) whose joint distribution is defined in Example 2.27 are dependent. Find $\text{COV}(X, Y)$.

Exercise 2.84 Find the correlation coefficient of N and J of Example 2.31.

Exercise 2.85 Let X and Y be independent random variables, $X \sim G(2, 100)$ and $W(1.5, 500)$. Find the variance of XY.

Exercise 2.86 Consider the trinomial distribution of Example 2.33.

(i) What is the probability that during 1 h of operation there will be no more than 20 errors?
(ii) What is the conditional distribution of wrong components, given that there are 15 mis-insertions in a given hour of operation?
(iii) Approximating the conditional distribution of (ii) by a Poisson distribution, compute the conditional probability of no more than 15 wrong components?

Exercise 2.87 In continuation of Example 2.34, compute the correlation between J_1 and J_2.

Exercise 2.88 In a bivariate normal distribution, the conditional variance of Y given X is 150 and the variance of Y is 200. What is the correlation ρ_{XY}?

Exercise 2.89 $n = 10$ electronic devices start to operate at the same time. The times till failure of these devices are independent random variables having an identical $E(100)$ distribution.

 (i) What is the expected value of the first failure?
(ii) What is the expected value of the last failure?

Exercise 2.90 A factory has $n = 10$ machines of a certain type. At each given day, the probability is $p = 0.95$ that a machine will be working. Let J denote the number of machines that work on a given day. The time it takes to produce an item on a given machine is $E(10)$, i.e., exponentially distributed with mean $\mu = 10$ [min]. The machines operate independently of each other. Let $X_{(1)}$ denote the minimal time for the first item to be produced. Determine

 (i) $P[J = k, X_{(1)} \leq x], k = 1, 2, \ldots$
(ii) $P[X_{(1)} \leq x \mid J \geq 1]$.

Notice that when $J = 0$ no machine is working. The probability of this event is $(0.05)^{10}$.

Exercise 2.91 Let $X_1, X_2, \ldots, X_{11}$ be a random sample of exponentially distributed random variables with p.d.f. $f(x) = \lambda e^{-\lambda x}, x \geq 0$.

 (i) What is the p.d.f. of the median $Me = X_{(6)}$?
(ii) What is the expected value of Me?

Exercise 2.92 Let X and Y be independent random variables having an $E(\beta)$ distribution. Let $T = X + Y$ and $W = X - Y$. Compute the variance of $T + \dfrac{1}{2}W$.

Exercise 2.93 Let X and Y be independent random variables having a common variance σ^2. What is the covariance $\mathrm{cov}(X, X + Y)$?

Exercise 2.94 Let (X, Y) have a bivariate normal distribution. What is the variance of $\alpha X + \beta Y$?

Exercise 2.95 Let X have a normal distribution $N(\mu, \sigma)$. Let $\Phi(z)$ be the standard normal c.d.f. Verify that $E\{\Phi(X)\} = P\{U < X\}$, where U is independent of X and $U \sim N(0, 1)$. Show that

$$E\{\Phi(X)\} = \Phi\left(\frac{\eta}{\sqrt{1 + \sigma^2}}\right).$$

Exercise 2.96 Let X have a normal distribution $N(\mu, \sigma)$. Show that

$$E\{\Phi^2(X)\} = \Phi_2\left(\frac{\mu}{\sqrt{1 + \sigma^2}}, \frac{\mu}{\sqrt{1 + \sigma^2}}; \frac{\sigma^2}{1 + \sigma^2}\right).$$

Exercise 2.97 X and Y are independent random variables having Poisson distributions, with means $\lambda_1 = 5$ and $\lambda_2 = 7$, respectively. Compute the probability that $X + Y$ is greater than 15.

Exercise 2.98 Let X_1 and X_2 be independent random variables having continuous distributions with p.d.f. $f_1(x)$ and $f_2(x)$, respectively. Let $Y = X_1 + X_2$. Show that the p.d.f. of Y is

$$g(y) = \int_{-\infty}^{\infty} f_1(x) f_2(y - x) \, dx.$$

[This integral transform is called the convolution of $f_1(x)$ with $f_2(x)$. The convolution operation is denoted by $f_1 * f_2$.]

Exercise 2.99 Let X_1 and X_2 be independent random variables having the uniform distributions on $(0, 1)$. Apply the convolution operation to find the p.d.f. of $Y = X_1 + X_2$.

Exercise 2.100 Let X_1 and X_2 be independent random variables having a common exponential distribution $E(1)$. Determine the p.d.f. of $U = X_1 - X_2$. [The distribution of U is called bi-exponential or Laplace and its p.d.f. is $f(u) = \dfrac{1}{2} e^{-|u|}$.]

Exercise 2.101 Apply the central limit theorem to approximate $P\{X_1 + \cdots + X_{20} \leq 50\}$, where $X_1, \cdots, X_{20}$ are independent random variables having a common mean $\mu = 2$ and a common standard deviation $\sigma = 10$.

Exercise 2.102 Let X have a binomial distribution $B(200, .15)$. Find the normal approximation to $\Pr\{25 < X < 35\}$.

Exercise 2.103 Let X have a Poisson distribution with mean $\lambda = 200$. Find, approximately, $\Pr\{190 < X < 210\}$.

Exercise 2.104 $X_1, X_2, \cdots, X_{200}$ are 200 independent random variables having a common beta distribution $B(3, 5)$. Approximate the probability $\Pr\{|\bar{X}_{200} - 0.375| < 0.2282\}$, where

$$\bar{X}_n = \frac{1}{n} \sum_{i=1}^{n} X_i, \quad n = 200.$$

Exercise 2.105 Use Python to compute the 0.95-quantiles of $t[10], t[15], t[20]$.

Exercise 2.106 Use Python to compute the 0.95-quantiles of $F[10, 30], F[15, 30], F[20, 30]$.

Exercise 2.107 Show that, for each $0 < \alpha < 1$, $t_{1-\alpha/2}^2[n] = F_{1-\alpha}[1, n]$.

Exercise 2.108 Verify the relationship

$$F_{1-\alpha}[\nu_1, \nu_2] = \frac{1}{F_\alpha[\nu_2, \nu_1]}, \quad 0 < \alpha < 1,$$

$\nu_1, \nu_2 = 1, 2, \cdots$.

Exercise 2.109 Verify the formula

$$V\{t[\nu]\} = \frac{\nu}{\nu - 2}, \quad \nu > 2.$$

Exercise 2.110 Find the expected value and variance of $F[3, 10]$.

Chapter 3
Statistical Inference and Bootstrapping

Preview In this chapter we introduce basic concepts and methods of statistical inference. The focus is on estimating the parameters of statistical distributions and testing hypotheses about them. Problems of testing if certain distributions fit observed data are also considered.

We begin with some basic problems of estimation theory. A statistical population is represented by the distribution function(s) of the observable random variable(s) associated with its elements. The actual distributions representing the population under consideration are generally unspecified or only partially specified. Based on some theoretical considerations, and/or practical experience, we often assume that a distribution belongs to a particular family such as normal, Poisson, Weibull, etc. Such assumptions are called the **statistical model**. If the model assumes a specific distribution with known parameters, there is no need to estimate the parameters. We may, however, use sample data to test whether the hypothesis concerning the specific distribution in the model is valid. This is a "**goodness of fit**" testing problem. If the model assumes only the family to which the distribution belongs, while the specific values of the parameters are unknown, the problem is that of estimating the unknown parameters. The present chapter presents the basic principles and methods of statistical estimation and testing hypotheses for infinite population models.

3.1 Sampling Characteristics of Estimators

The means and the variances of random samples vary randomly around the true values of the parameters. In practice we usually take one sample of data and then construct a single estimate for each population parameter. To illustrate the

Supplementary Information The online version contains supplementary material available at https://doi.org/10.1007/978-3-031-07566-7_3.

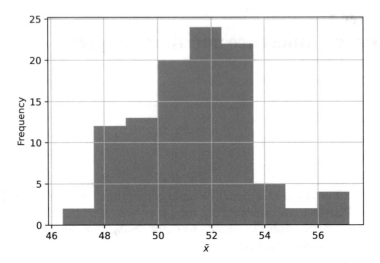

Fig. 3.1 Histogram of 100 sample means

concept of error in estimation, consider what happens if we take many samples from the same population. The collection of estimates (one from each sample) can itself be thought of as a sample taken from a hypothetical population of all possible estimates. The distribution of all possible estimates is called the **sampling distribution**. The sampling distributions of the estimates may be of a different type than the distribution of the original observations. In Figs. 3.1 and 3.2, we present the frequency distributions of $\bar{X}_{10}$ and of S^2_{10} for 100 random samples of size $n = 10$, drawn from the uniform distribution over the integers $\{1, \cdots, 100\}$.

We see in Fig. 3.1 that the frequency distribution of sample means does not resemble a uniform distribution but seems to be close to normal. Moreover, the spread of the sample means is from 46 to 57, rather than the original spread from 1 to 100. We have discussed in Chap. 2 the CLT which states that **when the sample size is large, the sampling distribution of the sample mean of a simple random sample, $\bar{X}_n$, for any population having a finite positive variance σ^2, is approximately normal with mean**

$$E\{\bar{X}\} = \mu \tag{3.1}$$

and variance

$$V\{\bar{X}_n\} = \frac{\sigma^2}{n}. \tag{3.2}$$

Notice that

$$\lim_{n \to \infty} V\{\bar{X}_n\} = 0.$$

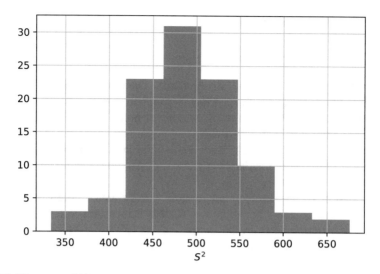

Fig. 3.2 Histogram of 100 sample variances

This means that the precision of the sample mean, as an estimator of the population mean μ, grows with the sample size.

Generally, if a function of the sample values $X_1, \cdots, X_n$, $\hat{\theta}(X_1, \cdots, X_n)$ is an estimator of a parameter θ of a distribution, then $\hat{\theta}_n$ is called an **unbiased estimator** if

$$E\{\hat{\theta}_n\} = \theta \quad \text{for all} \quad \theta. \tag{3.3}$$

Furthermore, $\hat{\theta}_n$ is called a **consistent estimator** of θ, if for any $\epsilon > 0$, $\lim_{n\to\infty} \Pr\{|\hat{\theta}_n - \theta| > \epsilon\} = 0$. Applying Chebyshev's inequality, we see that a sufficient condition for consistency is that $\lim_{n\to\infty} V\{\hat{\theta}_n\} = 0$. The sample mean is generally a consistent estimator. The standard deviation of the sampling distribution of $\hat{\theta}_n$ is called the **standard error** of $\hat{\theta}_n$, i.e., S.E. $\{\hat{\theta}_n\} = (V\{\hat{\theta}_n\})^{1/2}$.

3.2   Some Methods of Point Estimation

Consider a statistical model, which specifies the family $\mathcal{F}$ of the possible distributions of the observed V random variable. The family $\mathcal{F}$ is called a **parametric family** if the distributions in $\mathcal{F}$ are of the same functional type and differ only by the values of their parameters. For example, the family of all exponential distributions $E(\beta)$, when $0 < \beta < \infty$, is a parametric family. In this case we can write

$$\mathcal{F} = \{E(\beta) : 0 < \beta < \infty\}.$$

Another example of a parametric family is

$$\mathcal{F} = \{N(\mu, \sigma); -\infty < \mu < \infty, 0 < \sigma < \infty\},$$

which is the family of all normal distributions. The range Θ of the parameter(s) $\boldsymbol{\theta}$, is called the **parameter space**. Thus, a parametric statistical model specifies the parametric family $\mathcal{F}$. This specification gives both the functional form of the distribution and its parameter(s) space Θ.

We observe a random sample from the infinite population, which consists of the values of i.i.d. random variables $X_1, X_2, \cdots, X_n$, whose common distribution $F(x; \boldsymbol{\theta})$ is an element of $\mathcal{F}$.

A function of the observable random variables is called a **statistic**. A statistic cannot depend on unknown parameters. A statistic is thus a random variable, whose value can be determined from the sample values $(X_1, \cdots, X_n)$. In particular, a statistic $\hat{\theta}(X_1, \cdots, X_n)$, which yields values in the parameter space is called a **point estimator** of θ. If the distributions in F depend on several parameters we have to determine point estimators for each parameter, or for a function of the parameters. For example, the p-th quantile of a normal distribution is $\xi_p = \mu + z_p \sigma$, where μ and σ are the parameters and $z_p = \Phi^{-1}(p)$. This is a function of two parameters. An important problem in quality control is to estimate such quantiles. In this section we discuss a few methods of deriving point estimators.

3.2.1   Moment Equation Estimators

If $X_1, X_2, \cdots, X_n$ are i.i.d. random variables (a random sample), then the sample l-th moment $(l = 1, 2, \cdots)$ is

$$M_l = \frac{1}{n} \sum_{i=1}^{n} X_i^l. \tag{3.4}$$

The law of large numbers (strong) says that if $E\{|X|^l\} < \infty$, then M_l converges with probability one to the population l-th moment $\mu_l(F)$. Accordingly, we know that if the sample size n is large, then, with probability close to 1, M_l is close to $\mu_l(F)$. The method of moments, for parametric models, equates M_l to μ_l, which is a function of θ, and solves for θ. Generally, if $F(x; \boldsymbol{\theta})$ depends on k parameters $\theta_1, \theta_2, \cdots, \theta_k$, then we set up k equations

$$M_1 = \mu_1(\theta_1, \cdots, \theta_k),$$

$$M_2 = \mu_2(\theta_1, \cdots, \theta_k),$$

$$\vdots \tag{3.5}$$

$$M_k = \mu_k(\theta_1, \cdots, \theta_k),$$

and solve for $\theta_1, \cdots, \theta_k$. The solutions are functions of the sample statistics $M_1, \cdots, M_k$ and are therefore estimators. This method does not always yield simple or good estimators. We give now a few examples in which the estimators obtained by this method are reasonable.

Example 3.1   Consider the family $\mathcal{F}$ of Poisson distributions, i.e.,

$$\mathcal{F} = \{P(x; \theta); 0 < \theta < \infty\}.$$

The parameter space is $\Theta = (0, \infty)$. The distributions depend on one parameter, and

$$\mu_1(\theta) = E_\theta\{X\} = \theta.$$

Thus, the method of moments yields the estimator

$$\hat{\theta}_n = \bar{X}_n.$$

This is an unbiased estimator with $V\{\hat{\theta}_n\} = \dfrac{\theta}{n}$.   ∎

Example 3.2   Consider a random sample of $X_1, X_2, \cdots, X_n$ from a log-normal distribution $LN(\mu, \sigma)$. The distributions depend on $k = 2$ parameters.
   We have seen that

$$\mu_1(\mu, \sigma^2) = \exp\{\mu + \sigma^2/2\},$$

$$\mu_2(\mu, \sigma^2) = \exp\{2\mu + \sigma^2\}(e^{\sigma^2} - 1).$$

Thus, let $\theta_1 = \mu$, $\theta_2 = \sigma^2$ and set the equations

$$\exp\{\theta_1 + \theta_2/2\} = M_1$$

$$\exp\{2\theta_1 + \theta_2\}(e^{\theta_2} - 1) = M_2.$$

The solutions $\hat{\theta}_1$ and $\hat{\theta}_2$ of this system of equations are

$$\hat{\theta}_1 = \log M_1 - \frac{1}{2} \log \left(1 + \frac{M_2}{M_1^2} \right),$$

and

$$\hat{\theta}_2 = \log \left(1 + \frac{M_2}{M_1^2} \right).$$

The estimators obtained are **biased**, but we can show that they are **consistent**. Simple formulae for $V\{\hat{\theta}_1\}$, $V\{\hat{\theta}_2\}$ and $\text{cov}(\hat{\theta}_1, \hat{\theta}_2)$ do not exist. We can derive large sample approximations to these characteristics or approximate them by a method of resampling, called bootstrapping, which is discussed later in Sect. 3.10.   ∎

3.2.2   The Method of Least Squares

If $\mu = E\{X\}$, then the method of least squares chooses the estimator $\hat{\mu}$, which minimizes

$$Q(\mu) = \sum_{i=1}^{n} (X_i - \mu)^2. \tag{3.6}$$

It is immediate to show that the **least squares estimator** (LSE) is the sample mean, i.e.,

$$\hat{\mu} = \bar{X}_n.$$

Indeed, write

$$Q(\mu) = \sum_{i=1}^{n} (X_i - \bar{X}_n + \bar{X}_n - \mu)^2$$

$$= \sum_{i=1}^{n} (X_i - \bar{X}_n)^2 + n(\bar{X}_n - \mu)^2.$$

Thus, $Q(\hat{\mu}) \geq Q(\bar{X}_n)$ for all μ and $Q(\hat{\mu})$ is minimized only if $\hat{\mu} = \bar{X}_n$. This estimator is in a sense non-parametric. It is unbiased and consistent. Indeed,

$$V\{\hat{\mu}\} = \frac{\sigma^2}{n},$$

provided that $\sigma^2 < \infty$.

The LSE is more interesting in the case of linear regression (see Chap. 4).

In the simple linear regression case, we have n independent random variables $Y_1, \cdots, Y_n$, with equal variances, σ^2, but expected values which depend linearly on known regressors (predictors) $x_1, \cdots, x_n$. That is,

$$E\{Y_i\} = \beta_0 + \beta_1 x_i, \quad i = 1, \cdots, n. \tag{3.7}$$

The least squares estimators of the regression coefficients β_0 and β_1 are the values which minimize

$$Q(\beta_0, \beta_1) = \sum_{i=1}^{n} (Y_i - \beta_0 - \beta_1 x_i)^2. \tag{3.8}$$

These LSEs are

$$\hat{\beta}_0 = \bar{Y}_n - \hat{\beta}_1 \bar{x}_n, \tag{3.9}$$

and

$$\hat{\beta}_1 = \frac{\sum_{i=1}^{n} Y_i (x_i - \bar{x}_n)}{\sum_{i=1}^{n} (x_i - \bar{x}_n)^2}, \tag{3.10}$$

where $\bar{x}_n$ and $\bar{Y}_n$ are the sample means of the xs and the Ys, respectively. Thus, $\hat{\beta}_0$ and $\hat{\beta}_1$ are linear combinations of the Ys, with known coefficients. From the results of Sect. 2.8,

$$E\{\hat{\beta}_1\} = \sum_{i=1}^{n} \frac{(x_i - \bar{x}_n)}{SS_x} E\{Y_i\}$$

$$= \sum_{i=1}^{n} \frac{(x_i - \bar{x}_n)}{SS_x} (\beta_0 + \beta_1 x_i)$$

$$= \beta_0 \sum_{i=1}^{n} \frac{(x_i - \bar{x}_n)}{SS_x} + \beta_1 \sum_{i=1}^{n} \frac{(x_i - \bar{x}_n) x_i}{SS_x},$$

where $SS_x = \sum_{i=1}^{n} (x_i - \bar{x}_n)^2$. Furthermore,

$$\sum_{i=1}^{n} \frac{x_i - \bar{x}_n}{SS_x} = 0$$

and

$$\sum_{i=1}^{n} \frac{(x_i - \bar{x}_n)x_i}{SS_x} = 1.$$

Hence, $E\{\hat{\beta}_1\} = \beta_1$. Also,

$$\begin{aligned}
E\{\hat{\beta}_0\} &= E\{\bar{Y}_n\} - \bar{x}_n E\{\hat{\beta}_1\} \\
&= (\beta_0 + \beta_1 \bar{x}_n) - \beta_1 \bar{x}_n \\
&= \beta_0.
\end{aligned}$$

Thus, $\hat{\beta}_0$ and $\hat{\beta}_1$ are both **unbiased**. The variances of these LSE are given by

$$V\{\hat{\beta}_1\} = \frac{\sigma^2}{SS_x}, \tag{3.11}$$
$$V\{\hat{\beta}_0\} = \frac{\sigma^2}{n} + \frac{\sigma^2 \bar{x}_n^2}{SS_x},$$

and

$$\text{cov}(\hat{\beta}_0, \hat{\beta}_1) = -\frac{\sigma^2 \bar{x}_n}{SS_x}. \tag{3.12}$$

Thus, $\hat{\beta}_0$ and $\hat{\beta}_1$ are **not** independent. A hint for deriving these formulae is given in Exercise 3.8.

The correlation between $\hat{\beta}_0$ and $\hat{\beta}_1$ is

$$\rho = -\frac{\bar{x}_n}{\left(\frac{1}{n}\sum_{i=1}^{n} x_i^2\right)^{1/2}}. \tag{3.13}$$

3.2.3   Maximum Likelihood Estimators

Let $X_1, X_2, \cdots, X_n$ be i.i.d. random variables having a common distribution belonging to a parametric family $\mathcal{F}$. Let $f(x; \theta)$ be the p.d.f. of X, $\theta \in \Theta$. This is either a density function or a probability distribution function of a discrete random variable. Since $X_1, \cdots, X_n$ are independent, their joint p.d.f. is

$$f(x_1, \cdots, x_n; \theta) = \prod_{i=1}^{n} f(x_i; \theta).$$

The **likelihood function** of θ over Θ is defined as

$$L(\boldsymbol{\theta}; x_1, \cdots, x_n) = \prod_{i=1}^{n} f(x_i; \boldsymbol{\theta}). \tag{3.14}$$

The likelihood of $\boldsymbol{\theta}$ is thus the probability in the discrete case, or the joint density in the continuous case, of the observed sample values under $\boldsymbol{\theta}$. In the likelihood function $L(\boldsymbol{\theta}; x_1, \ldots, x_n)$, the sample values $(x_1, \ldots, x_n)$ are playing the role of parameters. A **maximum likelihood estimator** (MLE) of $\boldsymbol{\theta}$ is a point in the parameter space, $\hat{\theta}_n$, for which $L(\boldsymbol{\theta}; X_1, \cdots, X_n)$ is maximized. The notion of maximum is taken in a general sense. For example, the function

$$f(x; \lambda) = \begin{cases} \lambda e^{-\lambda x}, & x \geq 0 \\ 0, & x < 0 \end{cases}$$

as a function of λ, $0 < \lambda < \infty$, attains a maximum at $\lambda = \dfrac{1}{x}$.

On the other hand, the function

$$f(x; \theta) = \begin{cases} \dfrac{1}{\theta}, & 0 \leq x \leq \theta \\ 0, & \text{otherwise} \end{cases}$$

as a function of θ, over $(0, \infty)$ attains a lowest upper bound (supremum) at $\theta = x$, which is $\dfrac{1}{x}$. We say that it is maximized at $\theta = x$. Notice that it is equal to zero for $\theta < x$. We give a few examples.

Example 3.3 Suppose that $X_1, X_2, \cdots, X_n$ is a random sample from a normal distribution. Then, the likelihood function of (μ, σ^2) is

$$L(\mu, \sigma^2; X_1, \cdots, X_n) = \frac{1}{(2\pi)^{n/2}\sigma^n} \exp\left\{ -\frac{1}{2\sigma^2} \sum_{i=1}^{n}(X_i - \mu)^2 \right\}$$

$$= \frac{1}{(2\pi)^{n/2}(\sigma^2)^{n/2}} \exp\left\{ -\frac{1}{2\sigma^2} \sum_{i=1}^{n}(X_i - \bar{X}_n)^2 - \frac{n}{2\sigma^2}(\bar{X}_n - \mu)^2 \right\}.$$

Notice that the likelihood function of (μ, σ^2) depends on the sample variables only through the statistics $(\bar{X}_n, Q_n)$, where $Q_n = \sum_{i=1}^{n}(X_i - \bar{X}_n)^2$. These statistics are called the **likelihood** or **sufficient statistics**. To maximize the likelihood, we can maximize the log likelihood

$$l(\mu, \sigma^2; \bar{X}_n, Q_n) = -\frac{n}{2}\log(2\pi) - \frac{n}{2}\log(\sigma^2) - \frac{Q_n}{2\sigma^2} - \frac{n(\bar{X}_n - \mu)^2}{2\sigma^2}.$$

With respect to μ, we maximize by $\hat{\mu}_n = \bar{X}_n$. With respect to σ^2, differentiate

$$l(\hat{\mu}_n, \sigma^2; \bar{X}_n, Q_n) = -\frac{n}{2}\log(2\pi) - \frac{n}{2}\log(\sigma^2) - \frac{Q_n}{2\sigma^2}.$$

This is

$$\frac{\partial}{\partial \sigma^2}\log(\hat{\mu}, \sigma^2; \bar{X}_n, Q_n) = -\frac{n}{2\sigma^2} + \frac{Q_n}{2\sigma^4}.$$

Equating the derivative to zero and solving yields the MLE

$$\hat{\sigma}_n^2 = \frac{Q_n}{n}.$$

Thus, the MLEs are $\hat{\mu}_n = \bar{X}_n$ and

$$\hat{\sigma}_n^2 = \frac{n-1}{n}S_n^2.$$

$\hat{\sigma}_n^2$ is biased, but the bias goes to zero as $n \to \infty$.   ∎

Example 3.4 Let X have a negative binomial distribution NB(k, p). Suppose that k is known and $0 < p < 1$. The likelihood function of p is

$$L(p; X, k) = \binom{X+k-1}{k-1}p^k(1-p)^X.$$

Thus, the log likelihood is

$$l(p; X, k) = \log\binom{X+k-1}{k-1} + k\log p + X\log(1-p).$$

The MLE of p is

$$\hat{p} = \frac{k}{X+k}.$$

We can show that $\hat{p}$ has a positive bias, i.e., $E\{\hat{p}\} > p$. For large values of k, the bias is approximately

$$\text{Bias}(\hat{p}; k) = E\{\hat{p}; k\} - p$$

$$\cong \frac{3p(1-p)}{2k}, \quad \text{large } k.$$

The variance of $\hat{p}$ for large k is approximately $V\{\hat{p}; k\} \cong \frac{p^2(1-p)}{k}$. ∎

3.3 Comparison of Sample Estimates

3.3.1 Basic Concepts

Statistical hypotheses are statements concerning the parameters, or some characteristics, of the distribution representing a certain random variable (or variables) in a population. For example, consider a manufacturing process. The parameter of interest may be the proportion, p, of nonconforming items. If $p \leq p_0$, the process is considered to be acceptable. If $p > p_0$ the process should be corrected.

Suppose that 20 items are randomly selected from the process and inspected. Let X be the number of nonconforming items in the sample. Then X has a binomial distribution $B(20, p)$. On the basis of the observed value of X, we have to decide whether the process should be stopped for adjustment. In the statistical formulation of the problem, we are testing the hypothesis

$$H_0 : p \leq p_0,$$

against the hypothesis

$$H_1 : p > p_0.$$

The hypothesis H_0 is called the **null hypothesis**, while H_1 is called the **alternative hypothesis**. Only when the data provides significant evidence that the null hypothesis is wrong do we reject it in favor of the alternative. It may not be justifiable to disrupt a production process unless we have ample evidence that the proportion of nonconforming items is too high. It is important to distinguish between **statistical significance** and **practical or technological significance**. The statistical level of significance is the probability of rejecting H_0 when it is true. If we reject H_0 at a low level of significance, the probability of committing an error is small, and we are confident that our conclusion is correct. Rejecting H_0 might not be technologically significant, if the true value of p is not greater than $p_0 + \delta$, where δ is some acceptable level of indifference. If $p_0 < p < p_0 + \delta$, H_1 is true, but there is no technological significance to the difference $p - p_0$.

To construct a statistical test procedure based on a **test statistic**, X, consider first all possible values that could be observed. In our example X can assume the values $0, 1, 2, \cdots, 20$. Determine a **critical region** or **rejection region**, so that whenever the observed value of X belongs to this region, the null hypothesis H_0 is rejected. For example, if we were testing $H_0 : P \leq 0.10$ against $H_1 : P > 0.10$, we might reject H_0 if $X > 4$. The complement of this region, $X \leq 3$, is called the **acceptance region**.

There are two possible errors that can be committed. If the true proportion of nonconforming items, for example, were only 0.05 (unknown to us) and our sample happened to produce four items, we would incorrectly decide to reject H_0 and shut down the process that was performing acceptably. This is called a **type I error**. On the other hand, if the true proportion were 0.15 and only three nonconforming items were found in the sample, we would incorrectly allow the process to continue with more than 10% defectives (**a type II error**).

We denote the probability of committing a type I error by $\alpha(p)$, for $p \leq p_0$, and the probability of committing a type II error by $\beta(p)$, for $p > p_0$.

In most problems the critical region is constructed in such a way that the probability of committing a type I error will not exceed a preassigned value called the **significance level** of the test. Let α denote the significance level. In our example the significance level is

$$\alpha = \Pr\{X \geq 4; p = 0.1\} = 1 - B(3; 20, .1) = 0.133.$$

Notice that the significance level is computed with $p = 0.10$, which is the largest p value for which the null hypothesis is true.

To further evaluate the test procedure, we would like to know the probability of not rejecting (in practical terms "accepting") the null hypothesis for various values of p. Such a function is called the **operating characteristic function** and is denoted by $OC(p)$. The graph of $OC(p)$ vs. p is called the **OC curve**. Ideally we would like $OC(p) = 1$ whenever $H_0 : p \leq p_0$ is true and $OC(p) = 0$ when $H_1 : p > p_0$ is true. This, however, cannot be obtained when the decision is based on a random sample of items.

In our example we can compute the OC function as

$$OC(p) = \Pr\{X \leq 3; p\} = B(3; 20, p).$$

From Table 3.1 we find that

Table 3.1 The binomial c.d.f. $B(x; n, p)$, for $n = 20$, $p = 0.10(0.05)0.25$

x	$p = 0.10$	$p = 0.15$	$p = 0.20$	$p = 0.25$
0	0.1216	0.0388	0.0115	0.0032
1	0.3917	0.1756	0.0692	0.0243
2	0.6769	0.4049	0.2061	0.0913
3	0.8670	0.6477	0.4114	0.2252
4	0.9568	0.8298	0.6296	0.4148
5	0.9887	0.9327	0.8042	0.6172
6	0.9976	0.9781	0.9133	0.7858
7	0.9996	0.9941	0.9679	0.8982
8	0.9999	0.9987	0.9900	0.9591
9	1.0000	0.9998	0.9974	0.9861
10	1.0000	1.0000	0.9994	0.9961

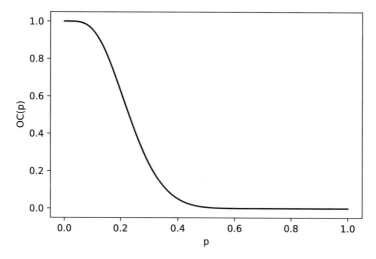

Fig. 3.3 The OC curve for testing $H_0 : p \leq 0.1$ Against $H_1 : p > 0.1$ with a sample of size $n = 20$ and rejection region $X \geq 4$

$$OC(0.10) = 0.8670$$
$$OC(0.15) = 0.6477$$
$$OC(0.20) = 0.4114$$
$$OC(0.25) = 0.2252.$$

Notice that the significance level α is the maximum probability of rejecting H_0 when it is true. Accordingly, $OC(p_0) = 1 - \alpha$. The OC curve for this example is shown in Fig. 3.3.

We see that as p grows, the value of $OC(p)$ decreases, since the probability of observing at least four nonconforming items out of 20 is growing with p.

Suppose that the significance level of the test is decreased, in order to reduce the probability of incorrectly interfering with a good process. We may choose the critical region to be $X \geq 5$. The new OC function is

$$OC(p) = \Pr\{X \leq 4; p\} = B(4; 20, p).$$

For this new critical region,

$$OC(0.10) = 0.9568,$$
$$OC(0.15) = 0.8298,$$
$$OC(0.20) = 0.6296,$$
$$OC(0.25) = 0.4148.$$

The new significance level is $\alpha = 1 - OC(0.1) = 0.0432$. Notice that, although we reduced the risk of committing a type I error, we increased the risk of committing a type II error. Only with a larger sample size can we reduce simultaneously the risks of both type I and II errors.

Instead of the OC function, one may consider the **power function**, for evaluating the sensitivity of a test procedure. The power function, denoted by $\psi(p)$, is the probability of **rejecting** the null hypothesis when the alternative is true. Thus, $\psi(p) = 1 - OC(p)$.

Finally, we consider an alternative method of performing a test. Rather than specifying in advance the desired significance level, say $\alpha = 0.05$, we can compute the probability of observing X_0 or more nonconforming items in a random sample if $p = p_0$. This probability is called the **attained significance level** or the P-**value** of the test. If the P-value is small, say ≤ 0.05, we consider the results to be **significant** and we reject the null hypothesis. For example, suppose we observed $X_0 = 6$ nonconforming items in a sample of size 20. The P-value is $\Pr\{X \geq 6; p = 0.10\} = 1 - B(5; 20, 0.10) = 0.0113$. This small probability suggests that we could reject H_0 in favor of H_1 without much of a risk.

The term P-value should not be confused with the parameter p of the binomial distribution.

3.3.2   Some Common One-Sample Tests of Hypotheses

3.3.2.1   The Z-Test: Testing the Mean of a Normal Distribution, σ^2 Known

One-Sided Test

The hypothesis for a one-sided test on the mean of a normal distribution is

$$H_0 : \mu \leq \mu_0,$$

against

$$H_1 : \mu > \mu_0,$$

where μ_0 is a specified value. Given a sample $X_1, \cdots, X_n$, we first compute the sample mean $\bar{X}_n$. Since large values of $\bar{X}_n$, relative to μ_0, would indicate that H_0 is possibly not true, the critical region should be of the form $\bar{X} \geq C$, where C is chosen so that the probability of committing a type I error is equal to α. (In many problems we use $\alpha = 0.01$ or 0.05 depending on the consequences of a type I error.) For convenience we use a modified form of the test statistic, given by the Z-statistic

$$Z = \sqrt{n}(\bar{X}_n - \mu_0)/\sigma. \tag{3.15}$$

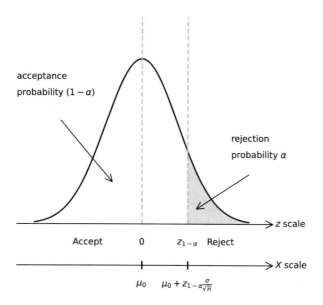

Fig. 3.4 Critical regions for the one-sided Z-test

The critical region, in terms of Z, is given by

$$\{Z : Z \geq z_{1-\alpha}\},$$

where $z_{1-\alpha}$ is the $1 - \alpha$ quantile of the standard normal distribution. This critical region is equivalent to the region

$$\{\bar{X}_n : \bar{X}_n \geq \mu_0 + z_{1-\alpha}\sigma/\sqrt{n}\}.$$

These regions are illustrated in Fig. 3.4.

The operating characteristic function of this test is given by

$$OC(\mu) = \Phi(z_{1-\alpha} - \delta\sqrt{n}), \tag{3.16}$$

where

$$\delta = (\mu - \mu_0)/\sigma. \tag{3.17}$$

Example 3.5 Suppose we are testing the hypothesis $H_0 : \mu \leq 5$, against $H_1 : \mu > 5$, with a sample of size $n = 100$ from a normal distribution with known standard deviation $\sigma = 0.2$. With a significance level of size $\alpha = 0.05$ we reject H_0 if

Table 3.2 OC values in the normal case

mu	$\delta\sqrt{n}$	z	OC(μ)
5.00	0.0	1.645	0.9500
5.01	0.5	1.145	0.8739
5.02	1.0	0.645	0.7405
5.03	1.5	0.145	0.5576
5.04	2.0	−0.355	0.3612
5.05	2.5	−0.855	0.1962

$$Z \geq z_{.95} = 1.645.$$

The values of the OC function are computed in the Table 3.2. In this table $z = z_{1-\alpha} - \delta\sqrt{n}$ and $\mathrm{OC}(\mu) = \Phi(z)$.   ∎

If the null hypothesis is $H_0 : \mu \geq \mu_0$ against the alternative $H_1 : \mu < \mu_0$, we reverse the direction of the test and reject H_0 if $Z \leq -z_{1-\alpha}$.

Two-Sided Test
The two-sided test has the form

$$H_0 : \mu = \mu_0$$

against

$$H_1 : \mu \neq \mu_0.$$

The corresponding critical region is given by

$$\{Z : Z \geq z_{1-\alpha/2}\} \cup \{Z : Z \leq -z_{1-\alpha/2}\}.$$

The operating characteristic function is

$$\mathrm{OC}(\mu) = \Phi(z_{1-\alpha/2} + \delta\sqrt{n}) - \Phi(-z_{1-\alpha/2} - \delta\sqrt{n}). \tag{3.18}$$

The P-value of the two-sided test can be determined in the following manner. First compute

$$|Z_0| = \sqrt{n}|\bar{X}_n - \mu_0|/\sigma,$$

and then compute the P-value

$$
\begin{aligned}
P &= \Pr\{Z \geq |Z_0|\} + P\{Z \leq -|Z_0|\} \\
&= 2(1 - \Phi(|Z_0|)).
\end{aligned}
\tag{3.19}
$$

3.3.2.2 The *t*-Test: Testing the Mean of a Normal Distribution, σ^2 Unknown

In this case, we replace σ in the above Z-test with the sample standard deviation, S, and $z_{1-\alpha}$ (or $z_{1-\alpha/2}$) with $t_{1-\alpha}[n-1]$ (or $t_{1-\alpha/2}[n-1]$). Thus, the critical region for the two-sided test becomes

$$\{t : |t| \geq t_{1-\alpha/2}[n-1]\},$$

where

$$t = (\bar{X}_n - \mu_0)\sqrt{n}/S. \tag{3.20}$$

The operating characteristic function of the one-sided test is given approximately by

$$OC(\mu) \cong 1 - \Phi\left(\frac{\delta\sqrt{n} - t_{1-\alpha}[n-1](1 - 1/8(n-1))}{(1 + t_{1-\alpha}^2[n-1]/2(n-1))^{1/2}}\right) \tag{3.21}$$

where $\delta = |\mu - \mu_0|/\sigma$. (This is a good approximation to the exact formula, which is based on the complicated non-central t-distribution.)

In Table 3.3, we present some numerical comparisons of the **power** of the one-sided test for the cases of σ^2 known and σ^2 unknown, when $n = 20$ and $\alpha = 0.05$. Notice that when σ is unknown, the power of the test is somewhat smaller than when it is known.

Example 3.6 The cooling system of a large computer consists of metal plates that are attached together, so as to create an internal cavity, allowing for the circulation of special-purpose cooling liquids. The metal plates are attached with steel pins that are designed to measure 0.5 mm in diameter. Experience with the process of manufacturing similar steel pins has shown that the diameters of the pins are normally distributed, with mean μ and standard deviation σ. The process is aimed at maintaining a mean of $\mu_0 = 0.5$ [mm]. For controlling this process, we want to test $H_0 : \mu = 0.5$ against $H_1 : \mu \neq 0.5$. If we have prior information that the process standard deviation is constant at $\sigma = 0.02$, we can use the Z-test to test the

Table 3.3 Power functions of Z- and *t*-tests

δ	σ known	σ unknown
0.0	0.050	0.050
0.1	0.116	0.111
0.2	0.226	0.214
0.3	0.381	0.359
0.4	0.557	0.527
0.5	0.723	0.691

above hypotheses. If we apply a significance level of $\alpha = 0.05$, then we will reject H_0 if $|Z| \geq z_{1-\alpha/2} = 1.96$.

Suppose that the following data were observed:

$$0.53, 0.54, 0.48, 0.50, 0.50, 0.49, 0.52.$$

The sample size is $n = 7$ with a sample mean of $\bar{X} = 0.509$. Therefore,

$$Z = |0.509 - 0.5|\sqrt{7}/.02 = 1.191.$$

Since this value of Z does not exceed the critical value of 1.96, do not reject the null hypothesis.

If there is no prior information about σ, use the sample standard deviation S and perform a t-test, and reject H_0 if $|t| > t_{1-\alpha/2}[6]$. In the present example $S = 0.022$, and $t = 1.082$. Since $|t| < t_{.975}[6] = 2.447$, we reach the same conclusion. ∎

3.3.2.3 The Chi-Squared Test: Testing the Variance of a Normal Distribution

Consider a one-sided test of the hypothesis

$$H_0 : \sigma^2 \leq \sigma_0^2,$$

against

$$H_1 : \sigma^2 > \sigma_0^2.$$

The test statistic corresponding to this hypothesis is

$$Q^2 = (n - 1)S^2/\sigma_0^2, \tag{3.22}$$

with a critical region

$$\{Q^2 : Q^2 \geq \chi_{1-\alpha}^2[n - 1]\}.$$

The operating characteristic function for this test is given by

$$\mathrm{OC}(\sigma^2) = \Pr\{\chi^2[n - 1] \leq \frac{\sigma_0^2}{\sigma^2}\chi_{1-\alpha}^2[n - 1]\}, \tag{3.23}$$

where $\chi^2[n - 1]$ is a chi-squared random variable with $n - 1$ degrees of freedom.

Example 3.7 Continuing the previous example, let us test the hypothesis

$$H_0 : \sigma^2 \leq 0.0004,$$

against

$$H_1 : \sigma^2 > 0.0004.$$

Since the sample standard deviation is $S = 0.022$, we find

$$Q^2 = (7 - 1)(0.022)^2/0.0004 = 7.26.$$

H_0 is rejected at level $\alpha = 0.05$ if

$$Q^2 \geq \chi^2_{0.95}[6] = 12.59.$$

Since $Q^2 < \chi^2_{.95}[6]$, H_0 is not rejected.   ■

Whenever n is odd, that is $n = 2m + 1$ $(m = 0, 1, \cdots)$, the c.d.f. of $\chi^2[n - 1]$ can be computed according to the formula

$$\Pr\{\chi^2[2m] \leq x\} = 1 - P\left(m - 1; \frac{x}{2}\right),$$

where $P(a; \lambda)$ is the c.d.f. of the Poisson distribution with mean λ. For example, if $n = 21$, $m = 10$ and $\chi^2_{.95}[20] = 31.41$. Thus, the value of the OC function at $\sigma^2 = 1.5\,\sigma_0^2$ is

$$OC(1.5\sigma_0^2) = \Pr\left\{\chi^2[20] \leq \frac{31.41}{1.5}\right\}$$

$$= 1 - P(9; 10.47) = 1 - .4007$$

$$= 0.5993.$$

If n is even, i.e., $n = 2m$, we can compute the OC values for $n = 2m - 1$ and for $n = 2m + 1$ and take the average of these OC values. This will yield a good approximation.

The power function of the test is obtained by subtracting the OC function from 1.

In Table 3.4 we present a few numerical values of the **power function** for $n = 20, 30, 40$ and for $\alpha = 0.05$. Here we have let $\rho = \sigma^2/\sigma_0^2$ and have used the values $\chi^2_{0.95}[19] = 30.1$, $\chi^2_{0.95}[29] = 42.6$, and $\chi^2_{0.95}[39] = 54.6$.

As illustrated in Table 3.4, the power function changes more rapidly as n grows.

Table 3.4 Power of the χ^2-test, $\alpha = 0.05$, $\rho = \sigma^2/\sigma_0^2$

	n		
ρ	20	30	40
1.00	0.050	0.050	0.050
1.25	0.193	0.236	0.279
1.50	0.391	0.497	0.589
1.75	0.576	0.712	0.809
2.00	0.719	0.848	0.920

3.3.2.4 Testing Hypotheses About the Success Probability, p, in Binomial Trials

Consider one-sided tests, for which

- The null hypothesis is $H_0 : p \leq p_0$.
- The alternative hypothesis is $H_1 : p > p_0$.
- The critical region is $\{X : X > c_\alpha(n, p_0)\}$,

where X is the number of successes among n trials and $c_\alpha(n, p_0)$ is the first value of k for which the binomial c.d.f., $B(k; n, p_0)$, exceeds $1 - \alpha$.

The operating characteristic function

$$OC(p) = B(c_\alpha(n, p_0); n, p). \tag{3.24}$$

Notice that $c_\alpha(n, p_0) = B^{-1}(1 - \alpha; n, p_0)$ is the $(1 - \alpha)$ quantile of the binomial distribution $B(n, p_0)$. In order to determine $c(n, p_0)$, one can use Python's `stats.binom(n, p0)` which, for $\alpha = 0.05$, $n = 20$ and $p_0 = 0.20$, gives

```
stats.binom(20, 0.2).ppf(0.95)
```

| 7.0

Table 3.5 is an output for the binomial distribution with $n = 20$ and $p = 0.2$. The smallest value of k for which $B(k; 20, .2) = \Pr\{X \leq k\} \geq 0.95$ is 7. Thus, we set $c_{0.05}(20, 0.20) = 7$. H_0 is rejected whenever $X > 7$. The level of significance of this test is actually 0.032, which is due to the discrete nature of the binomial distribution. The OC function of the test for $n = 20$ can be easily determined from the corresponding distribution of $B(20, p)$. For example, the $B(n, p)$ distribution for $n = 20$ and $p = 0.25$ is presented in Table 3.6.

We see that $B(7; 20, 0.25) = 0.8982$. Hence, the probability of accepting H_0 when $p = 0.25$ is $OC(0.25) = 0.8982$.

A large sample test in the binomial case can be based on the normal approximation to the binomial distribution. If the sample is indeed large, we can use the test statistic

$$Z = \frac{\hat{p} - p_0}{\sqrt{p_0 q_0}} \sqrt{n}, \tag{3.25}$$

Table 3.5 p.d.f. and c.d.f. of $B(20, .2)$

Binomial distribution: $n = 20$ $p = 0.2$		
a	$\Pr(X = a)$	$\Pr(X \leq a)$
0	0.0115	0.0115
1	0.0576	0.0692
2	0.1369	0.2061
3	0.2054	0.4114
4	0.2182	0.6296
5	0.1746	0.8042
6	0.1091	0.9133
7	0.0546	0.9679
8	0.0222	0.9900
9	0.0074	0.9974
10	0.0020	0.9994
11	0.0005	0.9999
12	0.0001	1.0000

Table 3.6 p.d.f. and c.d.f. of $B(20, .25)$

Binomial distribution: $n = 20$ $p = 0.25$		
a	$\Pr(X = a)$	$\Pr(X \leq a)$
0	0.0032	0.0032
1	0.0211	0.0243
2	0.0669	0.0913
3	0.1339	0.2252
4	0.1897	0.4148
5	0.2023	0.6172
6	0.1686	0.7858
7	0.1124	0.8982
8	0.0609	0.9591
9	0.0271	0.9861
10	0.0099	0.9961
11	0.0030	0.9991
12	0.0008	0.9998
13	0.0002	1.0000

with the critical region

$$\{Z : Z \geq z_{1-\alpha}\},$$

where $q_0 = 1 - p_0$. Here $\hat{p}$ is the sample proportion of successes. The operating characteristic function takes the form

$$OC(p) = 1 - \Phi\left(\frac{(p - p_0)\sqrt{n}}{\sqrt{pq}} - z_{1-\alpha}\sqrt{\frac{p_0 q_0}{pq}}\right), \quad (3.26)$$

where $q = 1 - p$ and $q_0 = 1 - p_0$.

For example, suppose that $n = 450$ and the hypotheses are $H_0 : p \leq .1$ against $H_1 : p > 0.1$. The critical region, for $\alpha = 0.05$, is

$$\{\hat{p} : \hat{p} \geq 0.10 + 1.645\sqrt{(0.1)(0.9)/450}\} = \{\hat{p} : \hat{p} \geq 0.1233\}.$$

Thus, H_0 is rejected whenever $\hat{p} \geq 0.1233$. The OC value of this test, at $p = 0.15$ is approximately

$$OC(0.15) \cong 1 - \Phi\left(\frac{0.05\sqrt{450}}{\sqrt{(0.15)(0.85)}} - 1.645\sqrt{\frac{(0.1)(0.9)}{(0.15)(0.85)}}\right)$$

$$= 1 - \Phi(2.970 - 1.382)$$

$$= 1 - 0.944 = 0.056.$$

The corresponding value of the power function is 0.949. Notice that the power of rejecting H_0 for H_1 when $p = 0.15$ is so high because of the large sample size.

3.4   Confidence Intervals

Confidence intervals for unknown parameters are intervals, determined around the sample estimates of the parameters, having the property that whatever the true value of the parameter is, in repetitive sampling a prescribed proportion of the intervals, say $1 - \alpha$, will contain the true value of the parameter. The prescribed proportion, $1 - \alpha$, is called the **confidence level** of the interval. In Fig. 3.5, we illustrate 50 simulated confidence intervals, which correspond to independent samples. All of these intervals are designed to estimate the mean of the population from which the samples were drawn. In this particular simulation, the population was normally distributed with mean $\mu = 10$. We see from the figure that 47 of these 50 random intervals cover the true value of μ.

If the sampling distribution of the estimator $\hat{\theta}_n$ is approximately normal, one can use, as a rule of thumb, the interval estimator with limits

$$\hat{\theta}_n \pm 2 \text{ S.E.}\{\hat{\theta}_n\}.$$

The confidence level of such an interval will be close to 0.95 for all θ.

Generally, if one has a powerful test procedure for testing the hypothesis $H_0 : \theta = \theta_0$ versus $H_1 : \theta \neq \theta_0$, one can obtain good confidence intervals for θ by the following method.

Let $T = T(\mathbf{X})$ be a test statistic for testing $H_0 : \theta = \theta_0$. Suppose that H_0 is rejected if $T \geq \bar{K}_\alpha(\theta_0)$ or if $T \leq \underline{K}_\alpha(\theta_0)$, where α is the significance level. The interval $(\underline{K}_\alpha(\theta_0), \bar{K}_\alpha(\theta_0))$ is the acceptance region for H_0. We can now consider the family of acceptance regions $\Theta = \{(\underline{K}_\alpha(\theta), \bar{K}_\alpha(\theta)), \; \theta \in \Theta\}$, where Θ is the

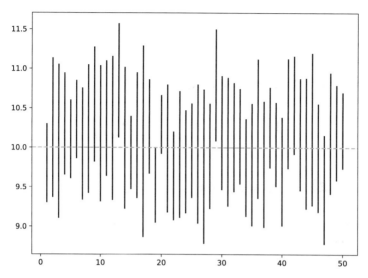

Fig. 3.5 Simulated confidence intervals for the mean of a normal distribution, samples of size $n = 10$ from $N(10, 1)$

parameter space. The interval $(L_\alpha(T), U_\alpha(T))$ defined as

$$L_\alpha(T) = \inf\{\theta : T \leq \bar{K}_\alpha(\theta)\}$$
$$U_\alpha(T) = \sup\{\theta : T \geq \underline{K}_\alpha(\theta)\} \tag{3.27}$$

is a confidence interval for θ at level of confidence $1 - \alpha$. Indeed, any hypothesis H_0 with $L_\alpha(T) < \theta_0 < U_\alpha(T)$ is accepted with the observed value of the test statistic. By construction, the probability of accepting such hypothesis is $1 - \alpha$, that is, if θ_0 is the true value of θ, the probability that H_0 is accepted is $(1 - \alpha)$. But H_0 is accepted if, and only if, θ_0 is covered by the interval $(L_\alpha(T), U_\alpha(T))$.

3.4.1 Confidence Intervals for μ; σ Known

For this case, the sample mean $\bar{X}$ is used as an estimator of μ, or as a test statistic for the hypothesis $H_0 : \mu = \mu_0$. H_0 is rejected, at level of significance α, if $\bar{X} \geq \mu_0 - z_{1-\alpha/2}\dfrac{\sigma}{\sqrt{n}}$ or $\bar{X} \leq \mu_0 + z_{1-\alpha/2}\dfrac{\sigma}{\sqrt{n}}$, where $z_{1-\alpha/2} = \Phi^{-1}(1 - \alpha/2)$. Thus, $\bar{K}_\alpha(\mu) = \mu + z_{1-\alpha/2}\dfrac{\sigma}{\sqrt{n}}$ and $\underline{K}_\alpha(\mu) = \mu - z_{1-\alpha/2}\dfrac{\sigma}{\sqrt{n}}$. The limits of the confidence interval are, accordingly, the roots μ of the equation

$$\bar{K}_\alpha(\mu) = \bar{X}$$

and

$$\underline{K}_\alpha(\mu) = \bar{X}.$$

These equations yield the confidence interval for μ,

$$\left(\bar{X} - z_{1-\alpha/2} \frac{\sigma}{\sqrt{n}}, \ \bar{X} + z_{1-\alpha/2} \frac{\sigma}{\sqrt{n}} \right). \tag{3.28}$$

3.4.2   Confidence Intervals for μ; σ Unknown

A confidence interval for μ, at level $1 - \alpha$, when σ is unknown, is obtained from the corresponding t-test. The confidence interval is

$$\left(\bar{X} - t_{1-\alpha/2}[n-1] \frac{S}{\sqrt{n}}, \ \bar{X} + t_{1-\alpha/2}[n-1] \frac{S}{\sqrt{n}} \right), \tag{3.29}$$

where $\bar{X}$ and S are the sample mean and standard deviation, respectively. $t_{1-\alpha/2}[n-1]$ is the $(1-\alpha/2)$-th quantile of the t-distribution with $n-1$ degrees of freedom.

3.4.3   Confidence Intervals for σ^2

We have seen that, in the normal case, the hypothesis $H_0 : \sigma = \sigma_0$, is rejected at level of significance α if

$$S^2 \geq \frac{\sigma_0^2}{n-1} \chi_{1-\alpha/2}^2[n-1]$$

or

$$S^2 \leq \frac{\sigma_0^2}{n-1} \chi_{\alpha/2}^2[n-1],$$

where S^2 is the sample variance and $\chi_{\alpha/2}^2[n-1]$ and $\chi_{1-\alpha/2}^2[n-1]$ are the $\alpha/2$-th and $(1-\alpha/2)$-th quantiles of χ^2, with $(n-1)$ degrees of freedom. The corresponding confidence interval for σ^2, at confidence level $(1-\alpha)$, is

$$\left(\frac{(n-1)S^2}{\chi_{1-\alpha}^2[n-1]}, \ \frac{(n-1)S^2}{\chi_{\alpha/2}^2[n-1]} \right), \tag{3.30}$$

Example 3.8 Consider a normal distribution with unknown mean μ and unknown standard deviation σ. Suppose that we draw a random sample of size $n = 16$ from this population and the sample values are

$$
\begin{array}{cccc}
16.16, & 9.33, & 12.96, & 11.49, \\
12.31, & 8.93, & 6.02, & 10.66, \\
7.75, & 15.55, & 3.58, & 11.34, \\
11.38, & 6.53, & 9.75, & 9.47.
\end{array}
$$

The mean and variance of this sample are $\bar{X} = 10.20$ and $S^2 = 10.977$. The sample standard deviation is $S = 3.313$. For a confidence level of $1 - \alpha = 0.95$, we find

$$t_{.975}[15] = 2.131,$$

$$\chi^2_{.975}[15] = 27.50,$$

$$\chi^2_{.025}[15] = 6.26.$$

Thus, the confidence interval for μ is $(8.435, 11.965)$. The confidence interval for σ^2 is $(5.987, 26.303)$. ∎

3.4.4 Confidence Intervals for p

Let X be the number of "success" in n independent trials, with unknown probability of "success," p. The sample proportion, $\hat{p} = X/n$, is an unbiased estimator of p. To construct a confidence interval for p, using $\hat{p}$, we must find limits $p_L(\hat{p})$ and $p_U(\hat{p})$ that satisfy

$$\Pr\{p_L(\hat{p}) < p < p_U(\hat{p})\} = 1 - \alpha.$$

The null hypothesis $H_0 : p = p_0$ is rejected if $\hat{p} \geq \bar{K}_\alpha(p_0)$ or $\hat{p} \leq \underline{K}_\alpha(p_0)$ where

$$\bar{K}_\alpha(p_0) = \frac{1}{n} B^{-1}(1 - \alpha/2; n, p_0)$$

and (3.31)

$$\underline{K}_\alpha(p_0) = \frac{1}{n} B^{-1}(\alpha/2; n, p_0).$$

$B^{-1}(\gamma; n, p)$ is the γ-th quantile of the binomial distribution $B(n, p)$. Thus, if $X = n\hat{p}$, the upper confidence limit for p, $p_U(\hat{p})$, is the largest value of p satisfying the equation

$$B(X; n, p) \geq \alpha/2.$$

The lower confidence limit for p is the smallest value of p satisfying

$$B(X; n, p) \leq 1 - \alpha/2.$$

Exact solutions to this equation can be obtained using tables of the binomial distribution. This method of searching for the solution in binomial tables is tedious. However, from the relationship between the F-distribution, the beta distribution, and the binomial distribution, the lower and upper limits are given by the formulae

$$p_L = \frac{X}{X + (n - X + 1)F_1} \tag{3.32}$$

and

$$p_U = \frac{(X + 1)F_2}{n - X + (X + 1)F_2}, \tag{3.33}$$

where

$$F_1 = F_{1-\alpha/2}[2(n - X + 1), 2X] \tag{3.34}$$

and

$$F_2 = F_{1-\alpha/2}[2(X + 1), 2(n - X)] \tag{3.35}$$

are the $(1 - \alpha/2)$-th quantiles of the F-distribution with the indicated degrees of freedom.

Example 3.9 Suppose that among $n = 30$ Bernoulli trials we find $X = 8$ successes. For level of confidence $1 - \alpha = 0.95$, the confidence limits are $p_L = 0.123$ and $p_U = 0.459$. Indeed,

$$B(7; 30, 0.123) = 0.975,$$

and

$$B(8; 30, 0.459) = 0.025.$$

Moreover,

$$F_1 = F_{.975}[46, 16] = 2.49$$

and

$$F_2 = F_{.975}[18, 44] = 2.07.$$

Hence,

$$p_L = 8/(8 + 23(2.49)) = 0.123$$

and

$$p_U = 9(2.07)/(22 + 9(2.07)) = 0.459.$$

∎

When the sample size n is large, we may use the normal approximation to the binomial distribution. This approximation yields the following formula for a $(1 - \alpha)$ confidence interval

$$\left(\hat{p} - z_{1-\alpha/2}\sqrt{\hat{p}\hat{q}/n}, \ \hat{p} + z_{1-\alpha/2}\sqrt{\hat{p}\hat{q}/n}\right), \tag{3.36}$$

where $\hat{q} = 1 - \hat{p}$. Applying this large sample approximation to our previous example, in which $n = 30$, we obtain the approximate 0.95-confidence interval $(0.108, 0.425)$. This interval is slightly different from the interval obtained with the exact formulae. This difference is due to the inaccuracy of the normal approximation.

It is sometimes reasonable to use only a one-sided confidence interval, for example, if $\hat{p}$ is the estimated proportion of nonconforming items in a population. Obviously, the true value of p is always greater than 0, and we may wish to determine only an upper confidence limit. In this case we apply the formula given earlier but replace $\alpha/2$ by α. For example, in the case of $n = 30$ and $X = 8$, the upper confidence limit for p, in a one-sided confidence interval, is

$$p_U = \frac{(X + 1)F_2}{n - X + (X + 1)F_2}$$

where $F_2 = F_{1-\alpha}[2(X + 1), 2(n - X)] = F_{.95}[18, 44] = 1.855$. Thus, the upper confidence limit of a 0.95 one-sided interval is $P_U = 0.431$. This limit is smaller than the upper limit of the two-sided interval.

3.5 Tolerance Intervals

Technological specifications for a given characteristic X may require that a specified proportion of elements of a statistical population satisfies certain constraints. For example, in the production of concrete, we may have the requirement that at least 90% of all concrete cubes, of a certain size, will have a compressive strength of at least 240 kg/cm^2. As another example, suppose that, in the production of washers, it is required that at least 99% of the washers produced will have a thickness between 0.121 and 0.129 inches. In both examples we want to be able to determine whether or not the requirements are satisfied. If the distributions of strength and thickness were completely known, we could determine if the requirements are met without data. However, if the distributions are not completely known, we can make these determinations only with a certain level of confidence and not with certainty.

3.5.1 Tolerance Intervals for the Normal Distributions

In order to construct tolerance intervals, we first consider what happens when the distribution of the characteristic X is completely known. Suppose, for example, that the compressive strength X of the concrete cubes is such that $Y = \ln X$ has a normal distribution with mean $\mu = 5.75$ and standard deviation $\sigma = 0.2$. The proportion of concrete cubes exceeding the specification of 240 kg/cm^2 is

$$\Pr\{X \geq 240\} = \Pr\{Y \geq \log 240\}$$
$$= 1 - \Phi((5.481 - 5.75)/0.2)$$
$$= \Phi(1.345) = 0.911$$

Since this probability is greater than the specified proportion of 90%, the requirement is satisfied.

We can also solve this problem by determining the compressive strength that is exceeded by 90% of the concrete cubes. Since 90% of the Y values are greater than the 0.1-th quantile of the $N(5.75, 0.04)$ distribution,

$$Y_{0.1} = \mu + z_{0.1}\sigma$$
$$= 5.75 - 1.28(0.2)$$
$$= 5.494.$$

Accordingly, 90% of the compressive strength values should exceed $e^{5.494} = 243.2$ kg/cm^2. Once again we see that the requirement is satisfied, since more than 90% of the cubes have strength values that exceed the specification of 240 kg/cm^2.

Notice that no sample values are required, since the distribution of X is known. Furthermore, we are **certain** that the requirement is met.

Consider the situation in which we have only partial information on the distribution of Y. Suppose we know that Y is normally distributed with standard deviation $\sigma = 0.2$, but the mean μ is unknown. The 0.1-th quantile of the distribution, $y_{0.1} = \mu + z_{0.1}\sigma$, cannot be determined exactly. Let $Y_1, \cdots, Y_n$ be a random sample from this distribution and let $\bar{Y}_n$ represent the sample mean. From the previous section, we know that

$$L(\bar{Y}_n) = \bar{Y}_n - z_{1-\alpha}\sigma/\sqrt{n}$$

is a $1 - \alpha$ lower confidence limit for the population mean, that is,

$$\Pr\{\bar{Y}_n - z_{1-\alpha}\sigma/\sqrt{n} < \mu\} = 1 - \alpha.$$

Substituting this lower bound for μ in the expression for the 0.1-th fractile, we obtain a **lower tolerance limit** for 90% of the log-compressive strengths, with confidence level $1 - \alpha$. More specifically, the lower tolerance limit at level of confidence $1 - \alpha$ is

$$L_{\alpha,.1}(\bar{Y}_n) = \bar{Y}_n - (z_{1-\alpha}/\sqrt{n} + z_{.9})\sigma.$$

In general we say that, **with confidence level of $1 - \alpha$, the proportion of population values exceeding the lower tolerance limit is at least $1 - \beta$**. This lower tolerance limit is

$$L_{\alpha,\beta}(\bar{Y}_n) = \bar{Y}_n - (z_{1-\alpha}/\sqrt{n} + z_{1-\beta})\sigma. \tag{3.37}$$

It can also be shown that the **upper tolerance limit** for a proportion $1 - \beta$ of the values, with confidence level $1 - \alpha$, is

$$U_{\alpha,\beta}(\bar{Y}_n) = \bar{Y}_n + (z_{1-\alpha}/\sqrt{n} + z_{1-\beta})\sigma \tag{3.38}$$

and a **tolerance interval** containing a proportion $1 - \beta$ of the values, with confidence $1 - \alpha$, is

$$(\bar{Y}_n - (z_{1-\alpha/2}/\sqrt{n} + z_{1-\beta/2})\sigma, \; \bar{Y}_n + (z_{1-\alpha/2}/\sqrt{n} + z_{1-\beta/2})\sigma).$$

When the standard deviation σ is unknown, we should use the sample standard deviation S to construct the tolerance limits and interval. The lower tolerance limits will be of the form $\bar{Y}_n - kS_n$ where the factor $k = k(\alpha, \beta, n)$ is determined so that with confidence level $1 - \alpha$ we can state that a proportion $1 - \beta$ of the population values will exceed this limit. The corresponding upper limit is given by $\bar{Y}_n + kS_n$ and the tolerance interval is given by

$$(\bar{Y}_n - k' S_n, \bar{Y}_n + k' S_n).$$

The "two-sided" factor $k' = k'(\alpha, \beta, n)$ is determined so that the interval will contain a proportion $1 - \beta$ of the population with confidence $1 - \alpha$. Approximate solutions, for large values of n, are given by

$$k(\alpha, \beta, n) \doteq t(\alpha, \beta, n) \tag{3.39}$$

and

$$k'(\alpha, \beta, n) \doteq t(\alpha/2, \beta/2, n), \tag{3.40}$$

where

$$t(a, b, n) = \frac{z_{1-b}}{1 - z_{1-a}^2/2n} + \frac{z_{1-a}(1 + z_b^2/2 - z_{1-a}^2/2n)^{1/2}}{\sqrt{n}(1 - z_{1-a}^2/2n)}. \tag{3.41}$$

Example 3.10 The following data represent a sample of 20 compressive strength measurements (kg/cm^2) of concrete cubes at age of 7 days.

349.09	308.88
238.45	196.20
385.59	318.99
330.00	257.63
388.63	299.04
348.43	321.47
339.85	297.10
348.20	218.23
361.45	286.23
357.33	316.69

Applying the transformation $Y = \ln X$, we find that $\bar{Y}_{20} = 5.732$ and $S_{20} = 0.184$. To obtain a lower tolerance limit for 90% of the log-compressive strengths with 95% confidence, we use the factor $k(0.05, 0.10, 20) = 2.548$. Thus, the lower tolerance limit for the transformed data is

$$\bar{Y}_{20} - k S_{20} = 5.732 - 2.548 \times 0.184 = 5.263,$$

and the corresponding lower tolerance limit for the compressive strength is

$$e^{5.263} = 193.09 \ [\text{kg/cm}^2].$$

■

Fig. 3.6 Normal Q-Q plot of simulated values from $N(10, 1)$

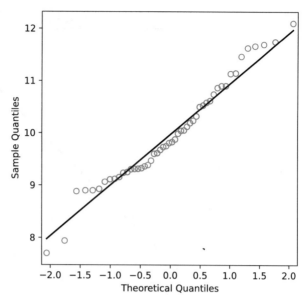

If the tolerance limits are within the specification range, we have a satisfactory production.

3.6 Testing for Normality with Probability Plots

It is often assumed that a sample is drawn from a population which has a normal distribution. It is, therefore, important to test the assumption of normality. We present here a simple test based on the **normal-scores** (NSCORES) of the sample values (Fig. 3.6). The normal scores corresponding to a sample $x_1, x_2, \cdots, x_n$ are obtained in the following manner. First, we let

$$r_i = \text{rank of } x_i, \quad i = 1, \cdots, n. \tag{3.42}$$

Here the rank of x_i is the position of x_i in a listing of the sample when it is arranged in increasing order. Thus, the rank of the smallest value is 1, that of the second smallest is 2, etc. We then let

$$p_i = (r_i - 3/8)/(n + 1/4), \quad i = 1, \cdots, n. \tag{3.43}$$

Then the normal score of x_i is

$$z_i = \Phi^{-1}(p_i),$$

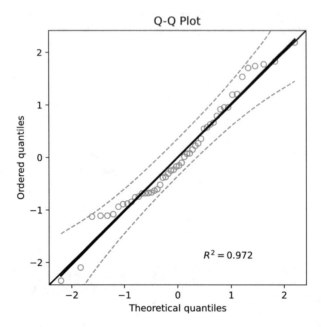

Fig. 3.7 Normal Q-Q plot of simulated values From $N(10, 1)$ with confidence intervals

i.e., the p_i-th quantile of the standard normal distribution. If the sample is drawn at random from a normal distribution $N(\mu, \sigma^2)$, the relationship between the normal scores, NSCORES, and x_i should be approximately linear. Accordingly, the correlation between $x_1, \cdots, x_n$ and their NSCORES should be close to 1 in large samples. The graphical display of the sample values versus their NSCORES is called a **normal Q-Q plot**.

In the following example, we provide a normal probability plotting of $n = 50$ values simulated from $N(10, 1)$, given in the previous section. If the simulation is good, and the sample is indeed generated from $N(10, 1)$, the X vs. NSCORES should be scattered randomly around the line $X = 10 + \text{NSCORES}$. We see in Fig. 3.7 that this is indeed the case. Also, the correlation between the x-values and their NSCORES is 0.976.

The linear regression of the x values on the NSCORES is

$$X = 10.043 + 0.953 * \text{NSCORES}.$$

We see that both the intercept and slope of the regression equation are close to the nominal values of μ and σ. Chapter 4 provides more details on linear regression, including testing statistical hypothesis on these coefficients.

In Table 3.7, we provide some critical values for testing whether the correlation between the sample values and their NSCORES is sufficiently close to 1. If the correlation is smaller than the critical value, an indication of non-normality has been

Table 3.7 Critical values for the correlation between sample values and their NSCORES (adapted from Ryan and Joiner (2000))

$n \setminus \alpha$	0.10	0.05	0.01
10	0.9347	0.9180	0.8804
15	0.9506	0.9383	0.9110
20	0.9600	0.9503	0.9290
30	0.9707	0.9639	0.9490
50	0.9807	0.9764	0.9664

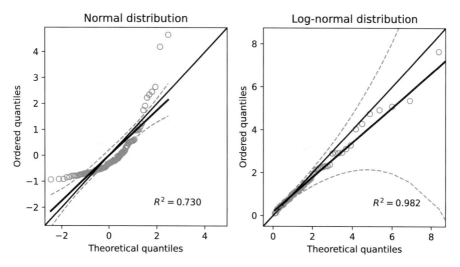

Fig. 3.8 Normal probability plot $n = 100$ random numbers generated from a log-normal distribution

established. In the example of Fig. 3.7, the correlation is $R^2 = 0.972$. This value is almost equal to the critical value for $\alpha = 0.05$ given in the following table. The hypothesis of normality is accepted.

In Python, we find implementations of Q-Q plots, for example, in the `scipy`, `statsmodels`, or `pingouin` packages. Figure 3.6 was created using the `statsmodels` package. The implementation in the `pingouin` package can display confidence intervals for the regression (see Fig. 3.8).

Both implementations can manage general distributions, and we will see these being used in the following example where we draw the sample from a non-normal distribution.

Example 3.11 Consider a sample of $n = 100$ observations from a log-normal distribution. The normal Q-Q plot of this sample is shown in Fig. 3.8 on the left. The correlation here is only 0.730. It is apparent that the relation between the NSCORES and the sample values is not linear. We reject the hypothesis that the sample has been generated from a normal distribution. If, on the other hand, we compare the distribution against a theoretical log-normal distribution, the correlation is 0.982 and we accept the hypothesis of log-normality.

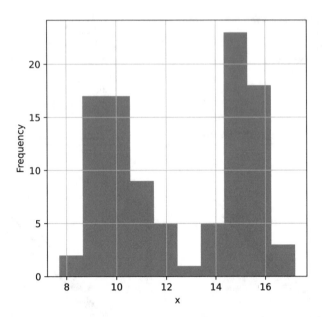

Fig. 3.9 Histogram of 100 random numbers, 50 generated from a $N(10, 1)$ and 50 from $N(15, 1)$

The code for generating a Q-Q plot to compare against a log-normal is shown here.

```
np.random.seed(1)
dist=stats.lognorm(s=0.1,loc=10)
x = dist.rvs(100)

fig, ax = plt.subplots(figsize=[5, 5])
stats.probplot(x, dist=stats.lognorm, sparams=[0.1, 10, 1], plot=ax)
plt.show()
```

■

Example 3.12 We consider here a sample of $n = 100$ values, with 50 of the values generated from $N(10, 1)$ and 50 from $N(15, 1)$. Thus, the sample represents a mixture of two normal distributions. The histogram is given in Fig. 3.9 and a normal probability plot in Fig. 3.10. The normal probability plot is definitely not linear. Although the correlation is 0.885, the hypothesis of a normal distribution is rejected.   ■

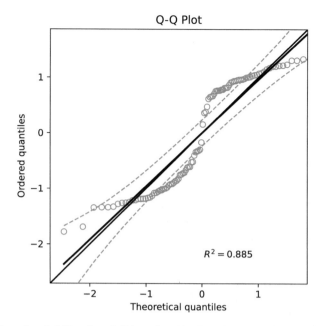

Fig. 3.10 Normal probability plot of 100 random Numbers Generated From a Mixture of Two Normal Distributions

3.7 Tests of Goodness of Fit

3.7.1 The Chi-Square Test (Large Samples)

The chi-square test is applied by comparing the observed frequency distribution of the sample to the expected one under the assumption of the model. More specifically, consider a (large) sample of size N. Let $\xi_0 < \xi_1 < \cdots < \xi_k$ be the limit points of k subintervals of the frequency distribution, and let f_i be the observed frequency in the i-th subinterval. If, according to the model, the c.d.f. is specified by the distribution function $F(x)$, then the expected frequency e_i in the i-th subinterval is

$$e_i = N(F(\xi_i) - F(\xi_{i-1})), \quad i = 1, \cdots, k.$$

The chi-square statistic is defined as

$$\chi^2 = \sum_{i=1}^{k} \frac{(f_i - e_i)^2}{e_i}.$$

We notice that

$$\sum_{i=1}^{k} f_i = \sum_{i=1}^{k} e_i = N,$$

and hence,

$$\chi^2 = \sum_{i=1}^{k} \frac{f_i^2}{e_i} - N.$$

The value of χ^2 is distributed approximately like $\chi^2[k-1]$. Thus, if $\chi^2 \geq \chi^2_{1-\alpha}[k-1]$, the distribution $F(x)$ does not fit the observed data.

Often, the c.d.f. $F(x)$ is specified by its family, e.g., normal or Poisson, but the values of the parameters have to be estimated from the sample. In this case, we reduce the number of degrees of freedom of χ^2 by the number of estimated parameters. For example, if $F(x)$ is $N(\mu, \sigma^2)$, where both μ and σ^2 are unknown, we use $N(\bar{X}, S^2)$ and compare χ^2 to $\chi^2_{1-\alpha}[k-3]$.

Example 3.13 In Sect. 3.1, we considered the sampling distribution of sample means from the uniform distribution over the integers $\{1, \cdots, 100\}$. The frequency distribution of the means of samples of size $n = 10$ is given in Fig. 3.1. We test here whether the model $N(50.5, 83.325)$ fits this data.

The observed and expected frequencies (for $N = 100$) are summarized in Table 3.8.

The sum of e_i here is 99.13, due to truncation of the tails of the normal distribution. The value of χ^2 is 12.86. The value of $\chi^2_{.95}[8]$ is 15.5. Thus, the deviation of the observed frequency distribution from the expected one is not significant at the $\alpha = 0.05$ level.   ∎

Example 3.14 We consider here a sample of 100 cycle times of a piston, which is described in detail in Chap. 2 in the Industrial Statistics book. We make a chi-

Table 3.8 Observed and expected frequencies of 100 sample means

Interval	f_i	e_i
27.5 – 32.5	3	1.84
32.5 – 37.5	11	5.28
37.5 – 42.5	12	11.32
42.5 – 47.5	11	18.08
47.4 – 52.5	19	21.55
52.5 – 57.5	24	19.70
57.5 – 62.5	14	12.73
62.5 – 67.5	4	6.30
67.5 – 72.5	2	2.33
Total	100	99.13

Table 3.9 Observed and expected frequencies of 100 cycle times

Lower limit	Upper limit	Observed frequency	Expected frequency
at or below	0.1050	7	6.1
0.1050	0.1100	9	7.7
0.1100	0.1150	17	12.6
0.1150	0.1200	12	16.8
0.1200	0.1250	18	18.1
0.1250	0.1300	11	15.9
0.1300	0.1350	12	11.4
at or above	0.1350	14	11.4

squared test whether the distribution of cycle times is normal. The estimated values of μ and σ are $\hat{\mu} = 0.1219$ and $\hat{\sigma} = 0.0109$.

In Table 3.9 we provide the observed and expected frequencies over $k = 8$ intervals.

The calculated value of χ^2 is 5.4036. We should consider the distribution of χ^2 with $k - 3 = 5$ degrees of freedom. The P value of the test is 0.37. The hypothesis of normality is not rejected. ∎

3.7.2 The Kolmogorov-Smirnov Test

The Kolmogorov-Smirnov (KS) test is a more accurate test of goodness of fit than the chi-squared test of the previous section.

Suppose that the hypothesis is that the sample comes from a specified distribution with c.d.f. $F_0(x)$. The test statistic compares the empirical distribution of the sample, $\hat{F}_n(x)$, to $F_0(x)$, and considers the maximal value, over all x values, that the distance $|\hat{F}_n(x) - F_0(x)|$ may assume. Let $x_{(1)} \leq x_{(2)} \leq \cdots \leq x_{(n)}$ be the ordered sample values. Notice that $\hat{F}_n(x_{(i)}) = \dfrac{i}{n}$. The KS test statistic can be computed according to the formula

$$D_n = \max_{1 \leq i \leq n} \left\{ \max \left\{ \frac{i}{n} - F_0(x_{(i)}), F_0(x_{(i)}) - \frac{i-1}{n} \right\} \right\} \tag{3.44}$$

We have shown earlier that $U = F(X)$ **has a uniform distribution on** $(0, 1)$.

Accordingly, if the null hypothesis is correct, $F_0(X_{(i)})$ is distributed like the i-th order statistic $U_{(i)}$ from a uniform distribution on $(0, 1)$, irrespective of the particular functional form of $F_0(x)$. The distribution of the KS test statistic, D_n, is therefore independent of $F_0(x)$, if the hypothesis, H, is correct. Tables of the critical values k_α and D_n are available. One can also estimate the value of k_α by the bootstrap method, discussed later.

Table 3.10 Some critical
values δ_α^*

α	0.15	0.10	0.05	0.025	0.01
δ_α^*	0.775	0.819	0.895	0.995	1.035

If $F_0(x)$ is a normal distribution, i.e., $F_0(x) = \Phi\left(\dfrac{x-\mu}{\sigma}\right)$, and if the mean μ and the Standard deviation, σ, are unknown, one can consider the test statistic

$$D_n^* = \max_{1 \le i \le n}\left\{\max\left\{\frac{i}{n} - \Phi\left(\frac{X_{(i)} - \bar{X}_n}{S_n}\right), \Phi\left(\frac{X_{(i)} - \bar{X}_n}{S_n}\right) - \frac{i-1}{n}\right\}\right\},$$

(3.45)

where $\bar{X}_n$ and S_n are substituted for the unknown μ and σ. The critical values k_α^* for D_n^* are given approximately by

$$k_\alpha^* = \delta_\alpha^* / \left(\sqrt{n} - 0.01 + \frac{0.85}{\sqrt{n}}\right),$$

(3.46)

where δ_α^* is given in Table 3.10.

To compute the Kolmogorov-Smirnov statistics in Python, use the function `scipy.stats.kstest`.

```
oturb = mistat.load_data('OTURB')

result = stats.kstest(oturb, 'norm',
          args=(np.mean(oturb), np.std(oturb, ddof=1)),
          alternative='two-sided')
```

For the data in Example 3.14 (file name **OTURB.csv**), we obtain $D_n^* = 0.1108$. According to Table 3.10, the critical value for $\alpha = 0.05$ is $k_{.05}^* = 0.895/(10 - 0.01 + 0.085) = 0.089$. Thus, the hypothesis of normality for the piston cycle time data is rejected at $\alpha = 0.05$.

3.8 Bayesian Decision Procedures

It is often the case that optimal decision depends on unknown parameters of statistical distributions. The Bayesian decision framework provides us the tools to integrate information that one may have on the unknown parameters with the information obtained from the observed sample in such a way that the expected loss due to erroneous decisions will be minimized. In order to illustrate an industrial decision problem of such nature, consider the following example.

Example 3.15 (Inventory Management) The following is the simplest inventory problem that is handled daily by organizations of all sizes worldwide. One such organization is Starbread Express that supplies bread to a large community in the Midwest. Every night, the shift manager has to decide how many loafs of bread, s, to bake for next day's consumption. Let X (a random variable) be the number of units demanded during the day. If a manufactured unit is left at the end of the day, we lose $\$\,c_1$ on that unit. On the other hand, if a unit is demanded and is not available, due to shortage, the loss is $\$\,c_2$. How many units, s, should be manufactured so that the total expected loss due to overproduction or to shortages will be minimized?

The loss at the end of the day is

$$L(s, X) = c_1(s - X)^+ + c_2(X - s)^+, \tag{3.47}$$

where $a^+ = \max(a, 0)$. The loss function $L(s, X)$ is a random variable. If the p.d.f. of X is $f(x)$, $x = 0, 1, \cdots$ then the **expected loss**, is a function of the quantity s, is

$$
\begin{aligned}
R(s) &= c_1 \sum_{x=0}^{s} f(x)(s - x) + c_2 \sum_{x=s+1}^{\infty} f(x)(x - s) \\
&= c_2 E\{X\} - (c_1 + c_2) \sum_{x=0}^{s} x f(x) \\
&\quad + s(c_1 + c_2) F(s) - c_2 s,
\end{aligned}
\tag{3.48}
$$

where $F(s)$ is the c.d.f. of X, at $X = s$, and $E\{X\}$ is the expected demand.

The **optimal** value of s, s^0, is the smallest integer s for which $R(s+1) - R(s) \geq 0$. Since, for $s = 0, 1, \cdots$

$$R(s + 1) - R(s) = (c_1 + c_2) F(s) - c_2,$$

we find that

$$s^0 = \textbf{smallest non-negative integer } s, \;\; \textbf{such that } F(s) \geq \frac{c_2}{c_1 + c_2}. \tag{3.49}$$

In other words, s^0 is the $c_2/(c_1 + c_2)$-th quantile of $F(x)$. We have seen that the optimal decision is a function of $F(x)$. If this distribution is unknown, or only partially known, one cannot determine the optimal value s^0.

After observing a large number, N, of X values, one can consider the empirical distribution, $F_N(x)$, of the demand and determine the level $S^0(F_N) = $ smallest s value such that $F_N(s) \geq \dfrac{c_2}{c_1 + c_2}$. The question is what to do when N is small. ∎

3.8.1 Prior and Posterior Distributions

We will focus attention here on parametric models. Let $f(x; \theta)$ denote the p.d.f. of some random variable X, which depends on a parameter θ. θ could be a vector of several real parameters, like in the case of a normal distribution. Let Θ denote the set of all possible parameters $\boldsymbol{\theta}$. Θ is called the **parameter space**. For example, the parameter space Θ of the family of normal distribution is the set $\Theta = \{(\mu, \sigma); -\infty < \mu < \infty, 0 < \sigma < \infty\}$. In the case of Poisson distributions,

$$\Theta = \{\lambda; 0 < \lambda < \infty\}.$$

In a Bayesian framework, we express our prior belief (based on prior information) on which $\boldsymbol{\theta}$ values are plausible, by a p.d.f. on Θ, which is called the **prior** probability density function. Let $h(\boldsymbol{\theta})$ denote the prior p.d.f. of $\boldsymbol{\theta}$. For example, suppose that X is a discrete random variable having a binomial distribution $B(n, \theta)$. n is known, but θ is unknown. The parameter space is $\Theta = \{\theta; 0 < \theta < 1\}$. Suppose we believe that θ is close to 0.8, with small dispersion around this value. In Fig. 3.11 we illustrate the p.d.f. of a beta distribution, Beta(80,20), whose functional form is

$$h(\theta; 80, 20) = \frac{99!}{79!19!}\theta^{79}(1 - \theta)^{19}, \quad 0 < \theta < 1.$$

If we wish, however, to give more weight to small values of θ, we can choose the Beta(8,2) as a prior density, i.e.,

$$h(\theta; 8, 2) = 72\theta^7(1 - \theta), \quad 0 < \theta < 1$$

(see Fig. 3.12).

The average p.d.f. of X, with respect to the prior p.d.f. $h(\boldsymbol{\theta})$, is called the **predictive p.d.f.** of X. This is given by

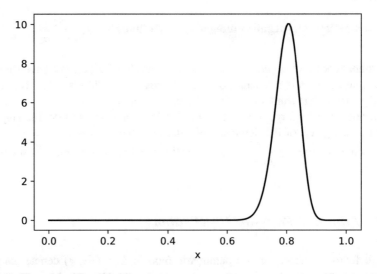

Fig. 3.11 The p.d.f. of Beta(80, 20)

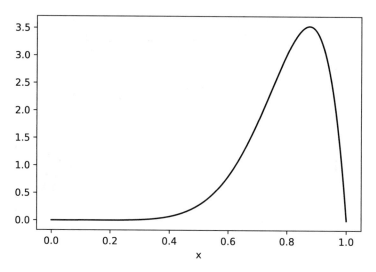

Fig. 3.12 The p.d.f. of Beta(8, 2)

$$f_h(x) = \int_{\Theta} f(x; \boldsymbol{\theta}) h(\boldsymbol{\theta}) \, d\boldsymbol{\theta}. \tag{3.50}$$

For the example above, the predictive p.d.f. is

$$
\begin{aligned}
f_h(x) &= 72 \binom{n}{x} \int_0^1 \theta^{7+x} (1-\theta)^{n-x+1} \, d\theta \\
&= 72 \binom{n}{x} \frac{(7+x)!(n+1-x)!}{(n+9-x)!}, \quad x = 0, 1, \cdots, n.
\end{aligned}
$$

Before taking observations on X, we use the predictive p.d.f. $f_h(x)$, to predict the possible outcomes of observations on X. After observing the outcome of X, say x, we convert the prior p.d.f. to a **posterior** p.d.f., by employing **Bayes' theorem**. If $h(\boldsymbol{\theta} \mid x)$ denotes the posterior p.d.f. of $\boldsymbol{\theta}$, given that $\{X = x\}$, Bayes' theorem yields

$$h(\boldsymbol{\theta} \mid x) = \frac{f(x \mid \boldsymbol{\theta}) h(\boldsymbol{\theta})}{f_h(x)}. \tag{3.51}$$

In the example above,

$$f(x \mid \theta) = \binom{n}{x} \theta^x (1-\theta)^{n-x}, \quad x = 0, 1, \cdots, n,$$

$$h(\theta) = 72\theta^7 (1-\theta), \quad 0 < \theta < 1,$$

and hence,

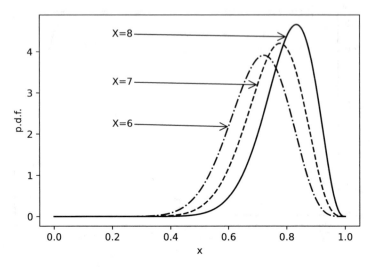

Fig. 3.13 The posterior p.d.f. of θ, $n = 10$, $X = 6, 7, 8$

$$h(\theta \mid x) = \frac{(n+9)!}{(7+x)!(n+1-x)!}\theta^{7+x}(1-\theta)^{n+1-x}, \quad 0 < \theta < 1.$$

This is again the p.d.f. of a beta distribution Beta($8 + x, n - x + 2$).

In Fig. 3.13 we present some of these posterior p.d.f. for the case of $n = 10$, $x = 6, 7, 8$. Notice that the posterior p.d.f. $h(\theta \mid x)$ is the **conditional** p.d.f. of θ, given $\{X = x\}$. If we observe a random sample of n independent and identically distributed (i.i.d.) random variables, and the observed values of $X_1, \cdots, X_n$ are $x_1, \cdots, x_n$, then the posterior p.d.f. of θ is

$$h(\theta \mid x_1, \cdots, x_n) = \frac{\prod_{i=1}^{n} f(x_i, \theta)h(\theta)}{f_h(x_1, \cdots, x_n)}, \tag{3.52}$$

where

$$f_h(x_1, \cdots, x_n) = \int_{\Theta} \prod_{i=1}^{n} f(x_i, \theta)h(\theta)\, d\theta \tag{3.53}$$

is the **joint predictive** p.d.f. of $X_1, \cdots, X_n$. If the i.i.d. random variables $X_1, X_2, \ldots$ are observed sequentially (time-wise), then the posterior p.d.f. of θ, given $x_1, \cdots, x_n, n \geq 2$ can be determined recursively, by the formula

$$H(\theta \mid x_1, \cdots, x_n) = \frac{f(x_n; \theta)h(\theta \mid x_1, \cdots, x_{n-1})}{\int_{\Theta} f(x_n; \theta')h(\theta' \mid x_1, \cdots, x_{n-1})\, d\theta'}.$$

The function

$$f_h(x_n \mid x_1, \cdots, x_{n-1}) = \int_\Theta f(x_n; \boldsymbol{\theta}) h(\boldsymbol{\theta} \mid x_1, \cdots, x_{n-1}) \, d\boldsymbol{\theta}$$

is called the **conditional predictive** p.d.f. of X_n, given $X_1 = x_1, \cdots, X_{n-1} = x_{n-1}$. Notice that

$$f_h(x_n \mid x_1, \cdots, x_{n-1}) = \frac{f_h(x_1, \cdots, x_n)}{f_h(x_1, \cdots, x_{n-1})}. \tag{3.54}$$

3.8.2 Bayesian Testing and Estimation

3.8.2.1 Bayesian Testing

We discuss here the problem of testing hypotheses as a Bayesian decision problem. Suppose that we consider a null hypothesis H_0 concerning a parameter θ of the p.d.f. of X. Suppose also that the parameter space Θ is partitioned to two sets Θ_0 and Θ_1. Θ_0 is the set of θ values corresponding to H_0, and Θ_1 is the complementary set of elements of Θ which are not in Θ_0. If $h(\theta)$ is a prior p.d.f. of θ, then the **prior probability** that H_0 is true is $\pi = \int_{\Theta_0} h(\theta) \, d\theta$. The prior probability that H_1 is true is $\bar{\pi} = 1 - \pi$.

The statistician has to make a decision whether H_0 or H_1 is true. Let $d(\pi)$ be a decision function, assuming the values 0 and 1, i.e.,

$$d(\pi) = \begin{cases} 0, & \text{decision to accept } H_0 \ (H_0 \text{ is true}) \\ 1, & \text{decision to reject } H_0 \ (H_1 \text{ is true}). \end{cases}$$

Let w be an indicator of the true situation, i.e.,

$$w = \begin{cases} 0, & \text{if } H_0 \text{ is true.} \\ 1, & \text{if } H_1 \text{ is true.} \end{cases}$$

We also impose a **loss function** for erroneous decision

$$L(d(\pi), w) = \begin{cases} 0, & \text{if } d(\pi) = w \\ r_0, & \text{if } d(\pi) = 0, \ w = 1 \\ r_1, & \text{if } d(\pi) = 1, \ w = 0, \end{cases} \tag{3.55}$$

where r_0 and r_1 are finite positive constants. The **prior risk** associated with the decision function $d(\pi)$ is

$$R(d(\pi), \pi) = d(\pi)r_1\pi + (1 - d(\pi))r_0(1 - \pi)$$
$$= r_0(1 - \pi) + d(\pi)[\pi(r_0 + r_1) - r_0]. \tag{3.56}$$

We wish to choose a decision function which **minimizes** the prior risk $R(d(\pi), \pi)$. Such a decision function is called the **Bayes decision function**, and the prior risk associated with the Bayes decision function is called the **Bayes risk**. According to the above formula of $R(d(\pi), \pi)$, we should choose $d(\pi)$ to be 1 if, and only if, $\pi(r_0 + r_1) - r_0 < 0$. Accordingly, the Bayes decision function is

$$d^0(\pi) = \begin{cases} 0, & \text{if } \pi \geq \dfrac{r_0}{r_0 + r_1} \\ 1, & \text{if } \pi < \dfrac{r_0}{r_0 + r_1} \end{cases} \tag{3.57}$$

Let $\pi^* = r_0/(r_0 + r_1)$, and define the indicator function

$$I(\pi; \pi^*) = \begin{cases} 1, & \text{if } \pi \geq \pi^* \\ 0, & \text{if } \pi < \pi^* \end{cases}$$

then, the Bayes risk is

$$R^0(\pi) = r_0(1 - \pi)I(\pi; \pi^*) + \pi r_1(1 - I(\pi; \pi^*)). \tag{3.58}$$

In Fig. 3.14 we present the graph of the Bayes risk function $R^0(\pi)$, for $r_0 = 1$ and $r_1 = 5$. We see that the function $R^0(\pi)$ attains its maximum at $\pi = \pi^*$. The maximal Bayes risk is $R^0(\pi^*) = r_0r_1/(r_0 + r_1) = 5/6$. If the value of π is close to π^*, the Bayes risk is close to $R^0(\pi^*)$.

The analysis above can be performed even before observations commenced. If π is close to 0 or 1, the Bayes risk $R^0(\pi)$ is small; we may reach decision concerning the hypotheses without even making observations. Recall that observations cost money, and it might not be justifiable to spend this money. On the other hand, if the cost of observations is negligible compared to the loss due to erroneous decision, it might be prudent to take as many observations as required to reduce the Bayes risk.

After observing a random sample, $x_1, \cdots, x_n$, we convert the prior p.d.f. of $\boldsymbol{\theta}$ to posterior and determine the posterior probability of H_0, namely,

$$\pi_n = \int_\Theta h(\boldsymbol{\theta} \mid x_1, \cdots, x_n) \, d\boldsymbol{\theta}.$$

The analysis then proceeds as before, replacing π with the posterior probability π_n.

Accordingly, the Bayes decision function is

$$d^0(x_1, \cdots, x_n) = \begin{cases} 0, & \text{if } \pi_n \geq \pi^* \\ 1, & \text{if } \pi_n < \pi^* \end{cases}$$

Fig. 3.14 The Bayes risk function

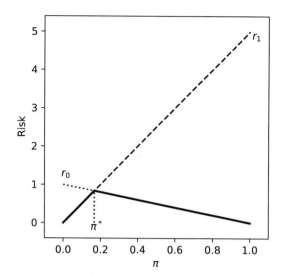

and the Bayes posterior risk is

$$R^0(\pi_n) = r_0(1 - \pi_n)I(\pi_n; \pi^*) + \pi_n r_1(1 - I(\pi_n; \pi^*)).$$

Under certain regularity conditions, $\lim_{n \to \infty} \pi_n = 1$ or 0, according to whether H_0 is true or false. We illustrate this with a simple example.

Example 3.16 Suppose that X has a normal distribution, with known $\sigma^2 = 1$. The mean μ is unknown. We wish to test $H_0 : \mu \le \mu_0$ against $H_1 : \mu > \mu_0$. Suppose that the prior distribution of μ is also normal, $N(\mu^*, \tau^2)$. The posterior distribution of μ, given $X_1, \cdots, X_n$, is normal with mean

$$E\{\mu \mid X_1, \cdots, X_n\} = \mu^* \frac{1}{(1 + n\tau^2)} + \frac{n\tau^2}{1 + n\tau^2} \bar{X}_n$$

and posterior variance

$$V\{\mu \mid X_1, \cdots, X_n\} = \frac{\tau^2}{1 + n\tau^2}.$$

Accordingly, the posterior probability of H_0 is

$$\pi_n = \Phi \left(\frac{\mu_0 - \dfrac{\mu^*}{1 + n\tau^2} - \dfrac{n\tau^2}{1 + n\tau^2} \bar{X}_n}{\sqrt{\dfrac{\tau^2}{1 + n\tau^2}}} \right).$$

According to the law of large numbers, $\bar{X}_n \to \mu$ (the true mean), as $n \to \infty$, with probability one. Hence,

$$
\lim_{n \to \infty} \pi_n = \begin{cases} 1, & \text{if } \mu < \mu_0 \\ \dfrac{1}{2}, & \text{if } \mu = \mu_0 \\ 0, & \text{if } \mu > \mu_0. \end{cases}
$$

Notice that the prior probability that $\mu = \mu_0$ is zero. Thus, if $\mu < \mu_0$ or $\mu > \mu_0$, $\lim_{n \to \infty} R^0(\pi_n) = 0$, with probability one, that is, if n is sufficiently large, the Bayes risk is, with probability close to one, smaller than some threshold r^*. This suggests to continue, step-wise or sequentially, collecting observations, until the Bayes risk $R^0(\pi_n)$ is, for the first time, smaller than r^*. At stopping, $\pi_n \geq 1 - \dfrac{r^*}{r_0}$ or $\pi_n \leq \dfrac{r^*}{r_1}$. We obviously choose $r^* < \dfrac{r_0 r_1}{r_0 + r_1}$. ∎

3.8.2.2   Bayesian Estimation

In an estimation problem, the decision function is an estimator $\hat{\theta}(x_1, \cdots, x_n)$, which yields a point in the parameter space Θ. Let $L(\hat{\theta}(x_1, \cdots, x_n), \theta)$ be a **loss function** which is non-negative, and $L(\theta, \theta) = 0$. The **posterior risk** of an estimator $\hat{\theta}(x_1, \cdots, x_n)$ is the expected loss, with respect to the posterior distribution of θ, given $(x_1, \cdots, x_n)$, i.e.,

$$
R_h(\hat{\theta}, \mathbf{x}_n) = \int_{\Theta} L(\hat{\theta}(\mathbf{x}_n), \theta) h(\theta \mid \mathbf{x}_n)\, d\theta, \tag{3.59}
$$

where $\mathbf{x}_n = (x_1, \cdots, x_n)$. We choose an estimator which **minimizes the posterior risk**. Such an estimator is called a **Bayes estimator** and designated by $\hat{\theta}_B(\mathbf{x}_n)$. We present here a few cases of importance.

Case A θ real, $L(\hat{\theta}, \theta) = (\hat{\theta} - \theta)^2$.
   In this case, the Bayes estimator of θ is posterior expectation of θ, i.e.,

$$
\hat{\theta}_B(\mathbf{x}_n) = E_h\{\theta \mid \mathbf{x}_n\}. \tag{3.60}
$$

The Bayes risk is the expected posterior variance, i.e.,

$$
R_h^0 = \int V_h\{\theta \mid \mathbf{x}_n\} f_h(x_1, \cdots, x_n)\, dx_1, \cdots,\ dx_n.
$$

Case B θ real, $L(\hat{\theta}, \theta) = c_1(\hat{\theta} - \theta)^+ + c_2(\theta - \hat{\theta})^+$, with $c_1, c_2 > 0$, and $(a)^+ = \max(a, 0)$.

As shown in inventory Example 3.15, at the beginning of the section, the Bayes estimator is

$$\hat{\theta}_B(\mathbf{x}_n) = \frac{c_2}{c_1 + c_2} \text{ th quantile of the posterior distribution of } \theta, \text{ given } \mathbf{x}_n.$$

When $c_1 = c_2$ we obtain the posterior median.

3.8.3 Credibility Intervals for Real Parameters

We restrict attention here to the case of a real parameter, θ. Given the values $x_1, \cdots, x_n$ of a random sample, let $h(\theta \mid \mathbf{x}_n)$ be the posterior p.d.f. of θ. An interval $C_{1-\alpha}(\mathbf{x}_n)$ such that

$$\int_{C_{1-\alpha}(\mathbf{x}_n)} h(\theta \mid \mathbf{x}_n)\, d\theta \geq 1 - \alpha \qquad (3.61)$$

is called a **credibility interval** for θ. A credibility interval $C_{1-\alpha}(\mathbf{x}_n)$ is called a **highest posterior density (HPD) interval** if for any $\theta \in C_{1-\alpha}(\mathbf{x}_n)$ and $\theta' \notin C_{1-\alpha}(\mathbf{x}_n)$, $h(\theta \mid \mathbf{x}_n) > h(\theta' \mid \mathbf{x}_n)$.

Example 3.17 Let $x_1, \cdots, x_n$ be the values of a random sample from a Poisson distribution $P(\lambda)$, $0 < \lambda < \infty$. We assign λ a Gamma distribution $G(\nu, \tau)$. The posterior p.d.f. of λ, given $\mathbf{x}_n = (x_1, \cdots, x_n)$, is

$$h(\lambda \mid \mathbf{x}_n) = \frac{(1 + n\tau)^{\nu + \Sigma x_i}}{\Gamma(\nu + \Sigma x_i)\tau^{\nu + \Sigma x_i}} \lambda^{\nu + \Sigma x_i - 1} e^{-\lambda \frac{1+n\tau}{\tau}}.$$

In other words, the posterior distribution is a gamma distribution $G\left(\nu + \sum_{i=1}^{n} x_i, \frac{\tau}{1+n\tau}\right)$. From the relationship between the gamma and the χ^2-distributions, we can express the limits of a credibility interval for λ, at level $(1 - \alpha)$ as

$$\frac{\tau}{2(1 + n\tau)} \chi^2_{\alpha/2}[\phi] \text{ and } \frac{\tau}{2(1 + n\tau)} \chi^2_{1-\alpha/2}[\phi]$$

where $\phi = 2\nu + 2 \sum_{i=1}^{n} x_i$. This interval is called an **equal tail credibility interval**. However, it is not an HPD credibility interval. In Fig. 3.15 we present the posterior density for the special case of $n = 10$, $\nu = 2$, $\tau = 1$, and $\sum_{i=1}^{10} x_i = 15$. For these values the limits of the credibility interval for λ, at level 0.95, are 0.9 and 2.364. As we see in Fig. 3.15, $h(0.9 \mid \mathbf{x}_n) > h(2.364 \mid \mathbf{x}_n)$. Thus, the equal-tail credibility interval is **not** an HPD interval. The limits of the HPD interval can be determined by

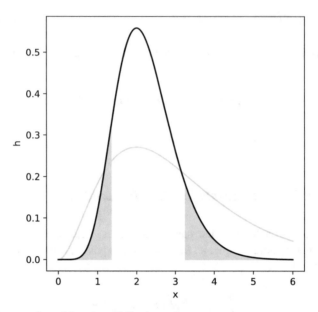

Fig. 3.15 The posterior p.d.f. and credibility intervals

trial and error. In the present case, they are approximately 0.86 and 2.29, as shown in Fig. 3.15.   ■

3.9   Random Sampling from Reference Distributions

We have seen in Sect. 1.4.1 an example of blemishes on ceramic plates. In that example (Table 1.2), the proportion of plates having **more** than one blemish is 0.23. Suppose that we decide to improve the manufacturing process and reduce this proportion. How can we test whether an alternative production process with new operating procedures and machine settings is indeed better so that the proportion of plates with more than one blemish is significantly smaller? The objective is to operate a process with a proportion of defective units (i.e., with more than one blemish) which is smaller than 0.10. After various technological modifications, we are ready to test whether the modified process conforms with the new requirement. Suppose that a random sample of ceramic plates is drawn from the modified manufacturing process. One has to test whether the proportion of defective plates in the sample is not significantly larger than 0.10. In the parametric model, it was assumed that the number of plates having more than one defect, in a random sample of n plates, has a binomial distribution $B(n, p)$. For testing $H_0 : p \leq 0.1$ a test was constructed based on the reference distribution $B(n, .1)$.

One can create, artificially on a computer, a population having ninety 0s and ten 1s. In this population the proportion of 1s is $p_0 = 0.10$. From this population one can draw a large number, M, of random samples with replacement (RSWR) of a given size n. In each sample, the sample mean $\bar{X}_n$ is the proportion of 1s in the sample. The sampling distribution of the M sample means is our **empirical reference distribution** for the hypothesis that the proportion of defective plates is $p \leq p_0$. We pick a value α close to zero and determine the $(1 - \alpha)$-th quantile of the empirical reference distribution. If the observed proportion in the real sample is greater than this quantile, the hypothesis $H : p \leq p_0$ is rejected.

Example 3.18 To illustrate, we created an empirical reference distribution of $M = 1000$ proportions of 1s in RSWR of size $n = 50$ using the following code.

```
random.seed(1)

# Create a population of 90 0s and 10 1s
population = [0] * 90
population.extend([1] * 10)

proportions = []
for m in range(1000):
    # sample 50 values from population using RSWR
    sample = random.choices(population, k=50)
    # keep the mean of the sample
    proportions.append(np.mean(sample))
```

It was assumed that population contained ninety 0s and ten 1s. We can get the frequency distribution of the calculated proportions using the Python `Counter` class.

```
from collections import Counter
from pprint import pprint

frequencies = Counter(proportions)
pprint(sorted(frequencies.items()))
```

```
[(0.0, 7),
 (0.02, 24),
 (0.04, 76),
 (0.06, 123),
 (0.08, 176),
 (0.1, 193),
 (0.12, 150),
 (0.14, 110),
 (0.16, 80),
 (0.18, 35),
 (0.2, 17),
 (0.22, 9)]
```

This frequency distribution represents the reference distribution. An outcome of such a simulation is given in Table 3.11.

For $\alpha = 0.05$, the 0.95-quantile of the empirical reference distribution is 0.15, since at least 50 out of 1000 observations are greater than 0.15. Thus, if in a real sample, of size $n = 50$, the defective proportion is greater than 0.15, the null hypothesis is rejected.   ∎

Table 3.11 Frequency distribution of $M = 1000$ means of RSWR from a set with ninety 0s and ten 1s

$\bar{x}$	f	$\bar{x}$	f
0.03	10	0.11	110
0.04	17	0.12	100
0.05	32	0.13	71
0.06	61	0.14	50
0.07	93	0.15	28
0.08	128	0.16	24
0.09	124	0.17	9
0.10	133	>0.17	10

Example 3.19 Consider a hypothesis on the length of aluminum pins (with cap), $H_0 : \mu \geq 60.1$ [mm]. We create now an empirical reference distribution for this hypothesis. In the dataset **ALMPIN.csv**, we have the actual sample values. The mean of the variable lenWcp is $\bar{X}_{70} = 60.028$. Since the hypothesis states that the process mean is $\mu \geq 60.1$, we transform the sample values to $Y = X - 60.028 + 60.1$. This transformed sample has a mean of 60.1. We now create a reference distribution of sample means, by drawing M RSWR of size $n = 70$ from the transformed sample. We can perform this using the following Python code.

```
random.seed(1)

# Load the dataset
X = mistat.load_data('ALMPIN')['lenWcp']
Y = X - np.mean(X) + 60.1

means = []
for m in range(1000):
  # sample 70 values from population using RSWR
  sample = random.choices(Y, k=70)
  # keep the mean of the sample
  sample_mean = np.mean(sample)
  means.append(0.005 * round(sample_mean/0.005))

quant_001 = np.quantile(means, 0.01)
```

After executing the code, we obtain an empirical reference distribution whose frequency distribution is given in Table 3.12.

In the dataset **ALMPIN.csv**, there are six variables, measuring various dimensions of aluminum pins. We calculate the transformed variable from the column lenWcp.

Since $\bar{X}_{70} = 60.028$ is smaller than 60.1, we consider as a test criterion the α-quantile of the reference distribution. If $\bar{X}_{70}$ is smaller than this quantile, we reject the hypothesis. For $\alpha = 0.01$, the 0.01-quantile in the above reference distribution is 60.08995. Accordingly we reject the hypothesis, since it is very implausible (less than one chance in a hundred) that $\mu \geq 60.1$. The estimated P-value is less than 10^{-3}, since the smallest value in the reference distribution is 60.08. ∎

Table 3.12 Frequency distribution of $\bar{X}_{70}$ from 1000 RSWR from lengthwcp

Midpoint	Count
60.080	1
60.085	9
60.090	78
60.095	230
60.100	350
60.105	239
60.110	79
60.115	12
60.120	2

3.10 Bootstrap Sampling

3.10.1 The Bootstrap Method

The bootstrap methodology was introduced in 1979 by B. Efron, as an elegant method of performing statistical inference by harnessing the power of the computer and without the need for extensive assumptions and intricate theory. Some of the ideas of statistical inference with the aid of computer sampling were presented in the previous sections. In the present section, we introduce the bootstrap method in more detail. In Chap. 7 we deal with predictive analytic models where the data is split into a training and a validation set. An extended approach to assessing predictive models is to apply cross-validation.

Given a sample of size n, $S_n = \{x_1, \cdots, x_n\}$, let t_n denote the value of some specified sample statistic T. The bootstrap method draws M random samples **with replacement (RSWR)** of size n from S_n. For each such sample, the statistic T is computed. Let $\{t_1^*, t_2^*, \cdots, t_M^*\}$ be the collection of these sample statistics. The distribution of these M values of T is called the **empirical bootstrap distribution** (EBD). It provides an approximation, if M is large, to the **bootstrap distribution** of all possible values of the statistic T that can be generated by repeatedly sampling from S_n.

General Properties of the (EBD)

1. The EBD is centered at the sample statistic t_n.
2. The mean of the EBD is an estimate of the mean of the sampling distribution of the statistic T, over all possible samples.
3. The standard deviation of the EBD is the **bootstrap estimate of the standard error of T**.
4. The $\alpha/2$-th and $(1 - \alpha/2)$-th quantiles of the EBD are **bootstrap confidence limits** for the parameter which are estimated by t_n, at level of confidence $(1 - \alpha)$.

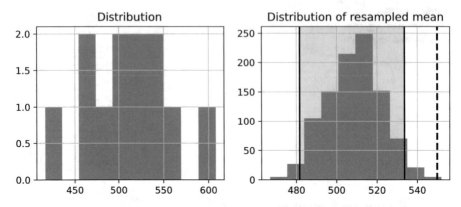

Fig. 3.16 Result of a bootstrap analysis of the ETCHRATE dataset

Example 3.20 We illustrate the bootstrap method with dataset **ETCHRATE.csv** in which we want to test if the sample is derived from a population with a specific mean. In principle, we could use Python code similar to the one we introduced in the previous section. However, there are packages that contain bootstrap implementations which provide additional functionality. Here, we use the `compute_bootci` method from the pingouin package.

```
etchrate = mistat.load_data('ETCHRATE')

B = pg.compute_bootci(etchrate, func=np.mean, n_boot=1000,
                      confidence=0.95, return_dist=True, seed=1)

ci, distribution = B
print(f' Mean: {np.mean(distribution)}')
print(f' 95%-CI: {ci[0]:.1f} - {ci[1]:.1f}')
```

```
Mean: 508.53783333333337
95%-CI: 481.7 - 533.7
```

Figure 3.16 shows the distribution of the data and the resampled means as histograms. The analysis returns the bootstrap confidence limits, at confidence level $(1-\alpha) = 0.95$. The bootstrap interval $(481.7, 533.7)$ is called a bootstrap confidence interval for μ. We see that this interval does not cover the tested mean of 550. ∎

3.10.2 Examining the Bootstrap Method

In the previous section, we introduced the bootstrap method as a computer intensive technique for making statistical inference. In this section some of the properties of the bootstrap methods are examined in light of the theory of sampling from finite populations. As we recall, the bootstrap method is based on drawing repeatedly M simple RSWR from the original sample.

Let $S_X = \{x_1, \cdots, x_n\}$ be the values of the n original observations on X. We can consider S_X as a finite population P of size n. Thus, the mean of this population, μ_n, is the sample mean $\bar{X}_n$, and the variance of this population is $\sigma_n^2 = \dfrac{n-1}{n} S_n^2$, where S_n^2 is the sample variance, $S_n^2 = \dfrac{1}{n-1} \sum_{i=1}^n (x_i - \bar{X}_n)^2$. Let $S_X^* = \{X_1^*, \cdots, X_n^*\}$ denote a simple RSWR from S_X. S_X^* is the bootstrap sample. Let $\bar{X}_n^*$ denote the mean of the bootstrap sample.

We have shown in Chap. 2 that the mean of a simple RSWR is an unbiased estimator of the corresponding sample mean. Thus,

$$E^*\{\bar{X}_n^*\} = \bar{X}_n, \tag{3.62}$$

where $E^*\{\cdot\}$ is the expected value with respect to the bootstrap sampling. Moreover, the bootstrap variance of $\bar{X}_n^*$ is

$$V^*\{\bar{X}_n^*\} = \frac{\dfrac{n-1}{n} S_n^2}{n} \tag{3.63}$$
$$= \frac{S_n^2}{n}\left(1 - \frac{1}{n}\right).$$

Thus, in large sample

$$V^*\{\bar{X}_n^*\} \cong \frac{S_n^2}{n}. \tag{3.64}$$

If the original sample S_X is a realization of n i.i.d. random variables, having a c.d.f. $F(x)$, with finite expected value μ_F and a finite variance σ_F^2, then, as shown in Sect. 4.8, the variance of $\bar{X}_n$ is σ_F^2/n. The sample variance S_n^2 is an unbiased estimator of σ_F^2. Thus, $\dfrac{S_n^2}{n}$ is an unbiased estimator of σ_F^2/n. Finally, the variance of the EBD of $\bar{X}_1^*, \cdots, \bar{X}_M^*$ obtained by repeating the bootstrap sampling M times independently is an unbiased estimator of $\dfrac{S_n^2}{n}\left(1 - \dfrac{1}{n}\right)$. Thus, the variance of the EBD is an approximation to the variance of $\bar{X}_n$.

We remark that this estimation problem is a simple one, and there is no need for bootstrapping in order to estimate the variance or standard error of the estimator $\bar{X}_n$.

3.10.3   Harnessing the Bootstrap Method

The effectiveness of the bootstrap method manifests itself when a formula for the variance of an estimator is hard to obtain. In Sect. 2.1.4, we provided a formula for the variance of the estimator S_n^2, in simple RSWR. By bootstrapping from the sample S_X, we obtain an EBD of S_n^{2*}. The variance of this EBD is an approximation to the true variance of S_n^2. Thus, for example, when $P = \{1, 2, \cdots, 100\}$, the true variance of S_n^2 is 31,131.2, while the bootstrap approximation, for a particular sample, is 33,642.9. Another sample will yield a different approximation. The approximation obtained by the bootstrap method becomes more precise as the sample size grows. For the above problem, if $n = 100$, $V\{S_n\} = 5693.47$ and the bootstrap approximation is distributed around this value.

The following are values of four approximations of $V\{S_n\}$ for $n = 100$, when $M = 100$. Each approximation is based on different random samples from P

$$6293.28, \quad 5592.07, \quad 5511.71, \quad 5965.89.$$

Each bootstrap approximation is an estimate of the true value of $V\{S_n\}$.

3.11   Bootstrap Testing of Hypotheses

In this section we present some of the theory and the methods of testing hypotheses by bootstrapping. Given a test statistic $\mathbf{T} = T(X_1, \cdots, X_n)$, the critical level for the test, k_α, is determined according to the distribution of $\mathbf{T}$ under the null hypothesis, which is the **reference distribution**.

The bootstrapping method, as explained before, is a randomization method which resamples the sample values and thus constructs a reference distribution for $\mathbf{T}$, independently of the unknown distribution F of $\mathbf{X}$. For each bootstrap sample, we compute the value of the test statistic $\mathbf{T}^* = T(x_1^*, \cdots, x_n^*)$. Let $\mathbf{T}_1^*, \cdots, \mathbf{T}_M^*$ be the M values of the test statistic obtained from the M samples from BP. Let $F_M^*(t)$ denote the empirical c.d.f. of these values. $F_M^*(t)$ is an estimator of the bootstrap distribution $F^*(t)$, from which we can estimate the critical value k^*. Specific procedures are given in the following subsections.

3.11.1   Bootstrap Testing and Confidence Intervals for the Mean

Suppose that $\{x_1, \cdots, x_n\}$ is a random sample from a parent population, having an unknown distribution, F, with mean μ and a finite variance σ^2.

We wish to test the hypothesis

$$H_0 : \mu \le \mu_0 \text{ against } H_1 : \mu > \mu_0.$$

Let $\bar{X}_n$ and S_n be the sample mean and sample standard deviation. Suppose that we draw from the original sample M bootstrap samples. Let $\bar{X}_1^*, \cdots, \bar{X}_M^*$ be the means of the bootstrap samples. Recall that, since the bootstrap samples are RSWR, $E^*\{\bar{X}_j^*\} = \bar{X}_n$ for $j = 1, \cdots, M$, where $E^*\{\cdot\}$ designates the expected value, with respect to the bootstrap sampling. Moreover, for large n,

$$\text{S.E.}^*\{\bar{X}_j^*\} \cong \frac{S_n}{\sqrt{n}}, \quad j = 1, \cdots, M.$$

Thus, if n is not too small, the central limit theorem implies that $F_M^*(\bar{X}^*)$ is approximately $\Phi\left(\dfrac{\bar{X}^* - \bar{X}_n}{S_n/\sqrt{n}}\right)$, i.e., the bootstrap means $\bar{X}_1^*, \cdots, \bar{X}_m^*$ are distributed approximately normally around $\mu^* = \bar{X}_n$. We wish to reject H_0 if $\bar{X}_n$ is significantly larger than μ_0. According to this normal approximation to $F_M^*(\bar{X}^*)$, we should reject H_0, at level of significance α, if $\dfrac{\mu_0 - \bar{X}_n}{S_n/\sqrt{n}} \le z_\alpha$ or $\bar{X}_n \ge \mu_0 + z_{1-\alpha}\dfrac{S_n}{\sqrt{n}}$. This is approximately the t-test of Sect. 3.3.2.2.

Notice that the reference distribution can be obtained from the EBD by subtracting $\Delta = \bar{X}_n - \mu_0$ from $\bar{X}_j^*$ ($j = 1, \ldots, M$). The reference distribution is centered at μ_0. The $(1 - \alpha/2)$-th quantile of the reference distribution is $\mu_0 + z_{1-\alpha/2}\dfrac{S_n}{\sqrt{n}}$. Thus, if $\bar{X}_n \ge \mu_0 + z_{1-\alpha/2}\dfrac{S_n}{\sqrt{n}}$, we reject the null hypothesis $H_0 : \mu \le \mu_0$.

If the sample size n is not large, it might not be justified to use the normal approximation. We use bootstrap procedures in the following sections.

3.11.2   Studentized Test for the Mean

A **studentized test statistic**, for testing the hypothesis $H_0 : \mu \le \mu_0$, is

$$t_n = \frac{\bar{X}_n - \mu_0}{S_n/\sqrt{n}}. \tag{3.65}$$

H is rejected if t_n is significantly greater than zero. To determine what is the rejection criterion, we construct an EBD by the following procedure:

1. Draw a RSWR, of size n, from the original sample.
2. Compute $\bar{X}_n^*$ and S_n^* of the bootstrap sample.
3. Compute the studentized statistic

$$t_n^* = \frac{\bar{X}_n^* - \bar{X}_n}{S_n^*/\sqrt{n}}. \tag{3.66}$$

4. Repeat this procedure M times.

Let t_p^* denote the p-th quantile of the EBD.

Case I $H : \mu \leq \mu_0$.
The hypothesis H is rejected if

$$t_n \geq t_{1-\alpha}^*.$$

Case II $H : \mu \geq \mu_0$.
We reject H if

$$t_n \leq t_\alpha^*.$$

Case III $H : \mu = \mu_0$.
We reject H if

$$|t_n| \geq t_{1-\alpha/2}^*.$$

The corresponding P^*-**levels** are:

For Case I: The proportions of t_n^* values greater than t_n.
For Case II: The proportions of t_n^* values smaller than t_n.
For Case III: The proportion of t_n^* values greater than $|t_n|$ or smaller than $-|t_n|$.

H is rejected if P^* is small.

Notice the difference in definition between t_n and t_n^*. t_n is centered around μ_0 while t_n^* around $\bar{X}_n$.

Example 3.21 In data file **HYBRID1.csv**, we find the resistance (in ohms) of Res3 in a hybrid microcircuit labeled hybrid 1 on $n = 32$ boards. The mean of Res 3 in hybrid 1 is $\bar{X}_{32} = 2143.4$. The question is whether Res 3 in hybrid 1 is significantly different from $\mu_0 = 2150$. We consider the hypothesis

$$H : \mu = 2150 \quad \text{(Case III)}.$$

With $M = 500$, we obtain with the Python commands below the following 0.95-confidence level bootstrap interval $(2111, 2176)$. We see that $\mu_0 = 2150$ is covered by this interval. We therefore infer that $\bar{X}_{32}$ is not significantly different than μ_0. The hypothesis H is **not** rejected. With `stats.ttest_1samp`, we see that the studentized difference between the sample mean $\bar{X}_{32}$ and μ_0 is $t_n = -0.374$. $M = 500$ bootstrap replicas yield the value $P^* = 0.692$. The hypothesis is **not** rejected.

The commands in Python are:

```
hybrid1 = mistat.load_data('HYBRID1')

ci = pg.compute_bootci(hybrid1, func='mean', n_boot=500,
                       confidence=0.95, method='per', seed=1)
print('bootstrap: ', ci)
print()

print(stats.ttest_1samp(hybrid1, 2150))
n = len(hybrid1) - 1
t_test_confinterval = stats.t.interval(0.95, n, loc=np.mean(hybrid1),
  scale=np.std(hybrid1)/np.sqrt(n))
print('t-test 95% conf-interval', t_test_confinterval)
print()

# get distribution of bootstrapped pvalues and determine percentage of
# values less than pvalue calculated from ttest_1samp
def stat_func(x):
    return stats.ttest_1samp(x, 2150).pvalue

ci, dist = pg.compute_bootci(hybrid1, func=stat_func, n_boot=500,
                       confidence=0.95, method='per',
                       seed=1, return_dist=True)
sum(dist < stats.ttest_1samp(hybrid1, 2150).pvalue) / len(dist)
```

```
bootstrap:  [2111.34 2175.69]

Ttest_1sampResult(statistic=-0.3743199001200656,
pvalue=0.7107146634282755)
t-test 95% conf-interval (2107.4796487616936, 2179.3328512383064)

0.692
```

■

3.11.3   Studentized Test for the Difference of Two Means

The problem is whether two population means μ_1 and μ_2 are the same. This problem is important in many branches of science and engineering, when two "treatments" are compared.

Suppose that one observes a random sample $X_1, \cdots, X_{n_1}$ from population 1 and another random sample $Y_1, \cdots, Y_{n_2}$ from population 2. Let $\bar{X}_{n_1}$, $\bar{Y}_{n_2}$, S_{n_1}, and S_{n_2} be the means and standard deviations of these two samples, respectively. Compute the studentized difference of the two sample means as

$$t = \frac{\bar{X}_{n_1} - \bar{Y}_{n_2} - \delta_0}{\left(\frac{S_{n_1}^2}{n_1} + \frac{S_{n_2}^2}{n_2}\right)^{1/2}}, \tag{3.67}$$

where $\delta = \mu_1 - \mu_2$. The question is whether this value is significantly different from zero. The hypothesis under consideration is

$$H : \mu_1 = \mu_2, \text{ or } \delta_0 = 0.$$

By the bootstrap method, we draw RSWR of size n_1 from the x-sample and an RSWR of size n_2 from the y-sample. Let $X_1^*, \cdots, X_{n_1}^*$ and $Y_1^*, \cdots, Y_{n_2}^*$ be these two bootstrap samples, with means and standard deviations $\bar{X}_{n_1}^*, \bar{Y}_{n_2}^*$ and $S_{n_1}^*, S_{n_2}^*$. We compute then the studentized difference

$$t^* = \frac{\bar{X}_{n_1}^* - \bar{Y}_{n_2}^* - (\bar{X}_{n_1} - \bar{Y}_{n_2})}{\left(\dfrac{S_{n_1}^{*2}}{n_1} + \dfrac{S_{n_2}^{*2}}{n_2} \right)^{1/2}}. \tag{3.68}$$

This procedure is repeated independently M times, to generate an EBD of $t_1^*, \cdots, t_M^*$.

Let $(D_{\alpha/2}^*, D_{1-\alpha/2}^*)$ be a $(1 - \alpha)$ level confidence interval for δ, based on the EBD. If $t_{\alpha/2}^*$ is the $\alpha/2$-quantile of t^* and $t_{1-\alpha/2}^*$ is its $(1 - \alpha/2)$-quantile, then

$$D_{\alpha/2}^* = (\bar{X}_{n_1} - \bar{Y}_{n_2}) + t_{\alpha/2}^* \left(\frac{S_{n_1}^2}{n_1} + \frac{S_{n_2}^2}{n_2} \right)^{1/2}$$

$$D_{1-\alpha/2}^* = (\bar{X}_{n_1} - \bar{Y}_{n_2}) + t_{1-\alpha/2}^* \left(\frac{S_{n_1}^2}{n_1} + \frac{S_{n_2}^2}{n_2} \right)^{1/2}. \tag{3.69}$$

If this interval does **not** cover the value $\delta_0 = 0$, we **reject** the hypothesis $H : \mu_1 = \mu_2$. The P^*-value of the test is the proportion of t_i^* values which are either smaller than $-|t|$ or greater than $|t|$.

Example 3.22 We compare the resistance coverage of Res 3 in hybrid 1 and hybrid 2. The data file **HYBRID2.csv** consists of two columns. The first represents the sample of $n_1 = 32$ observations on hybrid 1, and the second column consists of $n_2 = 32$ observations on hybrid 2. Using Python, we calculate $M = 500$ bootstrap samples of t^*.

```
random.seed(1)
hybrid2 = mistat.load_data('HYBRID2')
X = hybrid2['hyb1']
Y = hybrid2['hyb2']
Xbar = np.mean(X)
Ybar = np.mean(Y)
SX = np.std(X, ddof=1)
SY = np.std(Y, ddof=1)
print('Xbar {Xbar:.2f} / SX {SX:.3f}')
print('Ybar {Xbar:.2f} / SY {SX:.3f}')

def stat_func(x, y):
    return stats.ttest_ind(x, y, equal_var=False).statistic

tstar = []
for _ in range(500):
    Xstar = np.array(random.choices(X, k=len(X))) - Xbar
```

```
    Ystar = np.array(random.choices(Y, k=len(Y))) - Ybar
    tstar.append(stat_func(Xstar, Ystar))

# calculate confidence interval for t* and D*
alpha = 0.05
tstar_ci = np.quantile(tstar, [alpha/2, 1-alpha/2])
Dstar_ci = Xbar - Ybar + np.sqrt(SX**2/len(X) + SY**2/len(Y))*tstar_ci

print('tstar-CI', tstar_ci)
print('Dstar-CI', Dstar_ci)

t0 = stat_func(X, Y)
print(f't0 {t0:.3f}')
pstar = (sum(tstar < -abs(t0)) + sum(abs(t0) < tstar)) / len(tstar)
print(f'P*-value {pstar:.2f}')
```

```
Xbar {Xbar:.2f} / SX {SX:.3f}
Ybar {Xbar:.2f} / SY {SX:.3f}
tstar-CI [-2.26597133  1.99806399]
Dstar-CI [175.28976673 298.17679872]
t0 8.348
P*-value 0.00
```

```
ax = pd.Series(tstar).hist()
ax.axvline(t0, color='black', lw=2)
ax.axvline(-t0, color='black', lw=2)
ax.set_xlabel('t* values')
plt.show()
```

We see that $\bar{X}_{n_1} = 2143.41$, $\bar{Y}_{n_2} = 1902.81$, $S_{n_1} = 99.647$, and $S_{n_2} = 129.028$. The studentized difference between the means is $t = 8.348$. The bootstrap $(1 - \alpha)$-level confidence interval for $\delta / \left(\dfrac{S_{n_1}^2}{n_1} + \dfrac{S_{n_2}^2}{n_2} \right)^{1/2}$ is (-2.27, 2.00). The hypothesis that $\mu_1 = \mu_2$ or $\delta = 0$ is **rejected** with $P^* \approx 0$. In Fig. 3.17 we present the histogram of the EBD of t^*.

■

3.11.4 Bootstrap Tests and Confidence Intervals for the Variance

Let $S_1^{*2}, \cdots, S_M^{*2}$ be the variances of M bootstrap samples. These statistics are distributed around the sample variance S_n^2. Consider the problem of testing the hypotheses $H_0 : \sigma^2 \leq \sigma_0^2$ against $H_1 : \sigma^2 > \sigma_0^2$, where σ^2 is the variance of the parent population. As in Sect. 3.3.2.3, H_0 is rejected if S^2/σ_0^2 is sufficiently large.

Let $G_M^*(x)$ be the bootstrap empirical c.d.f. of $S_1^{*2}, \cdots, S_M^{*2}$. The bootstrap P^* value for testing H_0 is

$$P^* = 1 - G_M^* \left(\frac{S^4}{\sigma_0^2} \right). \tag{3.70}$$

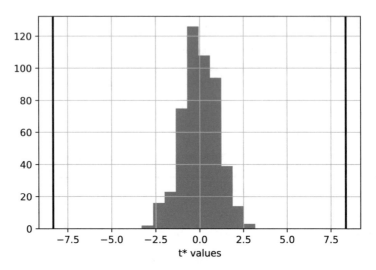

Fig. 3.17 Histogram of the EBD of $M = 500$ studentized differences

If P^* is sufficiently small, we reject H_0. For example, in a random sample of size
$n = 20$, the sample standard deviation is $S_{20} = 24.837$. Suppose that we wish
to test whether it is significantly larger than $\sigma_0 = 20$. We can create $M = 500$
bootstrapped samples using Python. The P^* value for testing the hypothesis H_0
is the proportion of bootstrap standard deviations greater than $S_{20}^2/\sigma_0 = 30.843$.
Running the program we obtain $P^* = 0.016$. The hypothesis H^0 is rejected. S_{20}
is significantly greater than $\sigma_0 = 20$. In a similar manner, we test the hypotheses
$H_0 : \sigma^2 \geq \sigma_0^2$ against $H_1 : \sigma^2 < \sigma_0^2$ or the two-sided hypothesis $H_0 : \sigma^2 = \sigma_0^2$
against $H_0 : \sigma^2 \neq \sigma_0^2$. Percentile bootstrap confidence limits for σ^2, at level $1 - \alpha$,
are given by $\dfrac{\alpha}{2}$-th and $\left(1 - \dfrac{\alpha}{2}\right)$-th quantiles of $G_M^*(x)$, or

$$S_{(j_{\alpha/2})}^{*2} \quad \text{and} \quad S_{(1+j_{1-\alpha/2})}^{*2}.$$

These bootstrap confidence limits for σ^2 at level 0.95, are 282.144 and 931.708.
The corresponding chi-squared confidence limits (see Sect. 3.4.3) are 356.77 and
1315.97. Another type of bootstrap confidence interval is given by the limits

$$\frac{S_n^4}{S_{(j_{1-\alpha/2})}^{*2}}, \quad \frac{S_n^4}{S_{(j_{\alpha/2})}^{*2}}.$$

These limits are similar to the chi-squared confidence interval limits but use the
quantiles of S_n^{*2}/S_n instead of those of $\chi^2[n - 1]$. For the sample of size $n = 20$
with $S_{20} = 24.837$, the above confidence interval for σ^2 is (408.43, 1348.74).

```
# create random sample with standard deviation 24.8
from statsmodels.distributions.empirical_distribution import ECDF
np.random.seed(seed=150)
X = stats.norm.rvs(scale=20, size=20)
S20 = np.std(X, ddof=1)
print(f'S20 = {S20:.3f}')

random.seed(1)
sigma0 = 20
GM = []
for _ in range(500):
    Xstar = random.choices(X, k=20)
    GM.append(np.std(Xstar, ddof=1)**2)

Pstar = sum(GM > (S20**2/sigma0)**2) / len(GM)
print(S20**2/sigma0)
print(f'Pstar = {Pstar:.3f}')

alpha=0.05
ci = np.quantile(GM, [alpha/2, 1-alpha/2])
print(f'0.95-ci [{ci[0]}, {ci[1]}]')

# chi2 confidence limit
n = 20
ci = (n-1)*S20**2 / stats.chi2.ppf([1-alpha/2, alpha/2], n-1)
print(f'chi2-ci [{ci[0]}, {ci[1]}]')

# bootstrapped based
GMquant = np.quantile(GM, [1-alpha/2, alpha/2])
ci = S20**4 / GMquant
print(f'boot-ci [{ci[0]}, {ci[1]}]')
```

```
S20 = 24.837
30.843882606534628
Pstar = 0.016
0.95-ci [282.1441534165316, 931.7079647027307]
chi2-ci [356.7685004417997, 1315.9662831350934]
boot-ci [408.4305942578152, 1348.7362154780665]
```

3.11.5 Comparing Statistics of Several Samples

It is often the case that we have to test whether the means or the variances of three or more populations are equal. In Chaps. 5, 6, and 7 in the Industrial Statistics book, we discuss the design and analysis of experiments where we study the effect of changing levels of different factors. Typically, we perform observations at different experimental conditions. The question is whether the observed differences between the means and variances of the samples observed under different factor level combinations are significant. The test statistic which we will introduce to test differences between means might be affected also by differences between variances. It is therefore prudent to test first whether the population variances are the same. If this hypothesis is rejected, one should not use the test for means, which is discussed below, but refer to a different type of analysis.

3.11.5.1 Comparing Variances of Several Samples

Suppose we have k samples, $k \geq 2$. Let $S_{n_1}^2, S_{n_2}^2, \cdots, S_{n_k}^2$ denote the variances of these samples. Let $S_{\max}^2 = \max\{S_{n_1}^2, \cdots, S_{n_k}^2\}$ and $S_{\min}^2 = \min\{S_{n_1}^2, \cdots, S_{n_k}^2\}$. The test statistic which we consider is the ratio of the maximal to the minimal variances, i.e.,

$$\tilde{F} = S_{\max}^2 / S_{\min}^2. \qquad (3.71)$$

The hypothesis under consideration is

$$H : \sigma_1^2 = \sigma_2^2 = \cdots = \sigma_k^2.$$

To test this hypothesis, we construct the following EBD.

- Step 1. Sample independently RSWR of sizes $n_1, \cdots, n_k$ respectively, from the given samples. Let $S_{n_1}^{*2}, \cdots, S_{n_k}^{*2}$ be the sample variances of these bootstrap samples.
- Step 2. Compute $W_i^{*2} = \frac{S_{n_i}^{*2}}{S_{n_i}^2}, i = 1, \cdots, k.$
- Step 3. Compute $\tilde{F}^* = \max_{1 \leq i \leq k}\{W_i^{*2}\} / \min_{1 \leq i \leq k}\{W_i^{*2}\}.$

Repeat these steps M times to obtain the EBD of $\tilde{F}_1^*, \cdots, \tilde{F}_M^*$.

Let $\tilde{F}_{1-\alpha}^*$ denote the $(1-\alpha)$-th quantile of this EBD distribution. The hypothesis H is **rejected** with level of significance α, if $\tilde{F} > \tilde{F}_{1-\alpha}^*$. The corresponding P^* level is the proportion of $\tilde{F}^*$ values which are greater than $\tilde{F}$.

Example 3.23 We compare now the variances of the resistance Res 3 in three hybrids. The data file is **HYBRID.csv**. In the present example, $n_1 = n_2 = n_3 = 32$. We find that $S_{n_1}^2 = 9929.54$, $S_{n_2}^2 = 16648.35$, and $S_{n_3}^2 = 21001.01$. The ratio of the maximal to minimal variance is $\tilde{F} = 2.11$. With $M = 500$ bootstrap samples, we find that $P^* = 0.544$. For $\alpha = 0.05$ we find that $\tilde{F}_{.95}^* = 2.389$. The sample $\tilde{F}$ is smaller than $\tilde{F}_{.95}^*$. The hypothesis of equal variances cannot be rejected at a level of significance of $\alpha = 0.05$.

In Python:

```
hybrid = mistat.load_data('HYBRID')

# variance for each column
S2 = hybrid.var(axis=0)
F0 = max(S2) / min(S2)
print('S2', S2)
print('F0', F0)

# Step 1: sample variances of bootstrapped samples for each column
B = {}
for seed, column in enumerate(hybrid.columns):
    B[column] = pg.compute_bootci(hybrid[column], func='var', n_boot=500,
                       confidence=0.95, seed=seed, return_dist=True)
Bt = pd.DataFrame({column: B[column][1] for column in hybrid.columns})
```

```
# Step 2: compute Wi
Wi = Bt / S2

# Step 3: compute F*
FBoot = Wi.max(axis=1) / Wi.min(axis=1)
FBoot95 = np.quantile(FBoot, 0.95)
print('FBoot 95%', FBoot95)
print('ratio', sum(FBoot >= F0)/len(FBoot))
```

```
S2 hyb1      9929.539315
  hyb2     16648.350806
  hyb3     21001.007056
dtype: float64
F0 2.1150031629110884
FBoot 95% 2.3889715227387565
ratio 0.102
```

∎

3.11.5.2   Comparing Several Means: The One-Way Analysis of Variance

The one-way analysis of variance (ANOVA) is a procedure of testing the equality of means, assuming that the variances of the populations are all equal. The hypothesis under test is

$$H : \mu_1 = \mu_2 \cdots = \mu_k.$$

Let $\bar{X}_{n_1}, S^2_{n_1}, \cdots, \bar{X}_{n_k}, S^2_{n_k}$ be the means and variances of the k samples. We compute the test statistic

$$F = \frac{\sum_{i=1}^{k} n_i (\bar{X}_{n_i} - \bar{\bar{X}})^2 / (k-1)}{\sum_{i=1}^{k} (n_i - 1) S^2_{n_i} / (N-k)}, \tag{3.72}$$

where

$$\bar{\bar{X}} = \frac{1}{N} \sum_{i=1}^{k} n_i \bar{X}_{n_i} \tag{3.73}$$

is the weighted average of the sample means, called the **grand mean**, and $N = \sum_{i=1}^{k} n_i$ is the total number of observations.

According to the bootstrap method, we repeat the following procedure M times:

- Step 1: Draw k RSWR of sizes $n_1, \cdots, n_k$ from the k given samples.
- Step 2: For each bootstrap sample, compute the mean and variance $\bar{X}^*_{n_i}$ and $S^{*2}_{n_i}$, $i = 1, \cdots, k$.
- Step 3: For each $i = 1, \cdots, k$ compute

$$\bar{Y}_i^* = \bar{X}_{n_i}^* - (\bar{X}_{n_i} - \bar{\bar{X}}).$$

[Notice that $\bar{\bar{Y}}^* = \frac{1}{N} \sum_{i=1}^{k} n_i \bar{Y}_i^* = \bar{\bar{X}}^*$, which is the grand mean of the k bootstrap samples.]

- Step 4: Compute

$$
\begin{aligned}
F^* &= \frac{\sum_{i=1}^{k} n_i (\bar{Y}_i^* - \bar{\bar{Y}}^*)^2 / (k-1)}{\sum_{i=1}^{k} (n_i - 1) S_{n_i}^{*2} / (N-k)} \\
&= \frac{\left[\sum_{i=1}^{k} n_i (\bar{X}_{n_i}^* - \bar{X}_{n_i})^2 - N(\bar{\bar{X}} - \bar{\bar{X}}^*)^2 \right] / (k-1)}{\sum_{i=1}^{k} (n_i - 1) S_{n_i}^{*2} / (N-k)}.
\end{aligned}
\tag{3.74}
$$

After M repetitions we obtain the EBD of $F_1^*, \cdots, F_M^*$.

Let $F_{1-\alpha}^*$ be the $(1-\alpha)$-th quantile of this EBD. The hypothesis H is **rejected**, at level of significance α, if $F > F_{1-\alpha}^*$. Alternatively H is rejected if the P^*-level is small, where

$$P^* = \text{proportion of } F^* \text{ values greater than } F$$

Example 3.24 Testing the equality of the means in the **HYBRID.csv** file, we obtain, using $M = 500$ bootstrap replicates, the following statistics:

$$\text{Hybrid1: } \bar{X}_{32} = 2143.406, \quad S_{32}^2 = 9929.539.$$

$$\text{Hybrid2: } \bar{X}_{32} = 1902.813, \quad S_{32}^2 = 16648.351.$$

$$\text{Hybrid3: } \bar{X}_{32} = 1850.344, \quad S_{32}^2 = 21001.007.$$

The test statistic is $F = 50.333$. The P^* level for this F is 0. Thus, the hypothesis H is **rejected**. The histogram of the EBD of the F^* values is presented in Fig. 3.18. In Python:

```
hybrid = mistat.load_data('HYBRID')
hybrid_long = pd.melt(hybrid, value_vars=hybrid.columns)

def test_statistic_F(samples):
    ''' Calculate test statistic F from samples '''
    k = len(samples)
    Ni = np.array([len(sample) for sample in samples])
    N = np.sum(Ni)
    XBni = np.array([np.mean(sample) for sample in samples])
    S2ni = np.array([np.var(sample, ddof=1) for sample in samples])
    XBB = np.sum(Ni * XBni) / N
    Sn = np.sum(Ni*(XBni - XBB)**2) / (k-1)
    Sd = np.sum((Ni-1)*S2ni) / (N-1)
    F0 = Sn / Sd
    return F0, XBni, XBB
```

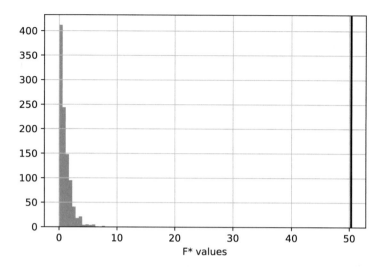

Fig. 3.18 Histogram of the EBD of $M = 500$ F^* values. The vertical line shows the observed F value

```
samples = [hybrid['hyb1'], hybrid['hyb2'], hybrid['hyb3']]
F0, XBni, XBB = test_statistic_F(samples)
DB = XBni - XBB
print(f'F = {F0:.3f}')

Ns = 1000
Fstar = []
for _ in range(Ns):
    Ysamples = []
    for sample, DBi in zip(samples, DB):
        Xstar = np.array(random.choices(sample, k=len(sample)))
        Ystar = Xstar - DBi
        Ysamples.append(Ystar)
    Fs = test_statistic_F(Ysamples)[0]
    Fstar.append(Fs)

ax = pd.Series(Fstar).hist(bins=14, color='grey')
ax.axvline(F0, color='black', lw=2)
ax.set_xlabel('F* values')
plt.show()
```

```
F = 50.333
```

Instead of using our own implementation of one-way ANOVA as the test statistic, we can also use one of several available implementations; here we use `stats.f_oneway`. Note that the F values are slightly different due to a more robust implementation used in the library implementation.

```
def test_statistic_F(samples):
    return stats.f_oneway(*samples).statistic

F0 = test_statistic_F(samples)
print(f'F = {F0:.3f}')

# Calculate sample shifts
Ni = np.array([len(sample) for sample in samples])
N = np.sum(Ni)
```

```
XBni = np.array([np.mean(sample) for sample in samples])
XBB = np.sum(Ni * XBni) / N
DB = XBni - XBB

Ns = 1000
Fstar = []
for _ in range(Ns):
    Ysamples = []
    for sample, DBi in zip(samples, DB):
        Xstar = np.array(random.choices(sample, k=len(sample)))
        Ysamples.append(Xstar - DBi)
    Fs = test_statistic_F(Ysamples)
    Fstar.append(Fs)
```

```
F = 49.274
```

∎

3.12   Bootstrap Tolerance Intervals

3.12.1   Bootstrap Tolerance Intervals for Bernoulli Samples

Trials (experiments) are called **Bernoulli** trials, if the results of the trials are either 0 or 1 (head or tail; good or defective, etc.); the trials are independently performed, and the probability for 1 in a trial is a fixed constant p, $0 < p < 1$. A random sample (RSWR) of size n, from a population of 0s and 1s, whose mean is p (proportion of 1's) will be called a **Bernoulli sample**. The number of 1s in such a sample has a binomial distribution. This is the sampling distribution of the number of 1s in all possible Bernoulli sample of size n and population mean p. p is the probability that in a random drawing of an element from the population, the outcome is 1.

Let X be the number of 1s in a RSWR of size n from such a population. If p is known, we can determine two integers $I_{\beta/2}(p)$ and $I_{1-\beta/2}(p)$ such that the proportion of Bernoulli samples for which $I_{\beta/2}(p) \leq X \leq I_{1-\beta/2}(p)$ is $(1 - \beta)$.

For example, if $n = 50$, if $p = 0.1$, $\beta = 0.05$ we obtain $I_{.025}(0.1) = 1$, and $I_{.975}(0.1) = 9$. Using Python:

```
stats.binom.ppf([0.025, 0.975], 50, 0.1)
```

```
array([1., 9.])
```

If p is unknown, and has to be estimated from a given Bernoulli sample of size n, we determine first the bootstrap $(1 - \alpha)$-level confidence interval for p. If the limits for this interval are $(p^*_{\alpha/2}, p^*_{1-\alpha/2})$, then the prediction interval $(I_{\beta/2}(p^*_{\alpha/2}), I_{1-\beta/2}(p^*_{1-\alpha/2}))$ is a bootstrap **tolerance interval** of confidence $(1-\alpha)$ and content $(1 - \beta)$.

Example 3.25   Consider the $n = 99$ electric voltage outputs of circuits, which is in data file **OELECT.csv**. Suppose that it is required that the output X will be between

216 and 224 volts. We create a Bernoulli sample in which we give a circuit the value 1 if its electric output is in the interval (216, 224) and the value 0 otherwise. This Bernoulli sample is stored in the variable elec_index.

The objective is to determine a (0.95, 0.95) tolerance interval for a future batch of $n = 100$ circuits from this production process. We create $M = 500$ RSWR from elec_index and determine for each sample the expected confidence interval. Due to limitations of the pg.compute_bootci implementation, we create samples for the upper and lower bound separately. In each case, we determine the 0.025 and 0.975 quartiles to finally derive the tolerance interval.

```
oelect = mistat.load_data('OELECT')

elec_index = np.array([1 if 216 <= value <= 224 else 0
                       for value in oelect])

def qbinomBoot(x, p):
    return stats.binom.ppf(p, 100, p=x.mean())

B_025 = pg.compute_bootci(elec_index, func=lambda x: qbinomBoot(x, p=0.025),
                          n_boot=500, seed=1, return_dist=True)
B_975 = pg.compute_bootci(elec_index, func=lambda x: qbinomBoot(x, p=0.975),
                          n_boot=500, seed=1, return_dist=True)
tol_int = [np.quantile(B_025[1], 0.025),np.quantile(B_975[1], 0.975)]
print(f'Tolerance interval ({tol_int[0]}, {tol_int[1]})')
```

```
Tolerance interval (49.0, 86.0)
```

The bootstrap tolerance interval is (49.0,86.0). In other words, with confidence level of 0.95, we predict that 95% of future batches of $n = 100$ circuits will have between 49.0 and 86.0 circuits which comply to the standard. The exact tolerance intervals are given by

$$
\begin{aligned}
\text{Lower} &= B^{-1}\left(\frac{\beta}{2}; n, \underline{p}_\alpha\right) \\
\text{Upper} &= B^{-1}\left(1 - \frac{\beta}{2}; n, \bar{p}_\alpha\right)
\end{aligned}
\tag{3.75}
$$

where $(\underline{p}_\alpha, \bar{p}_\alpha)$ is a $(1 - \alpha)$ confidence interval for p. In the present data, the 0.95-confidence interval for p is (0.585, 0.769). Thus, the (0.95,0.95) tolerance interval is (48,84), which is close to the bootstrap interval. ∎

3.12.2 Tolerance Interval for Continuous Variables

In a RSWR of size n, the p-th quantile, i.e., $X_{(np)}$, is an estimator of the p-th quantile of the distribution. Thus, we expect that the proportion of X-values in the population, falling in the interval $(X_{(n\beta/2)}, X_{(n(1-\beta/2))})$, is approximately $(1 - \beta)$ in large samples. As was explained in Chap. 1, $X_{(j)}, (j = 1, \cdots, n)$ is the j-th order statistic of the sample, and for $0 < p < 1$, $X_{(j.p)} = X_{(j)} + p(X_{(j+1)} - X_{(j)})$. By the bootstrap method, we generate M replicas of the statistics $X^*_{(n\beta/2)}$ and $X^*_{(n(1-\beta/2))}$. The $(1 - \alpha, 1 - \beta)$-tolerance interval is given by $(Y^*_{(M\alpha/2)}, Y^{**}_{(M(1-\alpha/2))})$, where $Y^*_{(M\alpha/2)}$ is the $\alpha/2$-quantile of the EBD of $X^*_{(n\beta/2)}$ and $Y^{**}_{(M(1-\alpha/2))}$ is the $(1 - \alpha/2)$-quantile of the EBD of $X^*_{(n(1-\beta/2))}$.

Example 3.26 Let us determine the $(0.95, 0.95)$ tolerance interval for samples of size $n = 100$, of piston cycle times. Use the sample in the data file **CYCLT.csv**. The original sample is of size $n_0 = 50$. Since future samples are of size $n = 100$, we draw from the original sample RSWR of size $n = 100$.

In Python, the calculations from Example 3.26 are straightforward:

```
cyclt = mistat.load_data('CYCLT')
cyclt = [*cyclt, *cyclt]  # create a dataset of size 100 by duplication

def getQuantile(x, p):
    return np.quantile(x, p)

B_025 = pg.compute_bootci(cyclt, func=lambda x: getQuantile(x, p=0.025),
        n_boot=500, seed=1, return_dist=True)
B_975 = pg.compute_bootci(cyclt, func=lambda x: getQuantile(x, p=0.975),
        n_boot=500, seed=1, return_dist=True)
print('0.025%', np.quantile(B_025[1], 0.025))
print('0.975%', np.quantile(B_975[1], 0.975))
```

```
0.025% 0.175
0.975% 1.141
```

The bootstrap $(0.95, 0.95)$ tolerance interval for $n = 100$ piston cycle times was estimated as $(0.175, 1.141)$. ∎

3.12.3 Distribution-Free Tolerance Intervals

The tolerance limits described above are based on the model of normal distribution. **Distribution-free** tolerance limits for $(1 - \beta)$ proportion of the population, at confidence level $(1 - \alpha)$, can be obtained for any model of continuous c.d.f. $F(x)$. As we will show below, if the sample size n is large enough, so that the following inequality is satisfied, i.e.,

$$\left(1 - \frac{\beta}{2}\right)^n - \frac{1}{2}(1 - \beta)^n \le \frac{\alpha}{2} \tag{3.76}$$

then the order statistics $X_{(1)}$ and $X_{(n)}$ are lower and upper tolerance limits. This is based on the following important property:

> **If X is a random variable having a continuous c.d.f. $F(x)$, then $U = F(x)$ has a uniform distribution on $(0, 1)$.**

Indeed

$$\Pr\{F(X) \le \eta\} = \Pr\{X \le F^{-1}(\eta)\}$$
$$= F(F^{-1}(\eta)) = \eta, \quad 0 < \eta < 1.$$

If $X_{(i)}$ is the i-th order statistic of a sample of n i.i.d. random variables having a common c.d.f. $F(x)$, then $U_{(i)} = F(X_{(i)})$ is the i-th order statistic of n i.i.d. random having a uniform distribution.

Now, the interval $(X_{(1)}, X_{(n)})$ contains at least a proportion $(1 - \beta)$ of the population if $X_{(1)} \le \xi_{\beta/2}$ and $X_{(n)} \ge \xi_{1-\beta/2}$, where $\xi_{\beta/2}$ and $\xi_{1-\beta/2}$ are the $\beta/2$ and $\left(1 - \dfrac{\beta}{2}\right)$ quantiles of $F(x)$.

Equivalently, $(X_{(1)}, X_{(n)})$ contains at least a proportion $(1 - \beta)$ if

$$U_{(1)} \le F(\xi_{\beta/2}) = \frac{\beta}{2}$$
$$U_{(n)} \ge F(\xi_{1-\beta/2}) = 1 - \beta/2.$$

By using the joint p.d.f. of $(U_{(1)}, U_{(n)})$, we show that

$$\Pr\left\{U_{(1)} \le \frac{\beta}{2}, U_{(n)} \ge 1 - \frac{\beta}{2}\right\} = 1 - 2\left(1 - \frac{\beta}{2}\right)^n + (1 - \beta)^n. \tag{3.77}$$

This probability is the confidence that the interval $(X_{(1)}, X_{(n)})$ covers the interval $(\xi_{\beta/2}, \xi_{1-\beta/2})$. By finding n which satisfies

$$1 - 2\left(1 - \frac{\beta}{2}\right)^n + (1 - \beta)^n \ge 1 - \alpha. \tag{3.78}$$

we can assure that the confidence level is at least $(1 - \alpha)$.

In Table 3.13 we give the values of n for some α and β values.

Table 3.13 can also be used to obtain the confidence level associated with fixed values of β and n. We see that with a sample of size 104, $(X_{(1)}, X_{(n)})$ is a tolerance interval for at least 90% of the population with approximately 99% confidence level or a tolerance interval for at least 95% of the population with slightly less than 90% confidence.

Other order statistics can be used to construct distribution-free tolerance intervals, that is, we can choose any integers j and k, where $1 \le j, k \le n/2$, and form

Table 3.13 Sample size required for $(X_{(1)}, X_{(n)})$ to be a $(1 - \alpha, 1 - \beta)$-level tolerance interval

β	α	n
0.10	0.10	58
	0.05	72
	0.01	104
0.05	0.10	118
	0.05	146
	0.01	210
0.01	0.10	593
	0.05	734
	0.01	1057

the interval $(X_{(j)}, X_{(n-k+1)})$. When $j > 1$ and $k > 1$, the interval will be shorter than the interval $(X_{(1)}, X_{(n)})$, but its confidence level will be reduced.

3.13   Non-Parametric Tests

Testing methods like the Z-tests, t-tests, etc. presented in this chapter were designed for specific distributions. The Z- and t-tests are based on the assumption that the parent population is normally distributed. What would be the effect on the characteristics of the test if this basic assumption is wrong? This is an important question, which deserves special investigation. We remark that if the population variance σ^2 is finite and the sample is large, then the t-test for the mean has approximately the required properties even if the parent population is not normal. In small samples, if it is doubtful whether the distribution of the parent population, we should perform a distribution free test or compute the P-value of the test statistic by the bootstrapping method. In the present section we present three non-parametric tests, namely, **sign test**, the **randomization test**, and the **Wilcoxon signed-rank test**.

3.13.1   The Sign Test

Suppose that $X_1, \cdots, X_n$ is a random sample from some **continuous** distribution, F, and has a positive p.d.f. throughout the range of X. Let ξ_p, for some $0 < p < 1$, be the p-th quantile of F. We wish to test the hypothesis that ξ_p does not exceed a specified value ξ^*, i.e.,

$$H_0 : \xi_p \leq \xi^*$$

against the alternative

$$H_1 : \xi_p > \xi^*.$$

If the null hypothesis H_0 is true, the probability of observing an X-value smaller than ξ^* is greater or equal to p; and if H_1 is true, then this probability is smaller than p. The sign test of H_0 versus H_1 reduces the problem to a test for p in a binomial model. The test statistic is $K_n = \#\{X_i \leq \xi^*\}$, i.e., the number of observed X-values in the sample which do not exceed ξ^*. K_n has a binomial distribution $B(n, \theta)$, irrespective of the parent distribution F. According to H_0, $\theta \geq p$, and according to H_1, $\theta < p$. The test proceeds then as in Sect. 3.3.2.

Example 3.27 We wish to test whether the median, $\xi_{.5}$, of the distribution of piston cycle times is greater than 0.50 [min]. The sample data is in file **CYCLT.csv**. The sample size is $n = 50$. Let $K_{50} = \sum_{i=1}^{50} I\{X_i \leq 0.5\}$. The null hypothesis is $H_0 : p \leq \frac{1}{2}$ vs. $H_1 : p > \frac{1}{2}$. From the sample values, we find $K_{50} = 24$. The P-value is $1 - B(23; 50, .5) = 0.664$. The null hypothesis H_0 is not rejected. The sample median is $M_e = 0.546$. This is however not significantly greater than 0.5. ∎

The sign test can be applied also to test whether tolerance specifications hold. Suppose that the standard specifications require that at least $(1 - \beta)$ proportion of products will have an X value in the interval (ξ^*, ξ^{**}). If we wish to test this, with level of significance α, we can determine the $(1 - \alpha, 1 - \beta)$ tolerance interval for X, based on the observed random sample, and accept the hypothesis

$$H_0 : \xi^* \leq \xi_{\beta/2} \text{ and } \xi_{1-\beta/2} \leq \xi^{**}$$

if the tolerance interval is included in (ξ^*, ξ^{**}).

We can also use the sign test. Given the random sample $X_1, \ldots, X_n$, we compute

$$K_n = \sum_{i=1}^{n} I\{\xi^* \leq X_i \leq \xi^{**}\}.$$

The null hypothesis H_0 above is equivalent to the hypothesis

$$H_0^* : p \geq 1 - \beta$$

in the binomial test. H_0^* is rejected, with level of significance α, if

$$K_n < B^{-1}(\alpha; n, 1 - \beta),$$

where $B^{-1}(\alpha; n, 1 - \beta)$ is the α-quantile of the binomial distribution $B(n, 1 - \beta)$.

Example 3.28 In Example 3.26 we have found that the bootstrap $(0.95, 0.95)$ tolerance interval for the **CYCLT.csv** sample is $(0.175, 1.141)$. Suppose that the specification requires that the piston cycle time in 95% of the cases will be in the interval $(0.2, 1.1)$ [min]. Can we accept the hypothesis

$$H_0^* : 0.2 \leq \xi_{.025} \text{ and } \xi_{.975} \leq 1.1$$

with level of significance $\alpha = 0.05$? For the data **CYCLT.csv** we find

$$K_{50} = \sum_{i=1}^{50} I\{.2 \leq X_i \leq 1.1\} = 41.$$

Also $B^{-1}(0.05; 50, 0.95) = 45$. Thus, since $K_{50} < 45$, H_0^* is rejected. This is in accord with the bootstrap tolerance interval, since $(0.175, 1.141)$ contains the interval $(0.2, 1.1)$. ∎

3.13.2 The Randomization Test

The randomization test described here can be applied to test whether two random samples come from the same distribution, F, without specifying the distribution F.

The null hypothesis, H_0, is that the two distributions, from which the samples are generated, are the same. The randomization test constructs a reference distribution for a specified test statistic, by randomly assigning to the observations the labels of the samples. For example, let us consider two samples, which are denoted by A_1 and A_2. Each sample is of size $n = 3$. Suppose that we observed

$$A_2 \ A_2 \ A_2 \ A_1 \ A_1 \ A_1$$
$$1.5 \ 1.1 \ 1.8 \ 0.75 \ 0.60 \ 0.80.$$

The sum of the values in A_2 is $T_2 = 4.4$ and that of A_1 is $T_1 = 2.15$. Is there an indication that the two samples are generated from different distributions? Let us consider the test statistic $D = (T_2 - T_1)/3$ and reject H_0 if D is sufficiently large. For the given samples, $D = 0.75$. We construct now the reference distribution for D under H_0.

There are $\binom{6}{3} = 20$ possible assignments of the letters A_1 and A_2 to the six values. Each such assignment yields a value for D. The reference distribution assigns each such value of D an equal probability of $1/20$. The 20 assignments of letters and the corresponding D values are given in Table 3.14.

Under the reference distribution, each one of these values of D is equally probable, and the P-value of the observed value of the observed D is $P = \dfrac{1}{20} = 0.05$. The null hypothesis is rejected at the $\alpha = 0.05$ level. If n is large, it becomes impractical to construct the reference distribution in this manner. For example, if $t = 2$ and $n_1 = n_2 = 10$, we have $\binom{20}{10} = 184,756$ assignments.

We can, however, estimate the P-value, by sampling, with replacement, from this reference distribution.

Table 3.14 Assignments for the randomized test

A_{ij}	Assignments									
0.75	1	1	1	1	1	1	1	1	1	1
0.6	1	1	1	1	2	2	2	2	2	2
0.8	1	2	2	2	1	1	1	2	2	2
1.5	2	1	2	2	1	2	2	1	1	2
1.1	2	2	1	2	2	1	2	1	2	1
1.8	2	2	2	1	2	2	1	2	1	1
D	0.750	0.283	0.550	0.083	0.150	0.417	−0.050	−0.050	−0.517	−0.250

A_{ij}	Assignments									
0.75	2	2	2	2	2	2	2	2	2	2
0.6	1	1	1	1	1	1	2	2	2	2
0.8	1	1	1	2	2	2	1	1	1	2
1.5	1	2	2	1	1	2	1	1	2	1
1.1	2	1	2	1	2	1	1	2	1	1
1.8	2	2	1	2	1	1	2	1	1	1
D	0.250	0.517	0.050	0.050	−0.417	−0.150	−0.083	−0.550	−0.283	−0.750

Example 3.29 File **OELECT.csv** contains $n_1 = 99$ random values of the output in volts of a rectifying circuit. File **OELECT1.csv** contains $n_2 = 25$ values of outputs of another rectifying circuit. The question is whether the differences between the means of these two samples is significant. Let $\bar{X}$ be the mean of **OELECT** and $\bar{Y}$ be that of **OELECT1**. We find that $D = \bar{X} - \bar{Y} = -10.7219$. In Python, we can use the function `randomizationTest` that is included in `mistat` package.

```
oelect = mistat.load_data('OELECT')
oelect1 = mistat.load_data('OELECT1')

_ = mistat.randomizationTest(oelect, oelect1, np.mean,
                        aggregate_stats=lambda x: x[0] - x[1],
                        n_boot=500, seed=1)
```

```
Original stat is -10.721980
Original stat is at quantile 1 of 501 (0.20%)
Distribution of bootstrap samples:
 min: -4.94, median: 0.16,  max: 4.55
```

The original mean -10.721 is the minimum, and the test rejects the hypothesis of equal means with a P-value $P = \dfrac{1}{501}$. ∎

3.13.3 The Wilcoxon Signed-Rank Test

In Sect. 3.13.1 we discussed the sign test. The Wilcoxon signed -rank (WSR) test is a modification of the sign test, which brings into consideration not only the signs of the sample values but also their magnitudes. We construct the test statistic in two steps. First, we rank the magnitudes (absolute values) of the sample values, giving the rank 1 to the value with smallest magnitude and the rank n to that with the maximal magnitude. In the second step, we sum the ranks multiplied by the signs of the values. For example, suppose that a sample of $n = 5$ is -1.22, $-.53$, 0.27, 2.25, 0.89. The ranks of the magnitudes of these values are, respectively, 4, 2, 1, 5, 3. The signed rank statistic is

$$W_5 = 0 \times 4 + 0 \times 2 + 1 + 5 + 3 = 9.$$

Here we assigned each negative value the weight 0 and each positive value the weight 1.

The WSR test can be used for a variety of testing problems. If we wish to test whether the distribution median, $\xi_{.5}$, is smaller or greater than some specified value ξ^*, we can use the statistics

$$W_n = \sum_{i=1}^{n} I\{X_i > \xi^*\} R_i, \tag{3.79}$$

where

$$I\{X_i > \xi^*\} = \begin{cases} 1, & \text{if } X_i > \xi^* \\ 0, & \text{otherwise.} \end{cases}$$

$R_i = \text{rank}(|X_i|)$.

The WSR test can be applied to test whether two random samples are generated from the same distribution against the alternative that one comes from a distribution having a larger location parameter (median) than the other. In this case we can give the weight 1 to elements of sample 1 and the weight 0 to the elements of sample 2. The ranks of the values are determined by combining the two samples. For example, consider two random samples $X_1, \ldots, X_5$ and $Y_1, \ldots, Y_5$ generated from $N(0, 1)$ and $N(2, 1)$. These are

X	0.188	0.353	$-$0.257	0.220	0.168
Y	1.240	1.821	2.500	2.319	2.190

The ranks of the magnitudes of these values are

X	2	5	4	3	1
Y	6	7	10	9	8

The value of the WSR statistic is

$$W_{10} = 6 + 7 + 10 + 9 + 8 = 40.$$

Notice that all the ranks of the Y values are greater than those of the X values. This yields a relatively large value of W_{10}. Under the null hypothesis that the two samples are from the same distribution, the probability that the sign of a given rank is 1 is $1/2$. Thus, the reference distribution, for testing the significance of W_n, is like that of

$$W_n^0 = \sum_{j=1}^{n} j B_j \left(1, \frac{1}{2}\right), \tag{3.80}$$

where $B_1 \left(1, \frac{1}{2}\right), \ldots, B_n \left(1, \frac{1}{2}\right)$ are mutually independent $B \left(1, \frac{1}{2}\right)$ random variables. The distribution of W_n^0 can be determined exactly. W_n^0 can assume the values $0, 1, \ldots, \dfrac{n(n+1)}{2}$ with probabilities which are the coefficients of the polynomial in t

$$P(t) = \frac{1}{2^n} \prod_{j=1}^{n} \left(1 + t^j\right).$$

These probabilities can be computed exactly. For large values of n, W_n^0 is approximately normal with mean

$$E\{W_n^0\} = \frac{1}{2} \sum_{j=1}^{n} j = \frac{n(n+1)}{4} \tag{3.81}$$

and variance

$$V\{W_n^0\} = \frac{1}{4} \sum_{j=1}^{n} j^2 = \frac{n(n+1)(2n+1)}{24}. \tag{3.82}$$

This can yield a large sample approximation to the P-value of the test. The WSR test, to test whether the median of a symmetric continuous distribution F is equal to ξ^*, can be performed in Python using the implementation in scipy.wilcoxon.

```
X = [0.188, 0.353, -0.257, 0.220, 0.168]
Y = [1.240, 1.821, 2.500, 2.319, 2.190]

print('Wilcoxon signed-rank test (unsuitable for ties)')
print(stats.ranksums(X, Y))
print('Mann-Whitney U test (suitable for ties)')
print(stats.mannwhitneyu(X, Y))
```

```
Wilcoxon signed-rank test (unsuitable for ties)
RanksumsResult(statistic=-2.6111648393354674,
pvalue=0.009023438818080326)
Mann-Whitney U test (suitable for ties)
MannwhitneyuResult(statistic=0.0, pvalue=0.007936507936507936)
```

3.14 Chapter Highlights

This chapter provides theoretical foundations for statistical inference. Inference on parameters of infinite populations is discussed using classical point estimation, confidence intervals, tolerance intervals, and hypothesis testing. Properties of point estimators such as moment equation estimators and maximum likelihood estimators are discussed in detail. Formulas for parametric confidence intervals and distribution-free tolerance intervals are provided. Statistical tests of hypothesis are presented with examples, including tests for normality with probability plots and the chi-square and Kolmogorov-Smirnov tests of goodness of fit. The chapter includes a section on Bayesian testing and estimation. Statistical inference is introduced by exploiting the power of the personal computer. Reference distributions are constructed through bootstrapping methods. Testing for statistical significance and the significance of least square methods in simple linear regression using bootstrapping is demonstrated. Industrial applications are used throughout with specially written software simulations. Through this analysis, confidence intervals and reference distributions are derived and used to test statistical hypothesis. Bootstrap analysis of variance is developed for testing the equality of several population means. Construction of tolerance intervals with bootstrapping is also presented. Three nonparametric procedures for testing are given: the sign test, randomization test, and Wilcoxon signed-rank test.

The main concepts and definitions introduced in this chapter include:

- Statistical inference
- Sampling distribution
- Unbiased estimators
- Consistent estimators
- Standard error
- Parameter space
- Statistic
- Point estimator
- Least squares estimators
- Maximum likelihood estimators
- Likelihood function
- Confidence intervals
- Tolerance intervals
- Testing statistical hypotheses
- Operating characteristic function

- Rejection region
- Acceptance region
- Type I error
- Type II error
- Power function
- OC curve
- Significance level
- P-value
- Normal scores
- Normal probability plot
- Chi-squared test
- Kolmogorov-Smirnov test
- Bayesian decision procedures
- Statistical inference
- The bootstrap method
- Sampling distribution of an estimate
- Reference distribution
- Bootstrap confidence intervals
- Bootstrap tolerance interval
- Bootstrap ANOVA
- Nonparametric tests

3.15 Exercises

Exercise 3.1 The consistency of the sample mean, $\bar{X}_n$, in RSWR, is guaranteed by the WLLN, whenever the mean exists. Let $M_l = \frac{1}{n} \sum_{i=1}^{n} X_i^l$ be the sample estimate of the l-th moment, which is assumed to exist ($l = 1, 2, \cdots$). Show that M_r is a consistent estimator of μ_r.

Exercise 3.2 Consider a population with mean μ and standard deviation $\sigma = 10.5$. Use the CLT to find approximately how large should the sample size, n, be so that $\Pr\{|\bar{X}_n - \mu| < 1\} = 0.95$.

Exercise 3.3 Let $X_1, \cdots, X_n$ be a random sample from a normal distribution $N(\mu, \sigma)$. What is the moment-equation estimator of the p-th quantile $\xi_p = \mu + z_p \sigma$?

Exercise 3.4 Let $(X_1, Y_1), \cdots, (X_n, Y_n)$ be a random sample from a bivariate normal distribution. What is the moment-equation estimator of the correlation ρ?

Exercise 3.5 Let $X_1, X_2, \ldots, X_n$ be a sample from a beta distribution Beta(ν_1, ν_2); $0 < \nu_1, \nu_2 < \infty$. Find the moment-equation estimators of ν_1 and ν_2.

Exercise 3.6 Let $\bar{Y}_1, \cdots, \bar{Y}_k$ be the means of k independent RSWR from normal distributions, $N(\mu, \sigma_i)$, $i = 1, \cdots, k$, with common means and variances σ_i^2 **known**. Let $n_1, \cdots, n_k$ be the sizes of these samples. Consider a weighted average $\bar{Y}_w = \dfrac{\sum_{i=1}^k w_i \bar{Y}_i}{\sum_{i=1}^k w_i}$, with $w_i > 0$. Show that for the estimator $\bar{Y}_w$ having smallest variance, the required weights are $w_i = \dfrac{n_i}{\sigma_i^2}$.

Exercise 3.7 Using the formula

$$\hat{\beta}_1 = \sum_{i=1}^n w_i Y_i,$$

with $w_i = \dfrac{x_i - \bar{x}_n}{SS_x}$, $i = 1, \ldots, n$, for the LSE of the slope β in a simple linear regression, derive the formula for $V\{\hat{\beta}_1\}$. We assume that $V\{Y_i\} = \sigma^2$ for all $i = 1, \cdots, n$. You can refer to Chap. 4 for a detailed exposition of linear regression.

Exercise 3.8 In continuation of the previous exercise, derive the formula for the variance of the LSE of the intercept β_0 and $\text{Cov}(\hat{\beta}_0, \hat{\beta}_1)$.

Exercise 3.9 Show that the correlation between the LSEs, $\hat{\beta}_0$ and $\hat{\beta}_1$, in the simple linear regression is

$$\rho = -\frac{\bar{x}_n}{\left(\dfrac{1}{n}\sum x_i^2\right)^{1/2}}.$$

Exercise 3.10 Let $X_1, \cdots, X_n$ be i.i.d. random variables having a Poisson distribution $P(\lambda)$, $0 < \lambda < \infty$. Show that the MLE of λ is the sample mean $\bar{X}_n$.

Exercise 3.11 Let $X_1, \cdots, X_n$ be i.i.d. random variables from a gamma distribution, $G(\nu, \beta)$, with **known** ν. Show that the MLE of β is $\hat{\beta}_n = \dfrac{1}{\nu}\bar{X}_n$, where $\bar{X}_n$ is the sample mean. What is the variance of $\hat{\beta}_n$?

Exercise 3.12 Consider Example 3.4. Let $X_1, \cdots, X_n$ be a random sample from a negative binomial distribution, N.B.$(2, p)$. Show that the MLE of p is

$$\hat{p}_n = \frac{2}{\bar{X}_n + 2},$$

where $\bar{X}_n$ is the sample mean.

(i) On the basis of the WLLN, show that $\hat{p}_n$ is a consistent estimator of p [Hint: $\bar{X}_n \to E\{X\} = (2 - p)/p$ in probability as $n \to \infty$].
(ii) Using the fact that if $X_1, \cdots, X_n$ are i.i.d. like NB(k, p), then $T_n = \sum_{i=1}^{n} X_i$ is distributed like N.B.(nk, p), and the results of Example 3.4 show that for large values of n,

$$\text{Bias}(\hat{p}_n) \cong \frac{3p(1 - p)}{4n} \quad \text{and}$$

$$V\{\hat{p}_n\} \cong \frac{p^2(1 - p)}{2n}.$$

Exercise 3.13 Let $X_1, \ldots, X_n$ be a random sample from a shifted exponential distribution

$$f(x; \mu, \beta) = \frac{1}{\beta} \exp\left\{-\frac{x - \mu}{\beta}\right\}, \quad x \geq \mu,$$

where $0 < \mu, \beta < \infty$.

(i) Show that the sample minimum $X_{(1)}$ is an MLE of μ.
(ii) Find the MLE of β.
(iii) What are the variances of these MLEs?

Exercise 3.14 We wish to test that the proportion of defective items in a given lot is smaller than $P_0 = 0.03$. The alternative is that $P > P_0$. A random sample of $n = 20$ is drawn from the lot **with** replacement (RSWR). The number of observed defective items in the sample is $X = 2$. Is there sufficient evidence to reject the null hypothesis that $P \leq P_0$?

Exercise 3.15 Compute and plot the operating characteristic curve OC(p), for binomial testing of $H_0 : P \leq P_0$ versus $H_1 : P > P_0$, when the hypothesis is accepted if two or less defective items are found in a RSWR of size $n = 30$.

Exercise 3.16 For testing the hypothesis $H_0 : P = 0.01$ versus $H_1 : P = 0.03$, concerning the parameter P of a binomial distribution, how large should the sample be, n, and what should be the critical value, k, if we wish that error probabilities will be $\alpha = 0.05$ and $\beta = 0.05$? [Use the normal approximation to the binomial.]

Exercise 3.17 As will be discussed in Chap. 2 in the Industrial Statistics book, the Shewhart 3-σ control charts for statistical process control provide repeated tests of the hypothesis that the process mean is equal to the nominal one, μ_0. If a sample mean $\bar{X}_n$ falls outside the limits $\mu_0 \pm 3\dfrac{\sigma}{\sqrt{n}}$, the hypothesis is rejected.

(i) What is the probability that $\bar{X}_n$ will fall outside the control limits when $\mu = \mu_0$?

(ii) What is the probability that when the process is in control, $\mu = \mu_0$, all sample means of 20 consecutive independent samples, will be within the control limits?

(iii) What is the probability that a sample mean will fall outside the control limits when μ changes from μ_0 to $\mu_1 = \mu_0 + 2\dfrac{\sigma}{\sqrt{n}}$?

(iv) What is the probability that a change from μ_0 to $\mu_1 = \mu_0 + 2\dfrac{\sigma}{\sqrt{n}}$ will not be detected by the next ten sample means?

Exercise 3.18 Consider the data in file **SOCELL.csv**. Use Python to test whether the mean ISC at time t_1 is significantly smaller than 4 (Amp). [Use 1-sample t-test.]

Exercise 3.19 Is the mean of ISC for time t_2 significantly larger than 4 (Amp)?

Exercise 3.20 Consider a one-sided t-test based on a sample of size $n = 30$, with $\alpha = 0.01$. Compute the OC(δ) as a function of $\delta = (\mu - \mu_0)/\sigma$, $\mu > \mu_0$.

Exercise 3.21 Compute the OC function for testing the hypothesis $H_0 : \sigma^2 \leq \sigma_0^2$ versus $H_1 : \sigma^2 > \sigma_0^2$, when $n = 31$ and $\alpha = 0.10$.

Exercise 3.22 Compute the OC function in testing $H_0 : p \leq p_0$ versus $H_1 : p > p_0$ in the binomial case, when $n = 100$ and $\alpha = 0.05$.

Exercise 3.23 Let $X_1, \ldots, X_n$ be a random sample from a normal distribution $N(\mu, \sigma)$. For testing $H_0 : \sigma^2 \leq \sigma_0^2$ against $H_1 : \sigma^2 > \sigma_0^2$, we use the test which rejects H_0 if $S_n^2 \geq \dfrac{\sigma_0^2}{n-1}\chi_{1-\alpha}^2[n-1]$, where S_n^2 is the sample variance. What is the power function of this test?

Exercise 3.24 Let $S_{n_1}^2$ and $S_{n_2}^2$ be the variances of two independent samples from normal distributions $N(\mu_i, \sigma_i)$, $i = 1, 2$. For testing $H_0 : \dfrac{\sigma_1^2}{\sigma_2^2} \leq 1$ against $H_1 : \dfrac{\sigma_1^2}{\sigma_2^2} > 1$, we use the F-test, which rejects H_0 when $F = \dfrac{S_{n_1}^2}{S_{n_2}^2} > F_{1-\alpha}[n_1-1, n_2-1]$. What is the power of this test, as a function of $\rho = \sigma_1^2/\sigma_2^2$?

Exercise 3.25 A random sample of size $n = 20$ from a normal distribution gave the following values: 20.74, 20.85, 20.54, 20.05, 20.08, 22.55, 19.61, 19.72, 20.34, 20.37, 22.69, 20.79, 21.76, 21.94, 20.31, 21.38, 20.42, 20.86, 18.80, 21.41. Compute

(i) Confidence interval for the mean μ, at level of confidence $1 - \alpha = 0.99$.
(ii) Confidence interval for the variance σ^2, at confidence level $1 - \alpha = 0.99$.
(iii) A confidence interval for σ, at level of confidence $1 - \alpha = 0.99$.

Exercise 3.26 Let C_1 be the event that a confidence interval for the mean, μ, covers it. Let C_2 be the event that a confidence interval for the standard deviation σ covers it. The probability that both μ **and** σ are simultaneously covered is

$$\Pr\{C_1 \cap C_2\} = 1 - \Pr\{\overline{C_1 \cap C_2}\}$$

$$= 1 - \Pr\{\bar{C}_1 \cup \bar{C}_2\} \geq 1 - \Pr\{\bar{C}_1\} - \Pr\{\bar{C}_2\}.$$

This inequality is called the **Bonferroni inequality**. Apply this inequality and the results of the previous exercise to determine the confidence interval for $\mu + 2\sigma$, at level of confidence not smaller than 0.98.

Exercise 3.27 Twenty independent trials yielded $X = 17$ successes. Assuming that the probability for success in each trial is the same, θ, determine the confidence interval for θ at level of confidence 0.95.

Exercise 3.28 Let $X_1, \cdots, X_n$ be a random sample from a Poisson distribution with mean λ. Let $T_n = \sum_{i=1}^{n} X_i$. Using the relationship between the Poisson and the gamma c.d.f., we can show that a confidence interval for the mean λ, at level $1 - \alpha$, has lower and upper limits, λ_L and λ_U, where

$$\lambda_L = \frac{1}{2n} \chi^2_{\alpha/2}[2T_n + 2], \quad \text{and}$$

$$\lambda_U = \frac{1}{2n} \chi^2_{1-\alpha/2}[2T_n + 2].$$

The following is a random sample of size $n = 10$ from a Poisson distribution 14, 16, 11, 19, 11, 9, 12, 15, 14, 13. Determine a confidence interval for λ at level of confidence 0.95. You can calculate the confidence intervals using either the exact value $\chi^2_p[\nu]$ or use an approximation. For large number of degrees of freedom, $\chi^2_p[\nu] \approx \nu + z_p \sqrt{2\nu}$, where z_p is the p-th quantile of the standard normal distribution.

Exercise 3.29 The mean of a random sample of size $n = 20$, from a normal distribution with $\sigma = 5$, is $\bar{Y}_{20} = 13.75$. Determine a $1 - \beta = 0.90$ content tolerance interval with confidence level $1 - \alpha = 0.95$.

Exercise 3.30 Use the **YARNSTRG.csv** data file to determine a $(0.95, 0.95)$ tolerance interval for log yarn strength. [Hint: notice that the interval is $\bar{Y}_{100} \pm k S_{100}$, where $k = t(0.025, 0.025, 100)$.]

Exercise 3.31 Use the minimum and maximum of the log yarn strength (see previous problem) to determine a distribution free tolerance interval. What are the values of α and β for your interval? How does it compare with the interval of the previous problem?

Exercise 3.32 Make a normal Q-Q plot to test, graphically, whether the ISC-t_1 of data file **SOCELL.csv**, is normally distributed.

Exercise 3.33 Using Python and data file **CAR.csv**.

 (i) Test graphically whether the turn diameter is normally distributed.
(ii) Test graphically whether the log (horsepower) is normally distributed.

Exercise 3.34 Use the **CAR.csv** file. Make a frequency distribution of turn diameter, with $k = 11$ intervals. Fit a normal distribution to the data and make a chi-squared test of the goodness of fit.

Exercise 3.35 Using Python and the **CAR.csv** data file, compute the K.S. test statistic D_n^* for the turn diameter variable, testing for normality. Compute k_α^* for $\alpha = 0.05$. Is D_n^* significant?

Exercise 3.36 The daily demand (loaves) for whole wheat bread at a certain bakery has a Poisson distribution with mean $\lambda = 100$. The loss to the bakery for undemanded unit at the end of the day is $C_1 = \$0.10$. On the other hand, the penalty for a shortage of a unit is $C_2 = \$0.20$. How many loaves of whole wheat bread should be baked every day?

Exercise 3.37 A random variable X has the binomial distribution $B(10, p)$. The parameter p has a beta prior distribution Beta(3, 7). What is the posterior distribution of p, given $X = 6$?

Exercise 3.38 In continuation to the previous exercise, find the posterior expectation and posterior standard deviation of p.

Exercise 3.39 A random variable X has a Poisson distribution with mean λ. The parameter λ has a gamma, $G(2, 50)$, prior distribution.

 (i) Find the posterior distribution of λ given $X = 82$.
(ii) Find the 0.025-th and 0.975-th quantiles of this posterior distribution.

Exercise 3.40 A random variable X has a Poisson distribution with mean which is either $\lambda_0 = 70$ or $\lambda_1 = 90$. The prior probability of λ_0 is $1/3$. The losses due to wrong actions are $r_1 = \$100$ and $r_2 = \$150$. Observing $X = 72$, which decision would you take?

Exercise 3.41 A random variable X is normally distributed, with mean μ and standard deviation $\sigma = 10$. The mean μ is assigned a prior normal distribution with mean $\mu_0 = 50$ and standard deviation $\tau = 5$. Determine a credibility interval for μ, at level 0.95. Is this credibility interval also a HPD interval?

Exercise 3.42 Read file **CAR.csv** in Python using `mistat.load_data`. There are five variables stored in columns. Write a function which samples 64 values from column `mpg` (MPG/City), with replacement and store in a variable. Let $k1$ be the mean of the sample. Execute this function $M = 200$ times to obtain a sampling distribution of the sample means. Check graphically whether this sampling distribution is approximately normal. Also check whether the standard deviation of the sampling distribution is approximately $S/8$, where S is the standard deviation of `mpg`.

Exercise 3.43 Read file **YARNSTRG.csv** using Python. Use bootstrap sampling $M = 500$ times, to obtain confidence intervals for the mean. Use samples of size $n = 30$. Check in what proportion of samples the confidence intervals cover the mean.

Exercise 3.44 The average turn diameter of 58 US-made cars, in data file **CAR.csv**, is $\bar{X} = 37.203$ [m]. Is this mean significantly larger than 37 [m]? In order to check this, use Python. After loading the data, you will need to filter the dataset to extract the data for the 58 US-made cars (origin = 1).

Write a function which samples with replacement from the turn column 58 values, and store them in a list. Repeat this 100 times. An estimate of the P-value is the proportion of means smaller than 36, greater than $2 \times 37.203 - 37 = 37.406$. What is your estimate of the P-value?

Exercise 3.45 You have to test whether the proportion of nonconforming units in a sample of size $n = 50$ from a production process is significantly greater than $p = 0.03$. Use Python to determine when should we reject the hypothesis that $p \leq 0.03$ with $\alpha = 0.05$.

Exercise 3.46 Generate 1000 bootstrap samples of the sample mean and sample standard deviation of the data in **CYCLT.csv** on 50 piston cycle times.

(i) Compute 95% confidence intervals for the sample mean and sample standard deviation.

(ii) Draw histograms of the EBD of the sample mean and sample standard deviation.

Exercise 3.47 Use Python to generate 1000 bootstrapped quartiles of the data in **CYCLT.csv**.

(i) Compute 95% confidence intervals for the first quartile, median, and third quartile.

(ii) Draw histograms of the bootstrap quartiles.

Exercise 3.48 Generate the EBD of size $M = 1000$, for the sample correlation ρ_{XY} between ISC-t1 and ISC-t2 in data file **SOCELL.csv**. Compute the bootstrap confidence interval for ρ_{XY}, at confidence level of 0.95.

Exercise 3.49 Generate the EBD of the regression coefficients (a, b) of miles per gallon/city, Y, versus horsepower, X, in data file **CAR.csv**. For each of the $M = 100$ bootstrap samples, run a simple regression with the `scipy` command `stats.linregress`. The result (e.g., called `result`) of this command contains the slope (b: `result.slope`) and the intercept (a: `result.intercept`).

 (i) Determine a bootstrap confidence interval for the intercept, at level 0.95.
 (ii) Determine a bootstrap confidence interval for the slope, at level 0.95.
(iii) Compare the bootstrap standard errors of intercept and slope to those obtained from the formulae of Sect. 4.3.2.1.

Exercise 3.50 Test the hypothesis that the data in **CYCLT.csv** comes from a distribution with mean $\mu_0 = 0.55$ s.

 (i) Calculate and compare the t-test P-value and the boostrapped P^*-value.
 (ii) Does the confidence interval derived in Exercise 3.46 include $\mu_0 = 0.55$?
(iii) Could we have guessed the answer of part (ii) after completing part (i)?

Exercise 3.51 Compare the variances of the two measurements recorded in data file **ALMPIN2.csv**

 (i) What is the P-value?
(ii) Draw box plots of the two measurements.

Exercise 3.52 Compare the means of the two measurements on the two variables Diameter1 and Diameter2 in **ALMPIN2.csv**. What is the bootstrap estimate of the P-values for the means and variances?

Exercise 3.53 Compare the variances of the gasoline consumption (MPG/City) of cars by origin. The data is saved in file **MPG.csv**. There are $k = 3$ samples of sizes $n_1 = 58$, $n_2 = 14$ and $n_3 = 37$. Do you accept the null hypothesis of equal variances?

Exercise 3.54 Test the equality of mean gas consumption (MPG/city) of cars by origin. The data file to use is **MPG.csv**. The sample sizes are $n_1 = 58$, $n_2 = 14$ and $n_3 = 37$. The number of samples is $k = 3$. Do you accept the null hypothesis of equal means using a bootstrap approach?

Exercise 3.55 Use Python to generate 50 random Bernoulli numbers, with $p = 0.2$. Use these numbers to obtain tolerance limits with $\alpha = 0.05$ and $\beta = 0.05$, for the number of nonconforming items in future batches of 50 items, when the process proportion defectives is $p = 0.2$. Repeat this for $p = 0.1$ and $p = 0.05$.

Exercise 3.56 Use Python to calculate a $(0.95, 0.95)$ tolerance interval for the piston cycle time from the data in **OTURB.csv**.

Exercise 3.57 Using the sign test, test the hypothesis that the median, $\xi_{.5}$, of the distribution of cycle time of the piston is not exceeding $\xi^* = 0.7$ [min]. The sample data is in file **CYCLT.csv**. Use $\alpha = 0.10$ for level of significance.

Exercise 3.58 Use the WSR test on the data of file **OELECT.csv** to test whether the median of the distribution $\xi_{.5} = 220$ [Volt].

Exercise 3.59 Apply the randomization test on the **CAR.csv** file to test whether the turn diameter of foreign cars, having four cylinders, is different from that of US-made cars with four cylinders.

Chapter 4
Variability in Several Dimensions and Regression Models

Preview When surveys or experiments are performed, measurements are usually taken on several characteristics of the observation elements in the sample. In such cases we have multivariate observations, and the statistical methods which are used to analyze the relationships between the values observed on different variables are called multivariate methods. In this chapter we introduce some of these methods. In particular, we focus attention on graphical methods, linear regression methods, and the analysis of contingency tables. The linear regression methods explore the linear relationship between a variable of interest and a set of variables, by which we try to predict the values of the variable of interest. Contingency tables analysis studies the association between qualitative (categorical) variables, on which we cannot apply the usual regression methods.

Several techniques for graphical analysis of data in several dimensions are introduced and demonstrated using case studies. These include matrix scatterplots, 3D-scatterplots, and multiple boxplots. Topics covered also include simple linear regression, multiple regression models, and contingency tables. Prediction intervals are constructed for currents of solar cells and resistances on hybrid circuits. Robust regression is used to analyze data on placement errors of components on circuit boards. A special section on indices of association for categorical variables includes an analysis of a customer satisfaction survey designed to identify the main components of customer satisfaction and dissatisfaction.

The analysis of variance (ANOVA) for testing the significance of differences between several sample means is introduced, as well as the method of multiple comparisons, which protects the overall level of significance. The comparisons of proportions for categorical data (binomial or multinomial) are also discussed.

Supplementary Information The online version contains supplementary material available at https://doi.org/10.1007/978-3-031-07566-7_4.

4.1   Graphical Display and Analysis

4.1.1   Scatterplots

Suppose we are given a data set consisting of N records (elements). Each record contains observed values on k variables. Some of these variables might be qualitative (categorical) and some quantitative. Scatterplots display the values of pairwise quantitative variables, in two-dimensional plots.

Example 4.1 Consider the data set **PLACE.csv**. The observations are the displacements (position errors) of electronic components on printed circuit boards. The data was collected by a large US manufacturer of automatic insertion machines used in mass production of electronic devices. The components are fed to the machine on reals. A robot arm picks the components and places them in a prescribed location on a printed circuit board. The placement of the component is controlled by a computer built into the insertion machine. There are 26 boards. Sixteen components are placed on each board. Each component has to be placed at a specific location (x, y) on a board and with correct orientation t. Due to mechanical and other design or environmental factors, some errors are committed in placement. It is interesting to analyze whether these errors are within the specified tolerances. There are $k = 4$ variables in the data set. The first one is categorical and gives the board number. The three other variables are continuous. The variable "x Dev" provides the error in placement along the x-axis of the system. The variable "y Dev" presents the error in placement along the y-axis. The variable "t Dev" is the error in angular orientation.

In Fig. 4.1 we present a scatterplot of y Dev versus x Dev of each record. The picture reveals immediately certain unexpected clustering of the data points. The

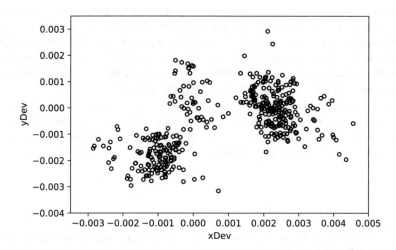

Fig. 4.1 Scatterplots of y Dev versus x Dev

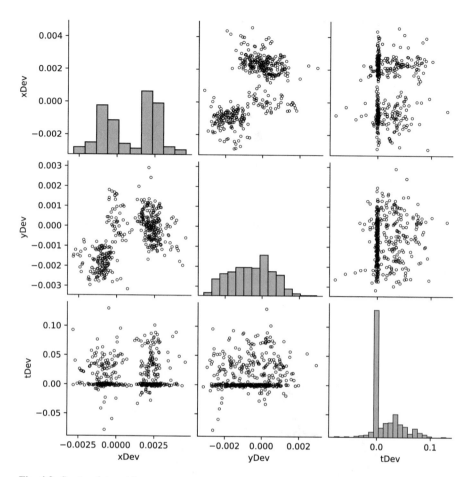

Fig. 4.2 Scatterplot matrix

y Dev of placements should not depend on their x Dev. The scatterplot of Fig. 4.1 shows three distinct clusters of points, which will be investigated later.

In a similar manner, we can plot the values of t Dev against those of x Dev or y Dev. This can be accomplished by performing what is called a **multiple scatterplot** or a **scatterplot matrix**. In Fig. 4.2 we present the scatterplot matrix of x Dev, y Dev, and t Dev.

The multiple (matrix) scatterplot gives us a general picture of the relationships between the three variables. Figure 4.1 is the middle left box in Fig. 4.2. Figure 4.2 directs us into further investigations. For example, we see in Fig. 4.2 that the variable t Dev has high concentration around zero with many observations to bigger than zero indicating a tilting of the components to the right. The frequency distribution of t Dev, which is presented in Fig. 4.3, reinforces this conclusion. Indeed, 50% of the t

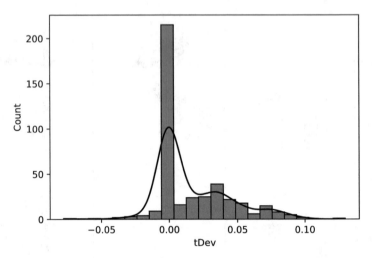

Fig. 4.3 Histogram of t Dev

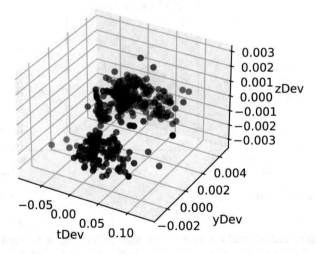

Fig. 4.4 3D-scatterplot

Dev values are close to zero. The other values tend to be positive. The histogram in Fig. 4.3 is skewed toward positive values.

An additional scatterplot can present the three-dimensional variability simultaneously. This graph is called a **3D-scatterplot**. In Fig. 4.4 we present this scatterplot for the three variables x Dev (X direction), y Dev (Y direction), and "t Dev" (Z direction). This plot expands the two-dimensional scatterplot by adding horizontally a third variable. ∎

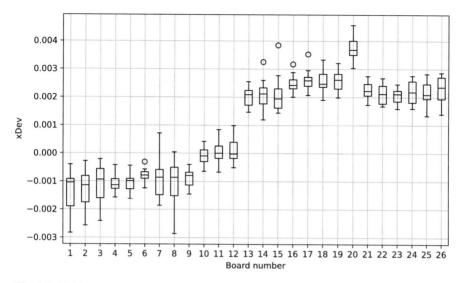

Fig. 4.5 Multiple boxplots of x Dev versus board number

4.1.2 Multiple Boxplots

Multiple boxplot or side-by-side boxplot is another graphical technique by which we present distributions of a quantitative variable at different categories of a categorical variable.

Example 4.2 Returning to the data set **PLACE.csv**, we wish to further investigate the apparent clusters, indicated in Fig. 4.1. As mentioned before, the data was collected in an experiment in which components were placed on 26 boards in a successive manner. The board number "board_n" is in the first column of the data set. We would like to examine whether the deviations in x, y, or θ tend to change with time. We can, for this purpose, plot the x Dev, y Dev, or t Dev against board_n. A more concise presentation is to graph multiple boxplots, by board number. In Fig. 4.5 we present these multiple boxplots of the x Dev against board number. We see in this figure an interesting picture. Boards 1–9 yield similar boxplots, while those of boards 10–12 are significantly above those of the first group, and those of boards 13–26 constitute a third group. These groups seem to be connected with the three clusters seen in Fig. 4.1. To verify it, we introduce a **code variable** to the data set, which assumes value 1 if board # ≤ 9, value 2 if $10 \leq$ board# ≤ 12, and value 3 if board # ≥ 13. We then plot again y-dev against x-dev, denoting the points in the scatterplot by the code variable symbols $\triangle$, $+$, and $\circ$.

In Fig. 4.6 we see this coded scatterplot. It is clear now that the three clusters are formed by these three groups of boards. The differences between these groups might be due to some deficiency in the placement machine, which caused the apparent time related drift in the errors. Other possible reasons could be the printed circuit board

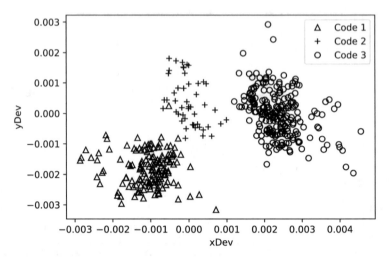

Fig. 4.6 Scatterplot of *y* Dev versus *x* Dev by code variables

composition or different batches of raw material, such as the glue used for placing the components. ■

4.2 Frequency Distributions in Several Dimensions

In Chap. 1 we studied how to construct frequency distributions of single variables, categorical or continuous. In the present section, we extend those concepts to several variables simultaneously. For the sake of simplification, we restrict the discussion to the case of two variables. The methods of this section can be generalized to a larger number of variables in a straightforward manner.

In order to enrich the examples, we introduce here two additional data sets. One is called **ALMPIN.csv**, and the other one is called **HADPAS.csv**. The **ALMPIN.csv** set consists of 70 records on 6 variables measured on aluminum pins used in airplanes. The aluminum pins are inserted with air-guns in pre-drilled holes in order to combine critical airplane parts such as wings, engine supports, and doors. Typical lot sizes consist of at least 1000 units providing a prime example of the discrete mass production operations in an industrial environment. The main role of the aluminum pins is to reliably secure the connection of two metal parts. The surface area where contact is established between the aluminum pins and the connected part determines the strength required to disconnect the part. A critical feature of the aluminum pin is that it fits perfectly to the pre-drilled holes. Parallelism of the aluminum pin is therefore essential, and the part diameter is measured in three different locations producing three measurements of the part width. Diameters 1, ,2 and 3 should be all equal. Any deviation indicates lack of parallelism and therefore potential reliability problems since the surface area with actual contact is not uniform. The measurements were taken in a computerized numerically controlled (CNC) metal cutting operation. The six variables are Diameter 1, Diameter 2, Diameter 3, Cap Diameter, LengthNocp, and LengthWcp. All the measurements are in millimeters. The first three variables give the pin diameter at three specified locations. Cap Diameter is the diameter of the cap on top of the pin. The last two variables are the length of the pin, without and with the cap, respectively.

Data set **HADPAS.csv** provides several resistance measurements (ohms) of five types of resistances (Res 3, Res 18, Res 14, Res 7, and Res 20), which are located in six hybrid micro circuits simultaneously manufactured on ceramic substrates. There are altogether 192 records for 32 ceramic plates.

4.2.1 Bivariate Joint Frequency Distributions

A joint frequency distribution is a function which provides the frequencies in the data set of elements (records) having values in specified intervals. More specifically, consider two variables X and Y. We assume that both variables are continuous. We partition the x-axis to k subintervals (ξ_{i-1}, ξ_i), $i = 1, \cdots, k_1$. We then partition the y-axis to k_2 subintervals (η_{j-1}, η_j), $j = 1, \cdots, k_2$. We denote by f_{ij} the number (count) of elements in the data set (sample) having x values in (ξ_{i-1}, ξ_i) and y values in (η_{j-1}, η_j), simultaneously. f_{ij} is called the **joint frequency** of the rectangle $(\xi_{i-1}, \xi_i) \times (\eta_{j-1}, \eta_j)$. If N denotes the total number of elements in the data set and then obviously

$$\sum_i \sum_j f_{ij} = N. \tag{4.1}$$

The frequencies f_{ij} can be represented in a table, called a table of the frequency distribution. The column totals provide the frequency distribution of the variable Lengthwcp. These row and column totals are called **marginal frequencies**. Generally, the marginal frequencies are

$$f_{i.} = \sum_{j=1}^{k_2} f_{ij}, \quad i = 1, \cdots, k_1 \tag{4.2}$$

and

$$f_{.j} = \sum_{i=1}^{k_1} f_{ij}, \quad j = 1, \cdots, k_2. \tag{4.3}$$

These are the sums of the frequencies in a given row or in a given column.

Table 4.1 Joint frequency distribution

LengthNocp	LengthWcp			
	59.9–60.0	60.0–60.1	60.1–60.2	Row total
49.8–49.9	16	17	0	33
49.9–50.0	5	27	2	34
50.0–50.1	0	0	3	3
Column total	21	44	5	70

Example 4.3 In Table 4.1 we present the joint frequency distribution of Length-Nocp and LengthWcp of the data set **ALMPIN.csv**. You can create this analysis in Python using a combination of binning and cross-tabulation.

```
almpin = mistat.load_data('ALMPIN')
binned_almpin = pd.DataFrame({
  'lenWcp': pd.cut(almpin['lenWcp'], bins=np.arange(59.9, 60.2, 0.1)),
  'lenNocp': pd.cut(almpin['lenNocp'], bins=np.arange(49.8, 50.1, 0.1)),
})
join_frequencies = pd.crosstab(binned_almpin['lenNocp'],
                               binned_almpin['lenWcp'])
print(join_frequencies)
```

```
lenWcp         (59.9, 60.0]   (60.0, 60.1]   (60.1, 60.2]
lenNocp
(49.8, 49.9]             16             17              0
(49.9, 50.0]              5             27              2
(50.0, 50.1]              0              0              3
```

The marginal frequencies can be obtained from the joint frequencies by summing the data along the rows or columns.

```
print('Row Totals', join_frequencies.sum(axis=1))
print('Column Totals', join_frequencies.sum(axis=0))
```

```
Row Totals lenNocp
(49.8, 49.9]    33
(49.9, 50.0]    34
(50.0, 50.1]     3
dtype: int64
Column Totals lenWcp
(59.9, 60.0]    21
(60.0, 60.1]    44
(60.1, 60.2]     5
dtype: int64
```

The row totals provide the frequency distribution of `lenNocp`.

We can visualize this table using a mosaic plot where the table entries are proportional to the size of the rectangles in the plot. A mosaic plot is an interesting way of visualizing the joint frequency distribution. Figures 4.7 presents a mosaic plot of the data in Table 4.1. It was created using the `mosaic` function in statsmodels.

Similar tabulation can be done of the frequency distributions of resistances, in data set **HADPAS.csv**. In Table 4.2 we provide the joint frequency distribution of Res 3 and Res 7.

■

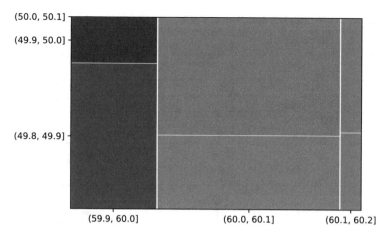

Fig. 4.7 Mosaic plot of data in Table 4.1

Table 4.2 Joint frequency distribution of Res 3 and Res 7 (in ohms)

	Res 7					
Res 3	1300–1500	1500–1700	1700–1900	1900–2100	2100–2300	Row total
1500–1700	1	13	1	0	0	15
1700–1900	0	15	31	1	0	47
1900–2100	0	1	44	40	2	87
2100–2300	0	0	5	31	6	42
2300–2500	0	0	0	0	1	1
Column total	1	29	81	72	9	192

The bivariate frequency distribution provides us also information on the association or dependence between the two variables. In Table 4.2 we see that resistance values of Res 3 tend to be similar to those of Res 7. For example, if the resistance value of Res 3 is in the interval (1500, 1700), 13 out of 15 resistance values of Res 7 are in the same interval. This association can be illustrated by plotting the box and whiskers plots of the variable Res 3 by the categories (intervals) of the variable Res 7. In order to obtain these plots, we partition first the 192 cases to five subgroups, according to the resistance values of Res 7. The single case having Res 7 in the interval (1300, 1500) belongs to subgroup 1. The 29 cases having Res 7 values in (1500, 1700) belong to subgroup 2 and so on. We can then perform an analysis by subgroups. Such an analysis yields Table 4.3. The Python commands for this analysis are:

Table 4.3 Means and standard deviations of Res 3

Subgroup	Interval of Res 7	Sample size	Mean	Standard deviation
1	1300–1500	1	1600.0	–
2	1500–1700	29	1718.9	80.27
3	1700–1900	81	1932.5	101.02
4	1900–2100	72	2076.5	93.53
5	2100–2300	9	2204.0	115.49

```
hadpas = mistat.load_data('HADPAS')
hadpas['res7fac'] = pd.cut(hadpas['res7'], bins=range(1300, 2500, 200))
hadpas['res3'].groupby(hadpas['res7fac']).agg(['count', 'mean', 'std'])
```

```
                count          mean         std
res7fac
(1300, 1500]        1   1600.000000         NaN
(1500, 1700]       29   1718.931034   80.266045
(1700, 1900]       81   1932.543210  101.018816
(1900, 2100]       72   2076.458333   93.530584
(2100, 2300]        9   2204.000000  115.489177
```

We see in Table 4.3 that the subgroup means grow steadily with the values of Res 7. The standard deviations do not change much. (There is no estimate of the standard deviation of subgroup 1.) A better picture of the dependence of Res 3 on the intervals of Res 7 is given by Fig. 4.8, in which the boxplots of the Res 3 values are presented by subgroup.

```
ax = hadpas.boxplot(column='res3', by='res7fac',
        color={'boxes':'grey', 'medians':'black', 'whiskers':'black'},
        patch_artist=True)
ax.set_title('')
ax.get_figure().suptitle('')
ax.set_xlabel('res7fac')
plt.show()
```

4.2.2 Conditional Distributions

Consider a population (or a sample) of elements. Each element assumes random values of two (or more) variables $X, Y, Z, \cdots$. The distribution of X, over elements whose Y value is restricted to a given interval (or set) A, is called the **conditional distribution of X, given Y is in** A. If the conditional distributions of X given Y is different from the marginal distribution of X, we say that the variables X and Y are **statistically dependent**. We will learn later how to test whether the differences between the conditional distributions and the marginal ones are significant and not due just to randomness in small samples.

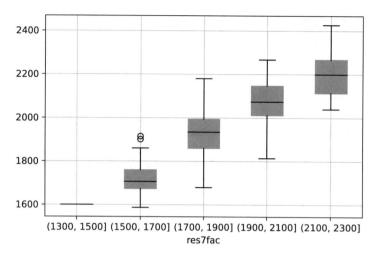

Fig. 4.8 Boxplots of Res 3 by intervals of Res 7

Table 4.4 Conditional and marginal frequency distributions of Res 3

	Res 7					
Res 3	1300–1500	1500–1700	1700–1900	1900–2100	2100–2300	Marginal distrib.
1500–1700	100.0	44.8	1.2	0.0	0.0	7.8
1700–1900	0.0	51.7	38.3	1.4	0.0	24.5
1900–2100	0.0	3.4	54.3	55.6	22.2	45.3
2100–2300	0.0	0.0	6.2	43.0	66.7	21.9
2300–2500	0.0	0.0	0.0	0.0	11.1	0.5
Column sums	100.0	100.0	100.0	100.0	100.0	100.0

Example 4.4 If we divide the frequencies in Table 4.2 by their column sums, we obtain the proportional frequency distributions of Res 3, given the intervals of Res 7. In Table 4.4 we compare these conditional frequency distributions, with the marginal frequency distribution of Res 3. We see in Table 4.4 that the proportional frequencies of the conditional distributions of Res 3 depend strongly on the intervals of Res 7 to which they are restricted.

∎

4.3   Correlation and Regression Analysis

In the previous sections, we presented various graphical procedures for analyzing multivariate data. In particular, we showed the multivariate scatterplots, three-

dimensional histograms, conditional boxplots, etc. In the present section, we start
with numerical analysis of multivariate data.

4.3.1   Covariances and Correlations

We introduce now a statistic which summarizes the simultaneous variability of two
variables. The statistic is called the **sample covariance**. It is a generalization of
the sample variance statistics, S_x^2, of one variable, X. We will denote the sample
covariance of two variables, X and Y, by S_{xy}. The formula of S_{xy} is

$$S_{xy} = \frac{1}{n-1} \sum_{i=1}^{n} (X_i - \bar{X})(Y_i - \bar{Y}), \tag{4.4}$$

where $\bar{X}$ and $\bar{Y}$ are the sample means of X and Y, respectively. Notice that S_{xx} is
the sample variance S_x^2 and S_{yy} is S_y^2. The sample covariance can assume positive
or negative values. If one of the variables, say X, assumes a constant value c, for all
X_i ($i = 1, \cdots, n$), then $S_{xy} = 0$. This can be immediately verified, since $\bar{X} = c$
and $X_i - \bar{X} = 0$ for all $i = 1, \cdots, n$.

It can be proven that, for any variables X and Y,

$$S_{xy}^2 \leq S_x^2 \cdot S_y^2. \tag{4.5}$$

This inequality is the celebrated **Schwarz inequality**. By dividing S_{xy} by $S_x \cdot S_y$, we
obtain a standardized index of dependence, which is called the **sample correlation**
(Pearson's product-moment correlation), namely:

$$R_{xy} = \frac{S_{xy}}{S_x \cdot S_y}. \tag{4.6}$$

From the Schwarz inequality, the sample correlation always assumes values between
-1 and $+1$. In Table 4.5 we present the sample covariances of the six variables
measured on the aluminum pins. Since $S_{xy} = S_{yx}$ (covariances and correlations are
symmetric statistics), it is sufficient to present the values at the bottom half of the
table (on and below the diagonal).

Example 4.5 In Tables 4.5 and 4.6, we present the sample covariances and sample
correlations in the data file **ALMPIN.csv**.

We see in Table 4.6 that the sample correlations between Diameter 1, Diameter
2, Diameter 3, and Cap Diameter are all greater than 0.9. As we see in Fig. 4.9
(the multivariate scatterplots), the points of these variables are scattered close to
straight lines. On the other hand, no clear relationship is evident between the first
four variables and the length of the pin (with or without the cap). The negative
correlations usually indicate that the points are scattered around a straight line

Table 4.5 Sample covariances of aluminum pins variables

	Y					
X	Diameter 1	Diameter 2	Diameter 3	Cap Diameter	Length Nocp	Length Wcp
Diameter 1	0.0270					
Diameter 2	0.0285	0.0329				
Diameter 3	0.0255	0.0286	0.0276			
Cap Diameter	0.0290	0.0314	0.0285	0.0358		
LengthNocp	−0.0139	−0.0177	−0.0120	−0.0110	0.1962	
LengthWcp	−0.0326	−0.0418	−0.0333	−0.0319	0.1503	0.2307

Table 4.6 Sample Correlations of Aluminum Pins Variables

	Y					
X	Diameter 1	Diameter 2	Diameter 3	Cap Diameter	Length Nocp	Length Wcp
Diameter 1	1.000					
Diameter 2	0.958	1.000				
Diameter 3	0.935	0.949	1.000			
Cap Diameter	0.933	0.914	0.908	1.000		
LengthNocp	−0.191	−0.220	−0.163	−0.132	1.000	
LengthWcp	−0.413	−0.480	−0.417	−0.351	0.707	1.000

having a negative slope. In the present case, it seems that the magnitude of these negative correlations is due to the one outlier (pin # 66). If we delete it from the data sets, the correlations are reduced in magnitude, as shown in Table 4.7.

We see that the correlations between the four diameter variables and the LengthNocp are much closer to zero after excluding the outlier. Moreover, the correlation with the Cap Diameter changed its sign. This shows that the sample correlation, as defined above, is sensitive to the influence of extreme observations (outliers). ∎

An important question to ask is, how **significant** is the value of the correlation statistic? In other words, what is the effect on the correlation of the random components of the measurements? If $X_i = \xi_i + e_i$, $i = 1, \cdots, n$, where ξ_i are deterministic components and e_i are random, and if $Y_i = \alpha + \beta \xi_i + f_i$, $i = 1, \cdots, n$, where α and β are constants and f_i are random components, how large could be the correlation between X and Y if $\beta = 0$?

Questions which deal with assessing the **significance** of the results will be discussed later.

4.3.2 Fitting Simple Regression Lines to Data

We have seen examples before in which the relationship between two variables X and Y is close to linear. This is the case when the (x, y) points scatter along a straight

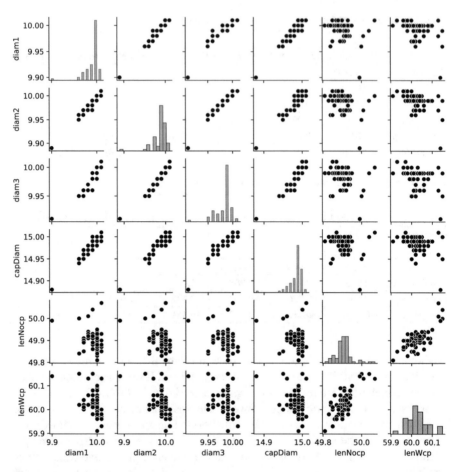

Fig. 4.9 Multiple scatterplots of the aluminum pins measurements

Table 4.7 Sample correlations of aluminum pins variables, after excluding outlying observation #66

X	Y					
	Diameter 1	Diameter 2	Diameter 3	Cap Diameter	Length Nocp	Length Wcp
Diameter 1	1.000					
Diameter 2	0.925	1.000				
Diameter 3	0.922	0.936	1.000			
Cap Diameter	0.876	0.848	0.876	1.000		
LengthNocp	−0.056	−0.103	−0.054	0.022	1.000	
LengthWcp	−0.313	−0.407	−0.328	−0.227	0.689	1.000

line. Suppose that we are given n pairs of observations $\{(x_i, y_i), i = 1, \cdots, n\}$. If the Y observations are related to those on X, according to the **linear model** :

$$y_i = \alpha + \beta x_i + e_i, \quad i = 1, \cdots, n \tag{4.7}$$

where α and β are constant coefficients and e_i are random components, with zero mean and constant variance, we say that Y relates to X according to a **simple linear regression**. The coefficients α and β are called the **regression coefficients**. α is the **intercept**, and β is the **slope coefficient**. Generally, the coefficients α and β are unknown. We fit to the data points a straight line, which is called the estimated regression line or prediction line.

4.3.2.1   The Least Squares Method

The most common method of fitting a regression line is the method of **least squares**.
   Suppose that $\hat{y} = a + bx$ is the straight line fitted to the data. The **principle** of **least squares** requires to determine estimates of α and β, a and b, which **minimize** the **sum of squares of residuals** around the line, i.e.:

$$SSE = \sum_{i=1}^{n}(y_i - a - bx_i)^2. \tag{4.8}$$

If we require that the regression line will pass through the point $(\bar{x}, \bar{y})$, where $\bar{x}$ and $\bar{y}$ are the sample means of the xs and ys, then

$$\bar{y} = a + b\bar{x},$$

or the coefficient a should be determined by the equation

$$a = \bar{y} - b\bar{x}. \tag{4.9}$$

Substituting this equation above, we obtain that

$$SSE = \sum_{i=1}^{n}(y_i - \bar{y} - b(x_i - \bar{x}))^2$$

$$= \sum_{i=1}^{n}(y_i - \bar{y})^2 - 2b \sum_{i=1}^{n}(x_i - \bar{x})(y_i - \bar{y}) + b^2 \sum_{i=1}^{n}(x_i - \bar{x})^2.$$

Dividing the two sides of the equation by $(n - 1)$ we obtain

$$\frac{SSE}{n - 1} = S_y^2 - 2bS_{xy} + b^2 S_x^2.$$

The coefficient b should be determined to minimize this quantity. One can write

$$
\frac{SSE}{n-1} = S_y^2 + S_x^2 \left(b^2 - 2b\frac{S_{xy}}{S_x^2} + \frac{S_{xy}^2}{S_x^4} \right) - \frac{S_{xy}^2}{S_x^2}
$$

$$
= S_y^2 (1 - R_{xy}^2) + S_x^2 \left(b - \frac{S_{xy}}{S_x^2} \right)^2 .
$$

It is now clear that the least squares estimate of β is

$$
b = \frac{S_{xy}}{S_x^2} = R_{xy}\frac{S_y}{S_x}. \tag{4.10}
$$

The value of $SSE/(n-1)$, corresponding to the least squares estimate, is

$$
S_{y|x}^2 = S_y^2 (1 - R_{xy}^2). \tag{4.11}
$$

$S_{y|x}^2$ is the sample variance of the residuals around the least squares regression line. By definition, $S_{y|x}^2 \geq 0$, and hence $R_{xy}^2 \leq 1$, or $-1 \leq R_{xy} \leq 1$. $R_{xy} = \pm 1$ only if $S_{y|x}^2 = 0$. This is the case when all the points (x_i, y_i), $i = 1, \cdots, n$, lie on a straight line. If $R_{xy} = 0$, then the slope of the regression line is $b = 0$ and $S_{y|x}^2 = S_y^2$.

Notice that

$$
R_{xy}^2 = \left(1 - \frac{S_{y|x}^2}{S_y^2} \right). \tag{4.12}
$$

Thus, R_{xy}^2 is the proportion of variability in Y, which is explainable by the linear relationship $\hat{y} = a + bx$. For this reason, R_{xy}^2 is also called the **coefficient of determination**. The coefficient of correlation (squared) measures the extent of linear relationship in the data. The linear regression line, or prediction line, could be used to predict the values of Y corresponding to X values, when R_{xy}^2 is not too small. To interpret the coefficient of determination—particularly when dealing with multiple regression models (see Sect. 4.4)—it is sometimes useful to consider an "adjusted" R^2. The adjustment accounts for the number of predictor or explanatory variables in the model and the sample size. In simple linear regression, we define

$$
R_{xy}^2(\text{adjusted}) = 1 - \left[(1 - R_{xy}^2)\frac{n-1}{n-2} \right]. \tag{4.13}
$$

Example 4.6 Telecommunication satellites are powered while in orbit by solar cells. Tadicell, a solar cell producer that supplies several satellite manufacturers, was requested to provide data on the degradation of its solar cells over time. Tadicell engineers performed a simulated experiment in which solar cells were subjected to

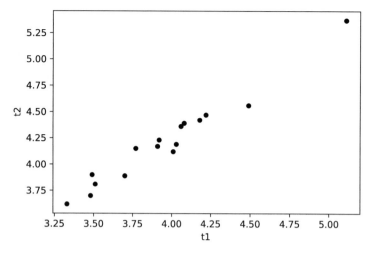

Fig. 4.10 Relationship of ISC values at t_1 and t_2

temperature and illumination changes similar to those in orbit and measured the short circuit current (ISC) (amperes) of solar cells at three different time periods, in order to determine their rate of degradation. In the following table, we present the ISC values of $n = 16$ solar cells, measured at three time epochs, 1 month apart. The data is given in file **SOCELL.csv**. In Fig. 4.10 we see the scatter of the ISC values at t_1 and at t_2

```
socell = mistat.load_data('SOCELL')
socell.plot.scatter(x='t1', y='t2', color='black')
plt.show()
```

We now make a regression analysis of ISC at time t_2, Y, versus ISC at time t_1, X. The computations can be easily performed in Python. There are several packages for linear regression. Here, we use `statsmodels` as it provides detailed information (Table 4.8).

```
socell = mistat.load_data('SOCELL')
model = smf.ols(formula='t2 ~ 1 + t1', data=socell).fit()
print(model.summary2())
```

```
                 Results: Ordinary least squares
=================================================================
Model:                OLS          Adj. R-squared:       0.957
Dependent Variable:   t2           AIC:                  -30.8366
Date:                 2022-05-18 21:13 BIC:              -29.2914
No. Observations:     16           Log-Likelihood:       17.418
Df Model:             1            F-statistic:          330.8
Df Residuals:         14           Prob (F-statistic):   3.88e-11
R-squared:            0.959        Scale:                0.0075846
-----------------------------------------------------------------
               Coef.    Std.Err.     t      P>|t|    [0.025   0.975]
-----------------------------------------------------------------
Intercept      0.5358    0.2031    2.6375   0.0195   0.1001   0.9715
```

Table 4.8 ISC values of solar cells at three time epochs

Cell	Time		
	t_1	t_2	t_3
1	4.18	4.42	4.55
2	3.48	3.70	3.86
3	4.08	4.39	4.45
4	4.03	4.19	4.28
5	3.77	4.15	4.22
6	4.01	4.12	4.16
7	4.49	4.56	4.52
8	3.70	3.89	3.99
9	5.11	5.37	5.44
10	3.51	3.81	3.76
11	3.92	4.23	4.14
12	3.33	3.62	3.66
13	4.06	4.36	4.43
14	4.22	4.47	4.45
15	3.91	4.17	4.14
16	3.49	3.90	3.81

```
t1              0.9287      0.0511   18.1890    0.0000    0.8192   1.0382
-----------------------------------------------------------------------
Omnibus:                    1.065       Durbin-Watson:          2.021
Prob(Omnibus):              0.587       Jarque-Bera (JB):       0.871
Skew:                      -0.322       Prob(JB):               0.647
Kurtosis:                   2.056       Condition No.:          39
=======================================================================
```

We see in the summary output that the least squares regression (prediction) line is $\hat{y} = 0.536 + 0.929x$. We read also that the coefficient of determination is $R_{xy}^2 = 0.959$. This means that only 4% of the variability in the ISC values, at time period t_2, are not explained by the linear regression on the ISC values at time t_1. Observation #9 is an "unusual observation." It has relatively much influence on the regression line, as can be seen in Fig. 4.10.

The model summary provides also additional analysis. The Stdev corresponding to the least squares regression coefficients is the square root of the variances of these estimates, which are given by the formulae:

$$S_a^2 = S_e^2 \left[\frac{1}{n} + \frac{\bar{x}^2}{\sum_{i=1}^n (x_i - \bar{x})^2} \right] \tag{4.14}$$

and

$$S_b^2 = S_e^2 / \sum_{i=1}^n (x_i - \bar{x})^2, \tag{4.15}$$

where

Table 4.9 Observed and predicted values of ISC at time t_2

i	y_i	$\hat{y}_i$	$\hat{e}_i$
1	4.42	4.419	0.0008
2	3.70	3.769	−0.0689
3	4.39	4.326	0.0637
4	4.19	4.280	−0.0899
5	4.15	4.038	0.1117
6	4.12	4.261	−0.1413
7	4.56	4.707	−0.1472
8	3.89	3.973	−0.0833
9	5.37	5.283	0.0868
10	3.81	3.797	0.0132
11	4.23	4.178	0.0523
12	3.62	3.630	−0.0096
13	4.36	4.308	0.0523
14	4.47	4.456	0.0136
15	4.17	4.168	0.0016
16	3.90	3.778	0.1218

$$S_e^2 = \frac{(1 - R_{xy}^2)}{n - 2} \sum_{i=1}^{n} (y_i - \bar{y})^2. \tag{4.16}$$

We see here that $S_e^2 = \frac{n-1}{n-2} S_{y|x}^2$. The reason for this modification is for testing purposes. The value of S_e^2 in the above analysis is 0.0076. The standard deviation of y is $S_y = 0.4175$. The standard deviation of the residuals around the regression line is $S_e = 0.08709$. This explains the high value of $R_{y|x}^2$.

In Table 4.9 we present the values of ISC at time t_2, y, and their predicted values, according to those at time t_1, $\hat{y}$. We present also a graph (Fig. 4.11) of the residuals, $\hat{e} = y - \hat{y}$, versus the predicted values $\hat{y}$. If the simple linear regression explains the variability adequately, the residuals should be randomly distributed around zero, without any additional relationship to the regression x.

In Fig. 4.11 we plot the residuals $\hat{e} = y - \hat{y}$, versus the predicted values $\hat{y}$, of the ISC values for time t_2. It seems that the residuals are randomly dispersed around zero. Later we will learn how to test whether this dispersion is indeed random. ∎

4.3.2.2 Regression and Prediction Intervals

Suppose that we wish to predict the possible outcomes of the Y for some specific value of X, say x_0. If the true regression coefficients α and β are known, then the predicted value of Y is $\alpha + \beta x_0$. However, when α and β are unknown, we predict the outcome at x_0 to be $\hat{y}(x_0) = a + bx_0$. We know, however, that the actual value of Y to be observed will not be exactly equal to $\hat{y}(x_0)$. We can determine a prediction

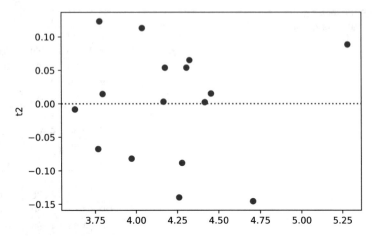

Fig. 4.11 Residual versus predicted ISC values

interval around $\hat{y}(x_0)$ such that the likelihood of obtaining a Y value within this interval will be high. Generally, the prediction interval limits, given by the formula

$$\hat{y}(x_0) \pm 3S_e^2 \cdot \left[1 + \frac{1}{n} + \frac{(x_0 - \bar{x})^2}{\sum_i (x_i - \bar{x})^2}\right]^{1/2}, \tag{4.17}$$

will yield good predictions. In Table 4.10 we present the 99% and 95% prediction intervals for the ISC values at time t_2, for selected ISC values at time t_1. In Fig. 4.12 we present the scatterplot, regression line, and prediction limits for Res 3 versus Res 7, of the **HADPAS.csv** set.

```
result = model.get_prediction(pd.DataFrame({'t1': [4.0,4.4,4.8,5.2]}))
columns = ['mean', 'obs_ci_lower', 'obs_ci_upper']
print(0.01)
print(result.summary_frame(alpha=0.01)[columns].round(3))
print(0.05)
print(result.summary_frame(alpha=0.05)[columns].round(3))
```

```
0.01
     mean   obs_ci_lower   obs_ci_upper
0   4.251        3.983          4.518
1   4.622        4.346          4.898
2   4.994        4.697          5.290
3   5.365        5.038          5.692
0.05
     mean   obs_ci_lower   obs_ci_upper
0   4.251        4.058          4.443
1   4.622        4.423          4.821
2   4.994        4.780          5.207
3   5.365        5.129          5.601
```

Table 4.10 Prediction intervals for ISC values at time t_2

x_0	$\hat{y}(x_0)$	Lower limit	Upper limit
4.0	4.251	3.983	4.518
4.4	4.622	4.346	4.898
4.8	4.993	4.697	5.290
5.2	5.364	5.038	5.692

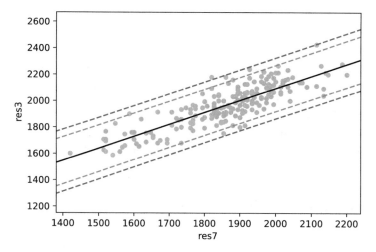

Fig. 4.12 Prediction intervals for Res 3 values, given the Res 7 values

4.4   Multiple Regression

In the present section, we generalize the regression to cases where the variability of a variable Y of interest can be explained, to a large extent, by the linear relationship between Y and k predicting or explaining variables $X_1, \ldots, X_k$. The number of explaining variables is $k \geq 2$. All the k variables $X_1, \ldots, X_k$ are continuous ones. The regression analysis of Y on several predictors is called **multiple regression**, and multiple regression analysis is an important statistical tool for exploring the relationship between the dependence of one variable Y on a set of other variables. Applications of multiple regression analysis can be found in all areas of science and engineering. This method plays an important role in the statistical planning and control of industrial processes.

The statistical linear model for multiple regression is

$$y_i = \beta_0 + \sum_{j=1}^{k} \beta_j x_{ij} + e_i, \quad i = 1, \cdots, n,$$

where $\beta_0, \beta_1, \cdots, \beta_k$ are the linear regression coefficients and e_i are random components. The commonly used method of estimating the regression coefficients, and testing their significance, is called **multiple regression analysis**. The method is based on the **principle of least squares**, according to which the regression coefficients are estimated by choosing $b_0, b_1, \cdots, b_k$ to minimize the sum of residuals

$$SSE = \sum_{i=1}^{n} (y_i - (b_0 + b_1 x_{i1} + \cdots + b_k x_{ik}))^2.$$

The first subsections of the present chapter are devoted to the methods of regression analysis, when both the regressant Y and the regressors $x_1, \ldots, x_k$ are quantitative variables. In Sect. 4.5 we present quantal response regression, in which the regressant is a qualitative (binary) variable and the regressors $x_1, \ldots, x_k$ are quantitative. In particular we present the logistic model and the logistic regression. In Sect. 4.6 we discuss the analysis of variance, for the comparison of sample means, when the regressant is quantitative, but the regressors are categorical variables.

4.4.1 Regression on Two Variables

The multiple regression linear model, in the case of two predictors, assumes the form

$$y_i = \beta_0 + \beta_1 x_{1i} + \beta_2 x_{2i} + e_i, \quad i = 1, \cdots, n. \tag{4.18}$$

$e_1, \ldots, e_n$ are independent r.v.s, with $E\{e_i\} = 0$ and $V\{e_i\} = \sigma^2, i = 1, \ldots, n$. The principle of least squares calls for the minimization of SSE. One can differentiate SSE with respect to the unknown parameters. This yields the least squares estimators, b_0, b_1, and b_2 of the regression coefficients, β_0, β_1, and β_2. The formula for these estimators are:

$$b_0 = \bar{Y} - b_1 \bar{X}_1 - b_2 \bar{X}_2; \tag{4.19}$$

and b_1 and b_2 are obtained by solving the set of linear equations

$$\left. \begin{array}{l} S_{x_1}^2 b_1 + S_{x_1 x_2} b_2 = S_{x_1 y} \\ S_{x_1 x_2} b_1 + S_{x_2}^2 b_2 = S_{x_2 y} \end{array} \right\} . \tag{4.20}$$

As before, $S_{x_1}^2$, $S_{x_1 x_2}$, $S_{x_2}^2$, $S_{x_1 y}$, and $S_{x_2 y}$ denote the sample variances and covariances of x_1, x_2, and y.

By simple substitution we obtain, for b_1 and b_2, the explicit formulae:

$$b_1 = \frac{S_{x_2}^2 S_{x_1 y} - S_{x_1 x_2} S_{x_2 y}}{S_{x_1}^2 S_{x_2}^2 - S_{x_1 x_2}^2}, \tag{4.21}$$

and

$$b_2 = \frac{S_{x_1}^2 S_{x_2 y} - S_{x_1 x_2} S_{x_1 y}}{S_{x_1}^2 S_{x_2}^2 - S_{x_1 x_2}^2}. \tag{4.22}$$

The values $\hat{y}_i = b_0 + b_1 x_{1i} + b_2 x_{2i}$ $(i = 1, \cdots, n)$ are called the **predicted values** of the regression, and the residuals around the regression plane are

$$\hat{e}_i = y_i - \hat{y}_i$$
$$= y_i - (b_0 + b_1 x_{1i} + b_2 x_{2i}), \quad i = 1, \cdots, n.$$

The mean square of the residuals around the regression plane is

$$S_{y|(x_1,x_2)}^2 = S_y^2 (1 - R_{y|(x_1,x_2)}^2), \tag{4.23}$$

where

$$R_{y|(x_1,x_2)}^2 = \frac{1}{S_y^2} (b_1 S_{x_1 y} + b_2 S_{x_2 y}), \tag{4.24}$$

is the **multiple squared correlation** (multiple-R^2), and S_y^2 is the sample variance of y. The interpretation of the multiple-R^2 is as before, i.e., the proportion of the variability of y which is **explainable** by the predictors (regressors) x_1 and x_2.

Example 4.7 We illustrate the fitting of a multiple regression on the following data, labeled **GASOL.csv**. The data set consists of 32 measurements of distillation properties of crude oils (see Daniel and Wood 1999). There are five variables, $x_1, \cdots, x_4$ and y. These are:

x1   Crude oil **gravity**, °API
x2   Crude oil **vapour** pressure, psi
astm   Crude oil **ASTM** 10% point, °F
endPt   Gasoline ASTM **endpoint**, °F
yield   **Yield** of gasoline (in percentage of crude oil)

The measurements of crude oil and gasoline volatility measure the temperatures at which a given amount of liquid has been vaporized.

The sample correlations between these five variables are :

We see that yield is highly correlated with endPt and to a lesser degree with astm (or x2).

The following is the statsmodels result of the regression of yield on astm and endPt:

	x2	astm	endPt	yield
x1	0.621	−0.700	−0.322	0.246
x2		−0.906	−0.298	0.384
astm			0.412	−0.315
endPt				0.712

```
gasol = mistat.load_data('GASOL')
# rename column 'yield' to 'Yield' as 'yield' is a special keyword in Python
gasol = gasol.rename(columns={'yield': 'Yield'})
model = smf.ols(formula='Yield ~ astm + endPt + 1', data=gasol).fit()
print(model.summary2())
```

```
                    Results: Ordinary least squares
=================================================================
Model:               OLS              Adj. R-squared:     0.949
Dependent Variable:  Yield            AIC:                150.3690
Date:                2022-05-18 21:13 BIC:                154.7662
No. Observations:    32               Log-Likelihood:     -72.184
Df Model:            2                F-statistic:        288.4
Df Residuals:        29               Prob (F-statistic): 7.26e-20
R-squared:           0.952            Scale:              5.8832
-----------------------------------------------------------------
             Coef.    Std.Err.    t     P>|t|    [0.025   0.975]
-----------------------------------------------------------------
Intercept   18.4676   3.0090   6.1374  0.0000  12.3135  24.6217
astm        -0.2093   0.0127 -16.4349  0.0000  -0.2354  -0.1833
endPt        0.1558   0.0069  22.7308  0.0000   0.1418   0.1698
-----------------------------------------------------------------
Omnibus:             1.604            Durbin-Watson:       1.076
Prob(Omnibus):       0.448            Jarque-Bera (JB):    1.055
Skew:                0.085            Prob(JB):            0.590
Kurtosis:            2.127            Condition No.:       2920
=================================================================
* The condition number is large (3e+03). This might indicate
strong multicollinearity or other numerical problems.
```

We compute now these estimates of the regression coefficients using the above formulae. The variances and covariances of endPt, astm, and yield are:

	astm	endPt	yield
astm	1409.355		
endPt	1079.565	4865.894	
yield	−126.808	532.188	114.970

```
# Covariance
gasol[['astm', 'endPt', 'Yield']].cov()
# Means
gasol[['astm', 'endPt', 'Yield']].mean()
```

The means of these variables are $\overline{\text{astm}} = 241.500$, $\overline{\text{endPt}} = 332.094$, and $\overline{\text{yield}} = 19.6594$. Thus, the least squares estimators of b_1 and b_2 are obtained by solving the equations

$$1409.355b_1 + 1079.565b_2 = -126.808$$
$$1079.565b_1 + 4865.894b_2 = 532.188.$$

The solution is

$$b_1 = -0.20933,$$

and

$$b_2 = 0.15581.$$

Finally, the estimate of β_0 is

$$b_0 = 19.6594 + 0.20933 \times 241.5 - 0.15581 \times 332.094$$
$$= 18.469.$$

These are the same results as in the Python output. Moreover, the multiple-R^2 is

$$R^2_{y|(\text{astm,endPt})} = \frac{1}{114.970}[0.20932 \times 126.808 + 0.15581 \times 532.88]$$
$$= 0.9530.$$

In addition,

$$S^2_{y|(\text{astm,endPt})} = S^2_y(1 - R^2_{y|(\text{astm,endPt})})$$
$$= 114.97(1 - .9530)$$
$$= 5.4036.$$

In Fig. 4.13 we present a scatterplot of the residuals $\hat{e}_i$ ($i = 1, \cdots, n$) against the predicted values $\hat{y}_i$ ($i = 1, \cdots, n$). This scatterplot does not reveal any pattern different than random. It can be concluded that the regression of y on x_3 and x_4 accounts for all the systematic variability in the yield, y. Indeed, $R^2 = .952$, and no more than 4.8% of the variability in y is unaccounted by the regression.   ■

The following are formulae for the variances of the least squares coefficients. First we convert $S^2_{y|(x_1,x_2)}$ to S^2_e, i.e.:

$$S^2_e = \frac{n-1}{n-3}S^2_{y|x}. \tag{4.25}$$

S^2_e is an unbiased estimator of σ^2. The variance formulae are:

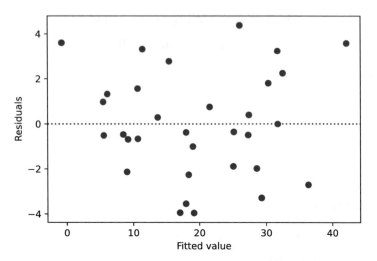

Fig. 4.13 Scatterplot of $\hat{e}$ versus $\hat{Y}$

$$S_{b_0}^2 = \frac{S_e^2}{n} + \bar{x}_1^2 S_{b_1}^2 + \bar{x}_2^2 S_{b_2}^2 + 2\bar{x}_1 \bar{x}_2 S_{b_1 b_2},$$

$$S_{b_1}^2 = \frac{S_e^2}{n-1} \cdot \frac{S_{x_2}^2}{D},$$

$$S_{b_2}^2 = \frac{S_e^2}{n-1} \cdot \frac{S_{x_1}^2}{D}, \tag{4.26}$$

$$S_{b_1 b_2} = -\frac{S_e^2}{n-1} \cdot \frac{S_{x_1 x_2}}{D},$$

where

$$D = S_{x_1}^2 S_{x_2}^2 - (S_{x_1 x_2})^2.$$

Example 4.8 Using the numerical Example 4.7 on the **GASOL.csv** data, we find that

$$S_e^2 = 5.8869,$$

$$D = 5,692,311.4,$$

$$S_{b_1}^2 = 0.0001624,$$

$$S_{b_2}^2 = 0.0000470,$$

$$S_{b_1,b_2} = -0.0000332,$$

and

$$S_{b_0}^2 = 9.056295.$$

The square roots of these variance estimates are the "std err" values printed in the Python output. The S_e value is shown in the regression summary output as `Scale`.
■

4.4.2 Partial Regression and Correlation

In performing the multiple least squares regression, one can study the effect of the predictors on the response in stages. This more pedestrian approach does not simultaneously provide all regression coefficients but studies the effect of predictors in more detail.

In Stage I we perform a simple linear regression of the yield y on one of the predictors, x_1 say. Let $a_0^{(1)}$ and $a_1^{(1)}$ be the intercept and slope coefficients of this simple linear regression. Let $\hat{\mathbf{e}}^{(1)}$ be the vector of residuals:

$$\hat{e}_i^{(1)} = y_i - (a_0^{(1)} + a_1^{(1)} x_{1i}), \quad i = 1, \cdots, n. \tag{4.27}$$

In Stage II we perform a simple linear regression of the second predictor, x_2, on the first predictor x_1. Let $c_0^{(2)}$ and $c_1^{(2)}$ be the intercept and slope coefficients of this regression. Let $\hat{\mathbf{e}}^{(2)}$ be the vector of residuals:

$$\hat{e}_i^{(2)} = x_{2i} - (c_0^{(2)} + c_1^{(2)} x_{1i}), \quad i = 1, \cdots, n. \tag{4.28}$$

In Stage III we perform a simple linear regression of $\hat{\mathbf{e}}^{(1)}$ on $\hat{\mathbf{e}}^{(2)}$. It can be shown that this linear regression must pass through the origin, i.e., it has a zero intercept. Let $d^{(3)}$ be the slope coefficient.

The simple linear regression of $\hat{\mathbf{e}}^{(1)}$ on $\hat{\mathbf{e}}^{(2)}$ is called the **partial regression**. The correlation between $\hat{\mathbf{e}}^{(1)}$ and $\hat{\mathbf{e}}^{(2)}$ is called the **partial correlation** of y and x_2, given x_1, and is denoted by $r_{yx_2 \cdot x_1}$.

From the regression coefficients obtained in the three stages, one can determine the multiple regression coefficients of y on x_1 and x_2, according to the formulae:

$$b_0 = a_0^{(1)} - d^{(3)} c_0^{(2)},$$

$$b_1 = a_1^{(1)} - d^{(3)} c_1^{(2)}, \tag{4.29}$$

$$b_2 = d^{(3)}.$$

Example 4.9 For the **GASOL** data, let us determine the multiple regression of the yield (y) on the astm (x_3) and the endPt (x_4) in stages.

```
stage1 = smf.ols(formula='Yield ~ 1 + astm', data=gasol).fit()
print(stage1.params)
print('R2(y, astm)', stage1.rsquared)
```

```
Intercept    41.388571
astm          -0.089976
dtype: float64
R2(y, astm) 0.09924028202169999
```

In Stage I, the simple linear regression of y on astm is

$$\hat{y} = 41.4 - 0.08998 \cdot x_3.$$

The residuals of this regression are $\hat{\mathbf{e}}^{(1)}$. Also $R^2_{y\,x_3} = 0.099$.

```
stage2 = smf.ols(formula='endPt ~ 1 + astm', data=gasol).fit()
print(stage2.params)
print('R2(endPt, astm)', stage2.rsquared)
```

```
Intercept    147.104971
astm           0.765999
dtype: float64
R2(endPt, astm) 0.16994727072324556
```

In Stage II, the simple linear regression of endPt on astm is

$$\hat{x}_4 = 147 + 0.766 \cdot x_3.$$

The residuals of this regression are $\hat{\mathbf{e}}^{(2)}$. In Fig. 4.14 we see the scatterplot of $\hat{\mathbf{e}}^{(1)}$ versus $\hat{\mathbf{e}}^{(2)}$. The partial correlation is $r_{yx_4 \cdot x_3} = 0.973$. This high partial correlation means that, after adjusting the variability of y for the variability of x_3, and the variability of x_4 for that of x_3, the adjusted x_4 values, namely, $\hat{e}_i^{(2)}$ ($i = 1, \cdots, n$), are still good predictors for the adjusted y values, namely, $\hat{e}_i^{(1)}$ ($i = 1, \cdots, n$).

```
residuals = pd.DataFrame({
  'e1': stage1.resid,
  'e2': stage2.resid,
})
print(np.corrcoef(stage1.resid, stage2.resid))

# use -1 in the formula to fix intercept to 0
stage3 = smf.ols(formula='e1 ~ e2 - 1', data=residuals).fit()
print(stage3.params)
print('R2(e1, e2)', stage3.rsquared)
```

```
[[1.          0.97306538]
 [0.97306538 1.         ]]
e2    0.155813
dtype: float64
R2(e1, e2) 0.9468562311053541
```

The regression of $\hat{\mathbf{e}}^{(1)}$ on $\hat{\mathbf{e}}^{(2)}$, determined in Stage III, is

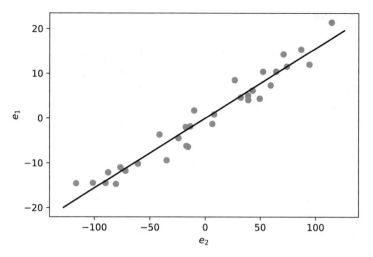

Fig. 4.14 Scatterplot of $\hat{e}_1$ versus $\hat{e}_2$

$$\hat{\hat{e}}^{(1)} = 0.156 \cdot \hat{e}^{(2)}.$$

We have found the following estimates:

$$a_0^{(1)} = 41.4, \quad a_1^{(1)} = -0.08998$$

$$c^{(2)} = 147.0, \quad c_1^{(2)} = 0.766$$

$$d^{(3)} = 0.156.$$

From the above formulae, we get

$$b_0 = 41.4 - 0.156 \times 147.0 = 18.468,$$

$$b_1 = -0.0900 - 0.156 \times 0.766 = -0.2095$$

$$b_2 = 0.156.$$

These values coincide with the previously determined coefficients. Finally, the relationship between the multiple and the partial correlations is

$$R^2_{y|(x_1,x_2)} = 1 - (1 - R^2_{yx_1})(1 - r^2_{yx_2 \cdot x_1}). \tag{4.30}$$

In the present example,

$$R^2_{y|(x_3,x_4)} = 1 - (1 - 0.099)(1 - .94673) = 0.9520.$$

■

4.4.3   Multiple Linear Regression

In the general case, we have k predictors ($k \geq 1$). Let (X) denote an array of n rows and $(k+1)$ columns, in which the first column consists of value 1 in all entries, and the second to $(k+1)$st columns consist of the values of the predictors $x_1, \cdots, x_k$. (X) is called the predictors matrix. Let Y be an array of n rows and one column, consisting of the values of the regressant. The linear regression model can be written in matrix notation as

$$Y = (X)\boldsymbol{\beta} + \mathbf{e}, \tag{4.31}$$

where $\boldsymbol{\beta}' = (\beta_0, \beta_1, \cdots, \beta_k)$ is the vector of regression coefficients and $\mathbf{e}$ is a vector of random residuals.

The sum of squares of residuals can be written as

$$SSE = (Y - (X)\boldsymbol{\beta})'(Y - (X)\boldsymbol{\beta}) \tag{4.32}$$

$(\ )'$ denotes the transpose of the vector of residuals. Differentiating SSE partially with respect to the components of $\boldsymbol{\beta}$, and equating the partial derivatives to zero, we obtain a set of linear equations in the LSE $\mathbf{b}$, namely,

$$(X)'(X)\mathbf{b} = (X)'Y. \tag{4.33}$$

$(X)'$ is the transpose of the matrix (X). These linear equations are called the **normal equations**.

If we define the matrix,

$$B = [(X)'(X)]^{-1}(X)', \tag{4.34}$$

where $[\ ]^{-1}$ is the inverse of $[\ ]$, then the general formula of the least squares regression coefficient vector $\mathbf{b}' = (b_0, \cdots, b_k)$ is given in matrix notation as

$$\mathbf{b} = (B)Y. \tag{4.35}$$

The vector of predicted y values, or FITS, is given by $\hat{\mathbf{y}} = (H)\mathbf{y}$, where $(H) = (X)(B)$. The vector of residuals $\hat{\mathbf{e}} = \mathbf{y} - \hat{\mathbf{y}}$ is given by

$$\hat{\mathbf{e}} = (I - H)\mathbf{y}, \tag{4.36}$$

where (I) is the $n \times n$ identity matrix. The variance of $\hat{\mathbf{e}}$, around the regression surface, is

$$S_e^2 = \frac{1}{n-k-1} \sum_{i=1}^{n} \hat{e}_i^2$$

$$= \frac{1}{n-k-1} \mathbf{Y}'(I-H)\mathbf{Y}.$$

The sum of squares of $\hat{e}_i$ $(i = 1, \cdots, n)$ is divided by $(n-k-1)$ to attain an unbiased estimator of σ^2. The multiple-R^2 is given by

$$R_{y|(x)}^2 = \frac{1}{(n-1)S_y^2} (\mathbf{b}'(X)'\mathbf{Y} - n\bar{y}^2) \tag{4.37}$$

where $x_{i0} = 1$ for all $i = 1, \cdots, n$, and S_y^2 is the sample variance of y. Finally, an estimate of the variance-covariance matrix of the regression coefficients $b_0, \cdots, b_k$ is

$$(S_b) = S_e^2[(X)'(X)]^{-1}. \tag{4.38}$$

Example 4.10 We use again the **ALMPIN** data set and regress the Cap Diameter (y) on Diameter 1 (x_1), Diameter 2 (x_2), and Diameter 3 (x_3). The "stdev" of the regression coefficients is the square root of the diagonal elements of the (S_b) matrix. To see this we present first the inverse of the $(X)'(X)$ matrix, which is given by the following symmetric matrix :

$$[(X)'(X)]^{-1} = \begin{bmatrix} 5907.2 & -658.63 & 558.05 & -490.70 \\ \cdot & 695.58 & -448.15 & -181.94 \\ \cdot & \cdot & 739.77 & -347.38 \\ \cdot & \cdot & \cdot & 578.77 \end{bmatrix}$$

```
almpin = mistat.load_data('ALMPIN')
# create the X matrix
X = almpin[['diam1', 'diam2', 'diam3']]
X = np.hstack((np.ones((len(X), 1)), X))
# calculate the inverse of XtX
np.linalg.inv(np.matmul(X.transpose(), X))
```

```
array([[5907.19803233,  -658.62798437,   558.04560732,  -490.70308612],
       [-658.62798432,   695.57552698,  -448.14874864,  -181.93582393],
       [ 558.04560746,  -448.14874866,   739.77378683,  -347.37983108],
       [-490.7030863 ,  -181.93582392,  -347.37983108,   578.76516931]])
```

The value of S_e^2 is the square of the printed s value, i.e., $S_e^2 = 0.0000457$. Thus, the variances of the regression coefficients are:

$$S_{b_0}^2 = 0.0000457 \times 5907.2 = 0.2699590$$

$$S_{b_1}^2 = 0.0000457 \times 695.58 = 0.0317878$$

$$S_{b_2}^2 = 0.0000457 \times 739.77 = 0.0338077$$

$$S_{b_3}^2 = 0.0000457 \times 578.77 = 0.0264496.$$

Thus, S_{b_i} ($i = 0, \cdots 3$) are the "Stdev" in the printout. The t-ratios are given by

$$t_i = \frac{b_i}{S_{b_i}}, \quad i = 0, \cdots, 3.$$

The t-ratios should be large to be considered significant. The significance criterion is given by the P-value. Large value of P indicates that the regression coefficient is not significantly different from zero. In the above table, see that b_2 is not significant. Notice that Diameter 2 by itself, as the sole predictor of Cap Diameter, is very significant. This can be verified by running a simple regression of y on x_2. However, in the presence of x_1 and x_3, x_2 loses its significance. This analysis can be done in Python as shown below.

```
almpin = mistat.load_data('ALMPIN')
model = smf.ols('capDiam ~ 1 + diam1 + diam2 + diam3', data=almpin).fit()
print(model.summary2())
print()
print(sms.anova.anova_lm(model))
```

```
                   Results: Ordinary least squares
=================================================================
Model:               OLS              Adj. R-squared:    0.874
Dependent Variable:  capDiam          AIC:               -496.9961
Date:                2022-05-18 21:13 BIC:               -488.0021
No. Observations:    70               Log-Likelihood:    252.50
Df Model:            3                F-statistic:       159.9
Df Residuals:        66               Prob (F-statistic): 3.29e-30
R-squared:           0.879            Scale:             4.5707e-05
-----------------------------------------------------------------
              Coef.    Std.Err.    t      P>|t|    [0.025   0.975]
-----------------------------------------------------------------
Intercept     4.0411   0.5196    7.7771   0.0000   3.0037   5.0786
diam1         0.7555   0.1783    4.2371   0.0001   0.3995   1.1115
diam2         0.0173   0.1839    0.0939   0.9255  -0.3499   0.3844
diam3         0.3227   0.1626    1.9840   0.0514  -0.0020   0.6474
-----------------------------------------------------------------
Omnibus:             2.922            Durbin-Watson:     2.307
Prob(Omnibus):       0.232            Jarque-Bera (JB):  2.161
Skew:                0.266            Prob(JB):          0.339
Kurtosis:            2.323            Condition No.:     11328
=================================================================
* The condition number is large (1e+04). This might indicate
strong multicollinearity or other numerical problems.

            df     sum_sq    mean_sq          F         PR(>F)
diam1      1.0   0.021657   0.021657   473.822506   7.918133e-32
diam2      1.0   0.000084   0.000084     1.832486   1.804520e-01
diam3      1.0   0.000180   0.000180     3.936223   5.141819e-02
Residual  66.0   0.003017   0.000046         NaN            NaN
```

The "analysis of variance" table provides a global summary of the contribution of the various factors to the variability of y. The total sum of squares of y, around

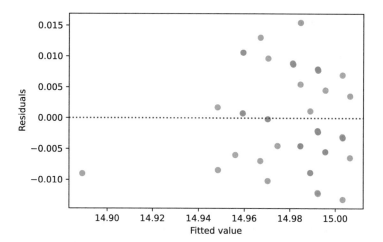

Fig. 4.15 A scatterplot of the residual versus the predicted values of CapDiam

its mean, is

$$\text{SST} = (n-1)S_y^2 = \sum_{i=1}^{n}(y_i - \bar{y})^2 = .024938.$$

This value of SST is partitioned into the sum of the variability explainable by the coefficients (SSR) and that due to the residuals around the regression (error, SSE). These are given by

$$\text{SSE} = (n-k-1)S_e^2$$

$$\text{SSR} = \text{SST} - \text{SSE}$$

$$= \mathbf{b}'(X)'\mathbf{y} - n\bar{\mathbf{Y}}_n^2$$

$$= \sum_{j=0}^{k} b_j \cdot \sum_{i=1}^{n} X_{ij} y_i - \left(\sum_{i=1}^{n} y(i)\right)^2 / n.$$

We present in Fig. 4.15 the scatterplot of the residuals, $\hat{e}_i$, versus the predicted values (FITS), $\hat{y}_i$. We see in this figure one point, corresponding to element # 66 in the data set, whose x-values have strong influence on the regression. ∎

The multiple regression can be used to test whether two or more simple linear regressions are parallel (same slopes) or have the same intercepts. We will show this by comparing two simple linear regressions.

Let $(x_i^{(1)}, Y_i^{(1)})$, $i = 1, \ldots, n$ be the data set of one simple linear regression of Y on x, and let $(x_j^{(2)}, Y_j^{(2)})$, $j = 1, \ldots, n_2$ be that of the second regression. By combining the data on the regression x from the two sets, we get the $\mathbf{x}$ vector

$$\mathbf{x} = (x_1^{(1)}, \ldots, x_{n_1}^{(1)}, x_1^{(2)}, \ldots, x_{n_2}^{(2)})'.$$

In a similar fashion, we combine the Y values and set

$$\mathbf{Y} = (Y_1^{(1)}, \ldots, Y_{n_1}^{(1)}, Y_1^{(2)}, \ldots, Y_{n_2}^{(2)})'.$$

Introduce a dummy variable z. The vector $\mathbf{z}$ has n_1 zeros at the beginning followed by n_2 ones. Consider now the multiple regression

$$\mathbf{Y} = b_0 \mathbf{1} + b_1 \mathbf{x} + b_2 \mathbf{z} + b_3 \mathbf{w} + \mathbf{e}, \tag{4.39}$$

where $\mathbf{1}$ is a vector of $(n_1 + n_2)$ ones and $\mathbf{w}$ is a vector of length $(n_1 + n_2)$ whose ith component is the product of the corresponding components of $\mathbf{x}$ and $\mathbf{z}$, i.e., $w_i = x_i z_i$ $(i = 1, \ldots, n_1 + n_2)$. Perform the regression analysis of $\mathbf{Y}$ on $(\mathbf{x}, \mathbf{z}, \mathbf{w})$. If b_2 is significantly different than 0, we conclude that the two simple regression lines have different intercepts. If b_3 is significantly different from zero, we conclude that the two lines have different slopes.

Example 4.11 In the present example, we compare the simple linear regressions of MPG/city (Y) on Turndiameter (x) of US-made cars and of Japanese cars. The data is in the file **CAR.csv**. The simple linear regression for US cars is

$$\hat{Y} = 49.0769 - 0.7565x$$

with $R^2 = 0.432$, $S_e = 2.735$ [56 degrees of freedom]. The simple linear regression for Japanese cars is

$$\hat{Y} = 42.0860 - 0.5743x,$$

with $R^2 = 0.0854$, $S_e = 3.268$ [35 degrees of freedom]. The combined multiple regression of $\mathbf{Y}$ on $\mathbf{x}, \mathbf{z}, \mathbf{w}$ yields the following table of P-values of the coefficients.

Coefficients:

	Value	Std. error	t − value	Pr(> $|t|$)
(Intercept)	49.0769	5.3023	9.2557	0.0000
mpgc	−0.7565	0.1420	−5.3266	0.0000
z	−6.9909	10.0122	−0.6982	0.4868
w	0.1823	0.2932	0.6217	0.5357

We see in this table that the P-values corresponding to $\mathbf{z}$ and $\mathbf{w}$ are 0.4868 and 0.5357, respectively. Accordingly, both b_2 and b_3 are **not** significantly different than zero. We can conclude that the two regression lines are not significantly different.

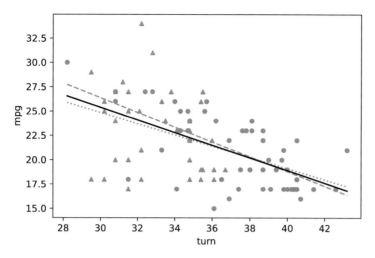

Fig. 4.16 Linear regression analysis for US (filled circle, dashed line) and Japanese cars (filled triangle, dotted line). The solid line is the linear regression of the combined data set

We can combine the data and have one regression line for both US and Japanese cars, namely:

$$\hat{Y} = 44.8152 - 0.6474x$$

with $R^2 = .3115$, $S_e = 3.337$ [93 degrees of freedom].

Here is how you can run this analysis in Python. Figure 4.16 compares the different regression equations.

```
# load dataset and split into data for US and Asia
car = mistat.load_data('CAR.csv')
car_US = car[car['origin'] == 1].copy()
car_Asia = car[car['origin'] == 3].copy()
# add the indicator variable z
car_US['z'] = 0
car_Asia['z'] = 1
# combine datasets and add variable w
car_combined = pd.concat([car_US, car_Asia])
car_combined['w'] = car_combined['z'] * car_combined['turn']

model_US = smf.ols('mpg ~ 1 + turn', data=car_US).fit()
model_Asia = smf.ols('mpg ~ 1 + turn', data=car_Asia).fit()
model_combined = smf.ols('mpg ~ 1 + turn+z+w', data=car_combined).fit()
model_simple = smf.ols('mpg ~ 1 + turn', data=car_combined).fit()
print('US\n', model_US.params)
print('Europe\n', model_Asia.params)
print(model_combined.summary2())
```

```
US
 Intercept    49.076922
turn         -0.756540
dtype: float64
Europe
 Intercept    42.086008
```

```
turn          -0.574288
dtype: float64
                Results: Ordinary least squares
==================================================================
Model:                OLS          Adj. R-squared:        0.298
Dependent Variable:   mpg          AIC:                   503.3067
Date:                 2022-05-18 21:13  BIC:               513.5223
No. Observations:     95           Log-Likelihood:        -247.65
Df Model:             3            F-statistic:           14.31
Df Residuals:         91           Prob (F-statistic):    1.03e-07
R-squared:            0.320        Scale:                 11.233
------------------------------------------------------------------
              Coef.   Std.Err.     t      P>|t|    [0.025    0.975]
------------------------------------------------------------------
Intercept    49.0769   5.3023   9.2557   0.0000   38.5445   59.6094
turn         -0.7565   0.1420  -5.3266   0.0000   -1.0387   -0.4744
z            -6.9909  10.0122  -0.6982   0.4868  -26.8790   12.8971
w             0.1823   0.2932   0.6217   0.5357   -0.4001    0.7646
------------------------------------------------------------------
Omnibus:               0.985       Durbin-Watson:          1.670
Prob(Omnibus):         0.611       Jarque-Bera (JB):       0.491
Skew:                 -0.059       Prob(JB):               0.782
Kurtosis:              3.332       Condition No.:          1164
==================================================================
* The condition number is large (1e+03). This might indicate
strong multicollinearity or other numerical problems.
```

```
# create visualization
ax = car_US.plot.scatter(x='turn', y='mpg', color='gray', marker='o')
car_Asia.plot.scatter(x='turn', y='mpg', ax=ax, color='gray', marker='^')

car_combined = car_combined.sort_values(['turn'])
ax.plot(car_combined['turn'], model_US.predict(car_combined),
        color='gray', linestyle='--')
ax.plot(car_combined['turn'], model_Asia.predict(car_combined),
        color='gray', linestyle=':')
ax.plot(car_combined['turn'], model_simple.predict(car_combined),
        color='black', linestyle='-')
plt.show()
```

■

4.4.4   Partial-F Tests and the Sequential SS

In the multiple regression analysis, a column typically entitled SEQ SS provides
a partition of the regression sum of squares, SSR, to additive components of
variance, each one with one degree of freedom. We have seen that the multiple-
R^2, $R^2_{y|(x_1,...,x_k)} = SSR/SST$ is the proportion of the total variability which is
explainable by the linear dependence of Y on all the k regressors. A simple linear
regression on the first variable x_1 yields a smaller $R^2_{y|x_1}$. The first component of
the SEQ SS is $SSR_{y|(x_1)} = SST \cdot R^2_{y|(x_1)}$. If we determine the multiple regression
of Y on x_1 and x_2, then $SSR_{y|(x_1,x_2)} = SST R^2_{y|(x_1,x_2)}$ is the amount of variability
explained by the linear relationship with the two variables. The difference

$$DSS_{x_2|x_1} = SST(R^2_{y|(x_1 x_2)} - R^2_{y|(x_1)}) \tag{4.40}$$

is the additional amount of variability explainable by x_2, after accounting for x_1. Generally, for $i = 2, \ldots, k$

$$DSS_{x_i|x_1 \ldots, x_{i-1}} = SST(R^2_{y|(x_1, \ldots, x_i)} - R^2_{y|(x_1, \ldots, x_{i-1})}) \tag{4.41}$$

is the additional contribution of the ith variable after controlling for the first $(i-1)$ variables.

Let

$$s^2_{e(i)} = \frac{SST}{n - i - 1}(1 - R^2_{y|(x_1, \ldots, x_i)}), \quad i = 1, \ldots, k \tag{4.42}$$

then

$$F^{(i)} = \frac{DSS_{x_i|x_1, \ldots, x_{i-1}}}{s^2_{e(i)}}, \quad i = 1, \ldots, k \tag{4.43}$$

is called the **partial-F** for testing the significance of the contribution of the variable x_i, after controlling for $x_1, \ldots, x_{i-1}$. If $F^{(i)}$ is greater than the $(1 - \alpha)$th quantile $F_{1-\alpha}[1, n - i - 1]$ of the F distribution, the additional contribution of X_i is significant. The partial-F test is used to assess whether the addition of the ith regression significantly improves the prediction of Y, given that the first $(i-1)$ regressors have already been included.

Example 4.12 In the previous example, we have examined the multiple regression of CapDiam, Y, on Diam1, Diam2, and Diam3 in the **ALMPIN.csv** file. We compute here the partial-F statistics corresponding to the SEQ SS values. We get the SEQ SS from calling anova_lm with the full model in the sum_sq column. The partial-F values are obtained by calling anova_lm with two models. It will return an analysis of the impact of changing the first to the second model. The reported F value is partial-F and Pr(>F) the P-value.

```
import warnings
almpin = mistat.load_data('ALMPIN')
model3 = smf.ols('capDiam ~ 1 + diam1+diam2+diam3', data=almpin).fit()
model2 = smf.ols('capDiam ~ 1 + diam1+diam2', data=almpin).fit()
model1 = smf.ols('capDiam ~ 1 + diam1', data=almpin).fit()
model0 = smf.ols('capDiam ~ 1', data=almpin).fit()

print('Full model\n', sms.anova.anova_lm(model))
print(f'SSE: diam1: {model1.ssr:.6f}')
print(f'     diam2: {model2.ssr:.6f}')
print(f'     diam3: {model3.ssr:.6f}')

# we capture a few irrelevant warnings here -
with warnings.catch_warnings():
    warnings.simplefilter("ignore")
    print('diam1:\n', sms.anova.anova_lm(model0, model1))
    print('diam2:\n', sms.anova.anova_lm(model1, model2))
    print('diam3:\n', sms.anova.anova_lm(model2, model3))
```

```
Full model
              df      sum_sq      mean_sq             F          PR(>F)
diam1       1.0    0.021657    0.021657    473.822506    7.918133e-32
diam2       1.0    0.000084    0.000084      1.832486    1.804520e-01
diam3       1.0    0.000180    0.000180      3.936223    5.141819e-02
Residual   66.0    0.003017    0.000046           NaN             NaN
SSE: diam1: 0.003280
     diam2: 0.003197
     diam3: 0.003017
diam1:
     df_resid          ssr   df_diff    ss_diff              F         Pr(>F)
0        69.0    0.024937       0.0        NaN            NaN            NaN
1        68.0    0.003280       1.0   0.021657    448.941202    1.153676e-31
diam2:
     df_resid          ssr   df_diff    ss_diff              F       Pr(>F)
0        68.0    0.003280       0.0        NaN            NaN          NaN
1        67.0    0.003197       1.0   0.000084      1.75555    0.189682
diam3:
     df_resid          ssr   df_diff    ss_diff              F       Pr(>F)
0        67.0    0.003197       0.0        NaN            NaN          NaN
1        66.0    0.003017       1.0    0.00018     3.936223    0.051418
```

Variable	SEQ SS	SSE	d.f.	Partial-F	P-value
Diam1	.021657	.003280	68	448.94	0
Diam2	.000084	.003197	67	1.75	0.190
Diam3	.000180	.003017	66	3.93	0.052

We see from these partial-F values, and their corresponding P-values, that after using Diam1 as a predictor, the additional contribution of Diam2 is insignificant. Diam3, however, in addition to the regressor Diam1, significantly decreases the variability which is left unexplained. ∎

The partial-F test is called sometimes **sequential-F test** (Draper and Smith 1998). We use the terminology Partial-F statistic because of the following relationship between the partial-F and the partial correlation. In Sect. 4.4.2 we defined the partial correlation $r_{yx_2 \cdot x_1}$, as the correlation between $\hat{\mathbf{e}}^{(1)}$, which is the vector of residuals around the regression of Y on x_1, and the vector of residuals $\hat{\mathbf{e}}^{(2)}$, of x_2 around its regression on x_1. Generally, suppose that we have determined the multiple regression of Y on $(x_1, \ldots, x_{i-1})$. Let $\hat{\mathbf{e}}(y \mid x_1, \ldots, x_{i-1})$ be the vector of residuals around this regression. Let $\hat{\mathbf{e}}(x_i \mid x_1, \ldots, x_{i-1})$ be the vector of residuals around the multiple regression of x_i on $x_1, \ldots, x_{i-1}$ ($i \geq 2$). **The correlation between $\hat{\mathbf{e}}(y \mid x_1, \ldots, x_{i-1})$ and $\hat{\mathbf{e}}(x_i \mid x_1, \ldots, x_{i-1})$ is the partial correlation between Y and x_i, given $x_1, \ldots, x_{i-1}$.** We denote this partial correlation by $r_{yx_i \cdot x_1, \ldots, x_{i-1}}$. The following relationship holds between the partial-F, $F^{(i)}$, and the partial correlation

$$F^{(i)} = (n - i - 1) \frac{r^2_{yx_i \cdot x_1, \ldots, x_{i-1}}}{1 - r^2_{yx_i \cdot x_1, \ldots, x_{i-1}}}, \quad i \geq 2. \tag{4.44}$$

This relationship is used to test whether $r_{yx_i \cdot x_1, \ldots, x_{i-1}}$ is significantly different than zero. $F^{(i)}$ should be larger than $F_{1-\alpha}[1, n - i - 1]$.

4.4.5 Model Construction: Step-Wise Regression

It is often the case that data can be collected on a large number of regressors, which might help us predict the outcomes of a certain variable, Y. However, the different regressors vary generally with respect to the amount of variability in Y which they can explain. Moreover, different regressors or predictors are sometimes highly correlated, and therefore not all of them might be needed to explain the variability in Y and to be used as predictors.

The following example is given by Draper and Smith (1998). The amount of steam [Pds] which is used monthly, Y, in a plant may depend on six regressors:

x_1 = Pounds of real fatty acid in storage per month

x_2 = Pounds of crude glycerine made in a month

x_3 = Monthly average wind velocity [miles/hour]

x_4 = Plant operating days per month

x_5 = Number of days per month with temperature below $32\,°F$

x_6 = Monthly average atmospheric temperature [F]

Are all these six regressors required to be able to predict Y? If not, which variables should be used? This is the problem of model construction.

There are several techniques for constructing a regression model. So far no implementation is available in any of the major Python packages. However, implementations are available in less popular packages. Here, we demonstrate how step-wise selection can be implemented in Python.

In the first step, we select the variable x_j ($j = 1, \ldots, k$) whose correlation with Y has maximal magnitude, provided it is significantly different than zero.

At each step the procedure computes a partial-F, or partial correlation, for each variable, x_l, which has not been selected in the previous steps. A variable having the **largest** significant partial-F is selected. The addition is followed by identifying the variable whose elimination has the **largest** partial-F. The procedure stops when no additional variables can be selected or removed. We illustrate the step-wise regression in the following example.

Example 4.13 In Example 4.7 we introduced the data file **GASOL.csv** and performed a multiple regression of Y on endPt and astm. In the following Python code, we derive a linear model which includes all variables which contribute significantly to the prediction. The code makes use of an implementation of step-wise regression in the mistat package.

```
gasol = mistat.load_data('GASOL')
gasol = gasol.rename(columns={'yield': 'Yield'})

outcome = 'Yield'
all_vars = set(gasol.columns)
all_vars.remove(outcome)

include, model = mistat.stepwise_regression(outcome, all_vars, gasol)

formula = ' + '.join(include)
formula = f'{outcome} ~ 1 + {formula}'
print()
print('Final model')
print(formula)
print(model.params)
```

```
Step 1 add - (F: 30.76)   endPt
Step 2 add - (F: 270.11)  astm endPt
Step 3 add - (F: 4.72)    astm endPt x1

Final model
Yield ~ 1 + x1 + astm + endPt
Intercept      4.032034
x1             0.221727
astm          -0.186571
endPt          0.156527
dtype: float64
```

The output shows at each step the variable whose partial-F value is maximal but greater than F_to_add, which is 4.00. Adding endPt (x_4) to the constant model has the largest partial-F statistic of 30.76 and is therefore added in Step 1. The fitted regression equation is

$$\hat{Y} = -16.662 + 0.1094x_4$$

with $R^2_{y|(x_4)} = .5063$. The partial-F for endPt is $F = 30.76$. Since this value is greater than F_to_remove, which is 4.00, endPt remains in the model.

In Step 2 the maximal partial correlation of Y and x_1, x_2, astm given endPt is that of astm, with a partial-$F = 270.11$. Variable astm (x_3) is selected, and the new regression equation is

$$\hat{Y} = 18.468 + .1558x_4 - .2090x_3,$$

with $R^2_{y|(x_4,x_3)} = .9521$. Since the partial-$F$ of x_4 is $(22.73)^2 = 516.6529$, the two variables remain in the model.

In Step 3 the variable x_1 is chosen. Since its partial-F is 4.72, it is included too. The final regression equation is

$$\hat{Y} = 4.032 + .1565x_4 - .1870x_3 + .2200x_1$$

with $R^2_{y|(x_4,x_3,x_1)} = .959$. Only 4.1% of the variability in Y is left unexplained. ■

4.4.6   Regression Diagnostics

As mentioned earlier, the least squares regression line is sensitive to extreme x or y values of the sample elements. Sometimes even one point may change the characteristics of the regression line substantially. We illustrate this in the following example.

Example 4.14 Consider again the SOCELL data. We have seen earlier that the regression line (L1) of ISC at time $t2$ on ISC at time $t1$ is $\hat{y} = 0.536 + 0.929x$, with $R^2 = .959$. In Fig. 4.17 we demonstrate how changes in the data can influence the regression.

The point having the largest x-value has a y-value of 5.37 (point #8).[1] If the y-value of this point is changed to 4.37, we obtain a different regression line (dashed), given by $\hat{y} = 2.04 + 0.532x$, with $R^2 = .668$. If on the other hand we change a point in the middle (point #5) of the possible x values at the same amount, the regression line changes only little (dotted). Here, we get $\hat{y} = 0.52 + 0.947x$, with $R^2 = .773$. ∎

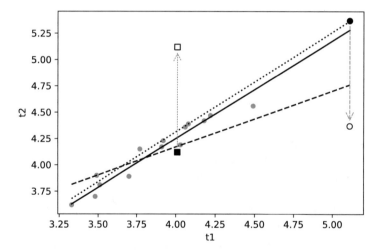

Fig. 4.17 Influence of data changes on linear regression for SOCELL data

---

[1] Note that this corresponds to row 9 of the data, as Python starts counting at 0.

In the present section, we present the diagnostic tools which are commonly used. The objective is to measure the degree of influence the points have on the regression line. In statsmodels, we get access to such information using the get_influence method.

```
socell = mistat.load_data('SOCELL')

model = smf.ols(formula='t2 ~ 1 + t1', data=socell).fit()
influence = model.get_influence()
# leverage: influence.hat_matrix_diag
# std. residuals: influence.resid_studentized
# Cook-s distance: influence.cooks_distance[0]
# DFIT: influence.dffits[0]
```

We start with the notion of the x-**leverage** of a point.

Consider the matrix (H) defined in Sect. 4.4.3. The vector of predicted values, $\hat{\mathbf{y}}$, is obtained as $(H)\mathbf{y}$. The x-**leverage** of the ith point is measured by the ith diagonal element of (H), which is

$$h_i = \mathbf{x}'_i((X)'(X))^{-1}\mathbf{x}_i, \quad i = 1, \cdots, n. \tag{4.45}$$

Here $\mathbf{x}'_i$ denotes the ith row of the predictors matrix (X), i.e.,

$$\mathbf{x}'_i = (1, x_{i1}, \cdots, x_{ik}).$$

In the special case of simple linear regression ($k = 1$), we obtain the formula

$$h_i = \frac{1}{n} + \frac{(x_i - \bar{x})^2}{\sum_{j=1}^{n}(x_j - \bar{x})^2}, \quad i = 1, \cdots, n. \tag{4.46}$$

Notice that $S_e\sqrt{h_i}$ is the **standard error** (square root of variance) **of the predicted value** $\hat{y}_i$. This interpretation holds also in the multiple regression case ($k > 1$). In Fig. 4.18 we present the x-leverage values of the various points in the **SOCELL** example.

From the above formula, we deduce that, when $k = 1$, $\sum_{i=1}^{n} h_i = 2$. Generally, for any k, $\sum_{i=1}^{n} h_i = k + 1$. Thus, the average x-leverage is $\bar{h} = \frac{k+1}{n}$. In the above solar cell example, the average x-leverage of the 16 points is $\frac{2}{16} = 0.125$. Point #8, (5.21,4.37), has a leverage value of $h_8 = .521$. This is indeed a high x-leverage. Point #5, in contrast, has a low leverage value of $h_5 = 0.064$.

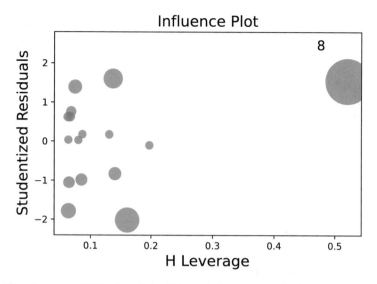

Fig. 4.18 x-Leverage of ISC values. Point #8 is marked as an unusual point

```
leverage = influence.hat_matrix_diag
print(f'average leverage: {np.mean(leverage):.3f}')
print(f'point #8: {leverage[8]:.3f}')
print(f'point #5: {leverage[5]:.3f}')
```

```
average leverage: 0.125
point #8: 0.521
point #5: 0.064
```

The standard error of the ith residual, $\hat{e}_i$, is given by

$$S\{\hat{e}_i\} = S_e\sqrt{1 - h_i}. \tag{4.47}$$

The **standardized residuals** are therefore given by

$$\hat{e}_i^* = \frac{\hat{e}_i}{S\{\hat{e}_i\}} = \frac{\hat{e}_i}{S_e\sqrt{1 - h_i}}, \tag{4.48}$$

$i = 1, \cdots, n$. There are several additional indices, which measure the effects of the points on the regression. We mention here two such measures, the **Cook distance** and the **fits distance**.

If we delete the ith point from the data set and recompute the regression, we obtain a vector of regression coefficients $\mathbf{b}^{(i)}$ and standard deviation of residuals $S_e^{(i)}$. The standardized difference

$$D_i = \frac{(\mathbf{b}^{(i)} - \mathbf{b})'((X)'(X))(\mathbf{b}^{(i)} - \mathbf{b})}{(k + 1)S_e} \tag{4.49}$$

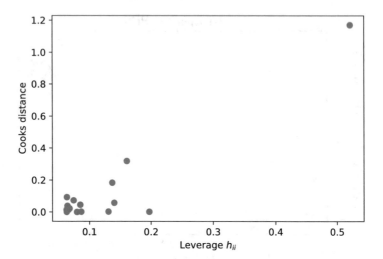

Fig. 4.19 Cook's distance for aluminum pins data

is the so-called Cook's distance. In Fig. 4.19 we present the Cook distance, for the **ALMPIN** data set.

The influence of the fitted values, denoted by DFIT, is defined as

$$\text{DFIT}_i = \frac{\hat{Y}_i - \hat{Y}_i^{(i)}}{S_e^{(i)} \sqrt{h_i}}, \quad i = 1, \cdots, n \tag{4.50}$$

where $\hat{Y}_i^{(i)} = b_0^{(i)} + \sum_{j=1}^{k} b_j^{(i)} x_{ij}$ are the predicted values of Y, at $(1, x_{i1}, \cdots, x_{ik})$, when the regression coefficients are $\mathbf{b}^{(i)}$. It is recommended to investigate observations with DFIT$_i$ larger than $2\sqrt{\frac{k}{n}}$ where k is the number of parameters. This metric is also calculated by `statsmodels`.

4.5   Quantal Response Analysis: Logistic Regression

We consider the case where the regressant Y is a binary random variable, and the regressors are quantitative. The distribution of Y at a given combination of x values $\mathbf{x} = (x_1, \ldots, x_k)$ is binomial $B(n, p(\mathbf{x}))$, where n is the number of identical and independent repetitions of the experiment at $\mathbf{x}$. $p(\mathbf{x}) = P\{Y = 1 \mid \mathbf{x}\}$. The question is how to model the function $p(\mathbf{x})$. An important class of models is the so-called quantal response models, according to which

$$p(\mathbf{x}) = F(\boldsymbol{\beta}'\mathbf{x}), \tag{4.51}$$

where $F(\cdot)$ is a c.d.f. and

$$\boldsymbol{\beta}'\mathbf{x} = \beta_0 + \beta_1 x_1 + \cdots + \beta_k x_k. \tag{4.52}$$

The **logistic regression** is a method of estimating the regression coefficients $\boldsymbol{\beta}$, in which

$$F(z) = \frac{e^z}{1 + e^z}, \quad -\infty < z < \infty, \tag{4.53}$$

is the logistic c.d.f.

The experiment is conducted at m different, and linearly independent, combinations of $\mathbf{x}$ values. Thus, let

$$(X) = (1, \mathbf{x}_1, \mathbf{x}_2, \ldots, \mathbf{x}_k)$$

be the predictors matrix of m rows and $(k + 1)$ columns. We assumed that $m > (k + 1)$ and the rank of (X) are $(k + 1)$. Let $\mathbf{x}^{(i)}$, $i = 1, \ldots, m$ denote the ith row vector of (X).

As mentioned above, we replicate the experiment at each $\mathbf{x}^{(i)}$ n times. Let $\hat{p}_i$ $(i = 1, \ldots, m)$ be the proportion of 1s observed at $\mathbf{x}^{(i)}$, i.e., $\hat{p}_{i,n} = \frac{1}{n} \sum_{j=1}^{n} Y_{ij}$, $i = 1, \ldots, m$; where $Y_{ij} = 0, 1$ is the observed value of the regressant at the jth replication $(j = 1, \ldots, n)$.

We have proven before that $E\{\hat{p}_{i,n}\} = p(\mathbf{x}^{(i)})$, and $V\{\hat{p}_{i,n}\} = \frac{1}{n} p(\mathbf{x}^{(i)})(1 - p(\mathbf{x}^{(i)}))$, $i = 1, \ldots, m$. Also, the estimators $\hat{p}_i$ $(i = 1, \ldots, m)$ are independent. According to the logistic model,

$$p(\mathbf{x}^{(i)}) = \frac{e^{\boldsymbol{\beta}'\mathbf{x}^{(i)}}}{1 + e^{\boldsymbol{\beta}'\mathbf{x}^{(i)}}}, \quad i = 1, \ldots, m. \tag{4.54}$$

The problem is to estimate the regression coefficients $\boldsymbol{\beta}$. Notice that the log-odds at $\mathbf{x}^{(i)}$ is

$$\log \frac{p(\mathbf{x}^{(i)})}{1 - p(\mathbf{x}^{(i)})} = \boldsymbol{\beta}'\mathbf{x}^{(i)}, \quad i = 1, \ldots, m. \tag{4.55}$$

Define $Y_{i,n} = \log \frac{\hat{p}_{i,n}}{1 - \hat{p}_{i,n}}$, $i = 1, \ldots, m$. $Y_{i,n}$ is finite if n is sufficiently large. Since $\hat{p}_{i,n} \to p(\mathbf{x}^{(i)})$ in probability, as $n \to \infty$ (WLLN), and since $\log \frac{x}{1-x}$ is a continuous function of x on $(0, 1)$, $Y_{i,n}$ is a consistent estimator of $\boldsymbol{\beta}'\mathbf{x}^{(i)}$. For large values of n, we can write the regression model

$$Y_{i,n} = \boldsymbol{\beta}'\mathbf{x}^{(i)} + e_{i,n} + e_{i,n}^*, \quad i = 1, \ldots, m \tag{4.56}$$

where

$$e_{i,n} = \frac{\hat{p}_{i,n} - p(\mathbf{x}^{(i)})}{p(\mathbf{x}^{(i)})(1 - p(\mathbf{x}^{(i)}))} \tag{4.57}$$

$e_{i,n}^*$ is a negligible remainder term if n is large. $e_{i,n}^* \to 0$ in probability at the rate of $\frac{1}{n}$. If we omit the remainder term $e_{i,n}^*$, we have the approximate regression model

$$Y_{i,n} \cong \boldsymbol{\beta}'\mathbf{x}^{(i)} + e_{i,n}, \quad i = 1, \ldots, m \tag{4.58}$$

where

$$E\{e_{i,n}\} = 0,$$

$$V\{e_{i,n}\} = \frac{1}{n} \cdot \frac{1}{p(\mathbf{x}^{(i)})(1 - p(\mathbf{x}^{(i)}))} \tag{4.59}$$

$$= \frac{\left(1 + e^{\boldsymbol{\beta}'\mathbf{x}^{(i)}}\right)^2}{n \cdot e^{\boldsymbol{\beta}'\mathbf{x}^{(i)}}},$$

$i = 1, \ldots, m$. The problem here is that $V\{e_{i,n}\}$ depends on the unknown $\boldsymbol{\beta}$ and varies from one $\mathbf{x}^{(i)}$ to another. An ordinary LSE of $\boldsymbol{\beta}$ is given by $\hat{\boldsymbol{\beta}} = [(X)'(X)]^{-1}(X)'Y$, where $\mathbf{Y}' = (Y_{1,n}, \ldots, Y_{m,n})$. Since the variances of $e_{i,n}$ are different, an estimator having smaller variances is the weighted LSE

$$\hat{\boldsymbol{\beta}}_w = [(X)'W(\boldsymbol{\beta})(X)]^{-1}(X)'W(\boldsymbol{\beta})Y, \tag{4.60}$$

where $W(\boldsymbol{\beta})$ is a diagonal matrix, whose ith term is

$$W_i(\boldsymbol{\beta}) = \frac{ne^{\boldsymbol{\beta}'\mathbf{x}^{(i)}}}{\left(1 + e^{\boldsymbol{\beta}'\mathbf{x}^{(i)}}\right)^2}, \quad i = 1, \ldots, m. \tag{4.61}$$

The problem is that the weights $W_i(\boldsymbol{\beta})$ depend on the unknown vector $\boldsymbol{\beta}$. An iterative approach to obtain $\hat{\boldsymbol{\beta}}_w$ is to substitute on the r.h.s. the value of $\hat{\boldsymbol{\beta}}$ obtained in the previous iteration, starting with the ordinary LSE, $\hat{\boldsymbol{\beta}}$. Other methods of estimating the coefficients $\boldsymbol{\beta}$ of the logistic regression are based on the maximum likelihood method. For additional information see Kotz et al. (1988), *Encyclopedia of Statistical Sciences* and Ruggeri et al. (2008), and *Encyclopedia of Statistics in Quality and Reliability*.

4.6   The Analysis of Variance: The Comparison of Means

4.6.1   The Statistical Model

When the regressors $x_1, x_2, \ldots, x_k$ are qualitative (categorical) variables and the variable of interest Y is quantitative, the previously discussed methods of multiple regression are invalid. The different values that the regressors obtain are different categories of the variables. For example, suppose that we study the relationship between film speed (Y) and the type of gelatine x used in the preparation of the chemical emulsion for coating the film, the regressor is a categorical variable. The values it obtains are the various types of gelatine, as classified according to manufacturers.

When we have k, $k \geq 1$, such qualitative variables, the combination of categorical levels of the k variables are called **treatment combinations** (a term introduced by experimentalists). Several observations, n_i, can be performed at the ith treatment combination. These observations are considered a random sample from the (infinite) population of all possible observations under the specified treatment combination. The statistical model for the jth observation is

$$Y_{ij} = \mu_i + e_{ij}, \quad i = 1, \ldots t \ \ j = 1, \ldots, n_i$$

where μ_i is the population mean for the ith treatment combination, t is the number of treatment combinations, and e_{ij} $(i = 1, \ldots, t; j = 1, \ldots, n_i)$ are assumed to be independent random variables (experimental errors) with $E\{e_{ij}\} = 0$ for all (i, j) and $v\{e_{ij}\} = \sigma^2$ for all (i, j). The comparison of the means μ_i $(i = 1, \ldots, t)$ provides information on the various effects of the different treatment combinations. The method used to do this analysis is called **analysis of variance** (ANOVA).

4.6.2   The One-Way Analysis of Variance (ANOVA)

In Sect. 3.11.5.1 we introduced the ANOVA F test statistics and presented the algorithm for bootstrap ANOVA for comparing the means of k populations. In the present section, we develop the rationale for the ANOVA. We assume here that the errors e_{ij} are independent and normally distributed. For the ith treatment combination (sample), let

$$\bar{Y}_i = \frac{1}{n_i} \sum_{j=1}^{n_i} Y_{ij}, \quad i = 1, \cdots, t \tag{4.62}$$

and

$$SSD_i = \sum_{j=1}^{n_i} (Y_{ij} - \bar{Y}_i)^2, \quad i = 1, \cdots, t. \tag{4.63}$$

Let $\bar{\bar{Y}} = \dfrac{1}{N} \sum_{i=1}^{t} n_i \bar{Y}_i$ be the grand mean of all the observations.

The one-way ANOVA is based on the following partition of the total sum of squares of deviations around $\bar{\bar{Y}}$:

$$\sum_{i=1}^{t} \sum_{j=1}^{n_i} (Y_{ij} - \bar{\bar{Y}})^2 = \sum_{i=1}^{t} SSD_i + \sum_{i=1}^{t} n_i (\bar{Y}_i - \bar{\bar{Y}})^2. \tag{4.64}$$

We denote the l.h.s. by SST and the r.h.s. by SSW and SSB, i.e.,

$$SST = SSW + SSB. \tag{4.65}$$

SST, SSW, and SSB are symmetric quadratic forms in deviations like $Y_{ij} - \bar{\bar{Y}}$, $\bar{Y}_{ij} - \bar{Y}_i$ and $\bar{Y}_i - \bar{\bar{Y}}$. Since $\sum_i \sum_j (Y_{ij} - \bar{\bar{Y}}) = 0$, only $N - 1$, linear functions $Y_{ij} - \bar{\bar{Y}} = \sum_{i'} \sum_{j'} c_{i'j'} Y_{i'j'}$, with

$$c_{i'j'} = \begin{cases} 1 - \dfrac{1}{N}, & i' = i, \ j' = j \\ -\dfrac{1}{N}, & \text{otherwise} \end{cases}$$

are linearly independent, where $N = \sum_{i=1}^{t} n_i$. For this reason we say that the quadratic form SST has $(N - 1)$ degrees of freedom (d.f.). Similarly, SSW has $(N - t)$ degrees of freedom, since $SSW = \sum_{i=1}^{t} SSD_i$, and the number of degrees of freedom of SSD_i is $(n_i - 1)$. Finally, SSB has $(t - 1)$ degrees of freedom. Notice that SSW is the total sum of squares of deviations **within** the t samples, and SSB is the sum of squares of deviations **between** the t sample means.

Dividing a quadratic form by its number of degrees of freedom, we obtain the mean squared statistic. We summarize all these statistics in a table called the **ANOVA table**. The ANOVA table for comparing t treatments is given in Table 4.11.

Generally, in an ANOVA table, D.F. designates degrees of freedom, S.S. designates the sum of squares of deviations, and M.S. designates the mean squared. In all tables,

Table 4.11 ANOVA table for one-way layout

Source of variation	D.F.	S.S.	M.S.
Between treatments	$t - 1$	SSB	MSB
Within treatments	$N - t$	SSW	MSW
Total (adjusted for mean)	$N - 1$	SST	–

$$M.S. = \frac{S.S.}{D.F.} \qquad (4.66)$$

We show now that

$$E\{MSW\} = \sigma^2. \qquad (4.67)$$

Indeed, according to the model, and since $\{Y_{ij}, j = 1, \cdots, n_i\}$ is a RSWR from the population corresponding to the ith treatment,

$$E\left\{\frac{SSD_i}{n_i - 1}\right\} = \sigma^2, \quad i = 1, \cdots, t.$$

Since $MSW = \sum_{i=1}^{t} \nu_i \left(\frac{SSD_i}{n_i - 1}\right)$, where $\nu_i = \frac{n_i - 1}{N - t}, i = 1, \cdots, t,$

$$E\{MSW\} = \sum_{i=1}^{t} \nu_i E\left\{\frac{SSD_i}{n_i - 1}\right\} = \sigma^2 \sum_{i=1}^{t} \nu_i$$

$$= \sigma^2.$$

Another important result is

$$E\{MSB\} = \sigma^2 + \frac{1}{t - 1} \sum_{i=1}^{t} n_i \tau_i^2, \qquad (4.68)$$

where $\tau_i = \mu_i - \bar{\mu}$ $(i = 1, \ldots, t)$ and $\bar{\mu} = \frac{1}{N} \sum_{i=1}^{t} n_i \mu_i$. Thus, under the null hypothesis $H_0 : \mu_1 = \ldots = \mu_t$, $E\{MSB\} = \sigma^2$. This motivates us to use, for testing H_0, the F statistic

$$F = \frac{MSB}{MSW}. \qquad (4.69)$$

H_0 is rejected, at the level of significance α, if

$$F > F_{1-\alpha}[t - 1, N - t].$$

Example 4.15 Three different vendors are considered for supplying cases for floppy disk drives. The question is whether the latch mechanism that opens and closes the disk loading slot is sufficiently reliable. In order to test the reliability of this latch, three independent samples of cases, each of size $n = 10$, were randomly selected from the production lots of these vendors. The testing was performed on a special apparatus that opens and closes a latch, until it breaks. The number of

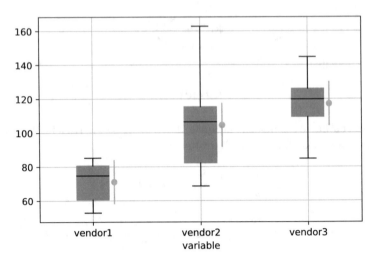

Fig. 4.20 Boxplots of Y by vendor. The errorbar next to each boxplot shows the linear regression result with the confidence interval

cycles is required until latch failure was recorded. In order to avoid uncontrollable environmental factors to bias the results, the order of testing of cases of different vendors was completely randomized. In file **VENDOR.csv**, we can find the results of this experiment, arranged in three columns. Column 1 represents the sample from vendor A_1; column 2 that of vendor A_2, and column 3 of vendor A_3. An ANOVA was performed. The analysis was done on $Y = $ (Number of Cycles)$^{1/2}$, in order to have data which is approximately normally distributed. The original data are expected to have positively skewed distribution, since it reflects the life length of the latch.

The following is the one-way ANOVA using the statsmodels package. We first load the data and convert it into a format where all values are in one column and the vendor indicated in second column. Figure 4.20 compares the distributions of values by vendor.

```
vendor = mistat.load_data('VENDOR')
vendor_long = pd.melt(vendor, value_vars=vendor.columns)
vendor_long['value'] = np.sqrt(vendor_long['value'])
```

```
                                0           1
variable[vendor1]     58.000928    84.123334
variable[vendor2]     91.546689   117.669096
variable[vendor3]    104.114252   130.236658
variable[vendor1]     13.061203
variable[vendor2]     13.061203
variable[vendor3]     13.061203
dtype: float64
```

Next, we create a regression model of the outcome as a function of the vendor. The anova_lm function returns the ANOVA analysis of the model.

```
model = smf.ols('value ~ variable', data=vendor_long).fit()
table = sm.stats.anova_lm(model, typ=1)
print(table)
```

	df	sum_sq	mean_sq	F	PR(>F)
variable	2.0	11365.667752	5682.833876	14.024312	0.000067
Residual	27.0	10940.751619	405.213023	NaN	NaN

We can also calculate confidence intervals for the reliability of the latches for each of the vendors. In this case, it is useful to exclude the intercept from the regression model.

```
model = smf.ols('value ~ -1 + variable', data=vendor_long).fit()
print(model.conf_int())
```

	0	1
variable[vendor1]	58.000928	84.123334
variable[vendor2]	91.546689	117.669096
variable[vendor3]	104.114252	130.236658

The ANOVA table shows that the F statistic is significantly large, having a P-value close to 0. The null hypothesis H_0 is rejected. The reliability of the latches from the three vendors is not the same (see Fig. 4.20). The 0.95 confidence intervals for the means show that vendors A_2 and A_3 manufacture latches with similar reliability. That of vendor A_1 is significantly lower. ∎

4.7 Simultaneous Confidence Intervals: Multiple Comparisons

Whenever the hypothesis of no difference between the treatment means is rejected, the question arises, which of the treatments have similar effects, and which ones differ significantly? In Example 4.15 we analyzed data on the strength of latches supplied by three different vendors. It was shown that the differences are very significant. We also saw that the latches from vendor A_1 were weaker from those of vendors A_2 and A_3, which were of similar strength. Generally, if there are t treatments, and the ANOVA shows that the differences between the treatment means are significant, we may have to perform up to $\binom{t}{2}$ comparisons, to rank the different treatments in terms of their effects.

If we compare the means of all pairs of treatments, we wish to determine $\binom{t}{2} = \frac{t(t-1)}{2}$ confidence intervals to the true differences between the treatment means. If each confidence interval has a confidence level $(1 - \alpha)$, the probability that **all** $\binom{t}{2}$ confidence intervals cover the true differences simultaneously is smaller than $(1 - \alpha)$. The simultaneous confidence level might be as low as $(1 - t\alpha)$.

There are different types of simultaneous confidence intervals. We present here the method of Scheffé, for simultaneous confidence intervals for any number of contrasts (Scheffé 1999). A **contrast** between t means $\bar{Y}_1, \cdots, \bar{Y}_t$ is a linear combination $\sum_{i=1}^{t} c_i \bar{Y}_i$, such that $\sum_{i=1}^{t} c_i = 0$. Thus, any difference between two means is a contrast, e.g., $\bar{Y}_2 - \bar{Y}_1$. Any second-order difference, e.g.,

$$(\bar{Y}_3 - \bar{Y}_2) - (\bar{Y}_2 - \bar{Y}_1) = \bar{Y}_3 - 2\bar{Y}_2 + \bar{Y}_1,$$

is a contrast. The space of all possible linear contrasts has dimension $(t - 1)$. For this reason, the coefficient we use, according to Scheffé's method, to obtain simultaneous confidence intervals of level $(1 - \alpha)$ is

$$S_\alpha = ((t - 1)F_{1-\alpha}[t - 1, t(n - 1)])^{1/2} \qquad (4.70)$$

where $F_{1-\alpha}[t - 1, t(n - 1)]$ is the $(1 - \alpha)$th quantile of the F distribution. It is assumed that all the t samples are of equal size n. Let $\hat{\sigma}_p^2$ denote the pooled estimator of σ^2, i.e.,

$$\hat{\sigma}_p^2 = \frac{1}{t(n - 1)} \sum_{i=1}^{t} SSD_i. \qquad (4.71)$$

Then the simultaneous confidence intervals for all contrasts of the form $\sum_{i=1}^{t} c_i \mu_i$ have limits

$$\sum_{i=1}^{t} c_i \bar{Y}_i \pm S_\alpha \frac{\hat{\sigma}_p}{\sqrt{n}} \left(\sum_{i=1}^{t} c_i^2 \right)^{1/2}. \qquad (4.72)$$

Example 4.16 In data file **HADPAS.csv**, we have the resistance values (ohms) of several resistors on six different hybrids at 32 cards. We analyze here the differences between the means of the $n = 32$ resistance values of resistor RES3, where the treatments are the $t = 6$ hybrids. The boxplots of the samples corresponding to the six hybrids are presented in Fig. 4.21.

In Table 4.12, we present the means and standard deviations of these six samples (treatments) (Fig. 4.21 and Table 4.12).

The pooled estimator of σ is

$$\hat{\sigma}_p = 133.74.$$

The Scheffé coefficient, for $\alpha = 0.05$, is

$$S_{.05} = (5F_{.95}[5, 186])^{1/2} = 3.332.$$

Upper and lower simultaneous confidence limits, with 0.95 level of significance, are obtained by adding to the differences between means

$$\pm S_\alpha \frac{\hat{\sigma}_p}{\sqrt{16}} = \pm 111.405.$$

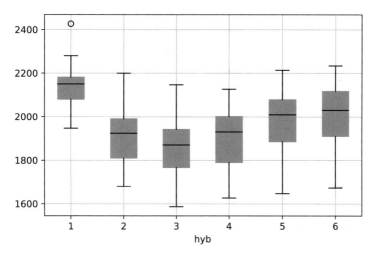

Fig. 4.21 Boxplots of resistance in six hybrids

Table 4.12 Means and std. of resistance RES3 by hybrid

Hybrid	$\bar{Y}$	S_y
1	2143.41	99.647
2	1902.81	129.028
3	1850.34	144.917
4	1900.41	136.490
5	1980.56	146.839
6	2013.91	139.816

Differences which are smaller in magnitude than 111.405 are considered insignificant.

Thus, if we order the sample means, we obtain:

Hybrid	Mean	Group mean
1	2143.41	2143.41
6	2013.91	1997.24
5	1980.56	
2	1902.81	
4	1900.41	1884.52
3	1850.34	

Thus, the difference between the means of Hybrid 1 and all the others are significant. The mean of Hybrid 6 is significantly different than those of 2, 4, and 3. The mean of Hybrid 5 is significantly larger than that of Hybrid 3. We suggest therefore the following **homogeneous group** of treatments (all treatments within the same homogeneous group have means which are not significantly different).

Homog group	Group mean
{1}	2143.41
{5,6}	1997.24
{2,3,4}	1884.52

Homog group	Means of groups
{1}	2143.41
{5, 6}	1997.24
{2, 3, 4}	1884.52

The difference between the means of $\{5, 6\}$ and $\{2, 3, 4\}$ is the contrast

$$-\frac{1}{3}\bar{Y}_2 - \frac{1}{3}\bar{Y}_3 - \frac{1}{3}\bar{Y}_4 + \frac{1}{2}\bar{Y}_5 + \frac{1}{2}\bar{Y}_6.$$

This contrast is significant, if it is greater than

$$S_\alpha \frac{\hat{\sigma}_p}{\sqrt{32}} \sqrt{\left(\frac{1}{2}\right)^2 + \left(\frac{1}{2}\right)^2 + \left(\frac{1}{3}\right)^2 + \left(\frac{1}{3}\right)^2 + \left(\frac{1}{3}\right)^2} = 71.912.$$

The above difference is thus significant. ∎

Tukey's Honest Significant Difference (HSD) test is a post hoc test commonly used to assess the significance of differences between pairs of group means. The assumptions in applying it are that the observations are independent and normality of distribution with homogeneity of variance. These are the same assumptions as for one-way ANOVA.

Example 4.17 We can apply the Tukey's HSD test on the **HADPAS.csv** data set from Example 4.16 using the function `pairwise_tukeyhsd` from the `statsmodels` package.

```
from statsmodels.stats.multicomp import pairwise_tukeyhsd
hadpas = mistat.load_data('HADPAS')
mod = pairwise_tukeyhsd(hadpas['res3'], hadpas['hyb'])
print(mod)
```

```
   Multiple Comparison of Means - Tukey HSD, FWER=0.05
=====================================================
group1 group2  meandiff  p-adj    lower      upper    reject
-----------------------------------------------------
     1      2 -240.5938  0.001  -336.8752  -144.3123    True
     1      3 -293.0625  0.001  -389.3439  -196.7811    True
     1      4    -243.0  0.001  -339.2814  -146.7186    True
     1      5 -162.8438  0.001  -259.1252   -66.5623    True
     1      6    -129.5  0.002  -225.7814   -33.2186    True
     2      3  -52.4688  0.6036 -148.7502    43.8127   False
     2      4   -2.4062     0.9  -98.6877    93.8752   False
```

```
2        5       77.75   0.189   -18.5314   174.0314   False
2        6   111.0938  0.0135    14.8123   207.3752    True
3        4   50.0625  0.6448    -46.2189   146.3439   False
3        5   130.2188  0.0019    33.9373   226.5002    True
3        6   163.5625   0.001     67.2811   259.8439    True
4        5   80.1562  0.1626    -16.1252   176.4377   False
4        6      113.5  0.0107     17.2186   209.7814    True
5        6    33.3438     0.9    -62.9377   129.6252   False
```

∎

4.8   Contingency Tables

4.8.1   The Structure of Contingency Tables

When the data is categorical, we generally summarize it in a table which presents the frequency of each category, by variable, in the data. Such a table is called a **contingency table**.

Example 4.18 Consider a test of a machine which inserts components into a board. The displacement errors of such a machine were analyzed in Example 4.1. In this test we perform a large number of insertions with $k = 9$ different components. The result of each trial (insertion) is either success (no insertion error) or failure (insertion error). In the present test, there are two categorical variables: component type and insertion result. The first variable has nine categories:

- $C1$: Diode
- $C2$: 1/2 watt canister
- $C3$: Jump wire
- $C4$: Small corning
- $C5$: Large corning
- $C6$: Small bullet
- $C7$: 1/8 watt dogbone
- $C8$: 1/4 watt dogbone
- $C9$: 1/2 watt dogbone

The second variable, insertion result, has two categories only (success, failure). Contingency Table 4.13 presents the frequencies of the various insertion results by component type.

Table 4.13 shows that the proportional frequency of errors in insertions is very small ($190/1, 056, 764 = 0.0001798$), which is about 180 FPM (failures per million). This may be judged to be in conformity with the industry standard. We see, however, that there are apparent differences between the failure proportions, by component types. In Fig. 4.22 we present the FPMs of the insertion failures, by

Table 4.13 Contingency table of insertion results by component type

| | Insertion result | | |
Component type	Failure	Success	Row total
$C1$	61	108,058	108,119
$C2$	34	136,606	136,640
$C3$	10	107,328	107,338
$C4$	23	105,042	105,065
$C5$	25	108,829	108,854
$C6$	9	96,864	96,873
$C7$	12	107,379	107,391
$C8$	3	105,851	105,854
$C9$	13	180,617	180,630
Column total	190	1,056,574	1,056,764

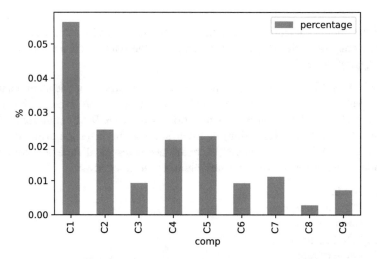

Fig. 4.22 Bar-chart of components error rates

component type. The largest one is that of $C1$ (diode), followed by components $\{C2, C4, C5\}$. Smaller proportions are those of $\{C3, C6, C9\}$. The smallest error rate is that of $C8$. The differences in the components error rates can be shown to be very significant. ∎

The structure of the contingency table might be considerably more complicated than that of Table 4.13. We illustrate here a contingency table with three variables.

Example 4.19 The data are the placement errors of an OMNI 4621 automatic insertion machine. The variables are:

  (i) Machine structure: Basic, EMI1, EMI2, EMI4
 (ii) Components: 805, 1206, SOT_23
(iii) Placement result: Error, No_Error

Table 4.14 Contingency table of placement errors

Comp. structure	805 Err	805 No Err	1206 Err	1206 No Err	SOT_23 Err	SOT_23 No Err	Total comp. Err	Total comp. No Err	Total rows
Basic	11	40,279	7	40,283	16	40,274	34	120,836	120,870
EMI1	11	25,423	8	25,426	2	25,432	21	76,281	76,302
EMI2	19	54,526	15	54,530	12	54,533	46	163,589	163,635
EMI4	14	25,194	4	25,204	5	25,203	23	75,601	75,624
Total	55	145,422	34	145,443	35	145,442	124	436,307	436,431

Table 4.15 Failure rates (FPM) by structure and component type

Structure	Component 805	1206	SOT_23
Basic	273	174	397
EMI1	433	315	79
EMI2	348	275	220
EMI4	555	159	198

The contingency table is given in Table 4.14 and summarizes the results of 436,431 placements.

We see in Table 4.14 that the total failure rate of this machine type is $124/436,307 = 284$ (FPM). The failure rates, by machine structure, in FPMs, are 281, 275, 281, and 304, respectively. The first three structural types have almost the same FPMs, while the fourth one is slightly larger. The component failure rates are 378, 234, and 241 FPM, respectively. It remains to check the failure rates according to structure × component. These are given in Table 4.15.

We see that the effect of the structure is different on different components. Again, one should test whether the observed differences are statistically significant or due only to chance variability. Methods for testing this will be discussed in Chapter 6 (Industrial Statistics book). ∎

The construction of contingency tables can be done from pandas DataFrames in Python. We illustrate this on the data in file **CAR.csv**. This file consists of information on 109 car models from 1989. The file contains 109 records on 5 variables: number of cylinders (4,6,8), origin (USA = 1, Europe = 2, Asia = 3), turn diameter [meters], horsepower, and number of miles per gallon in city driving. One variable, origin, is categorical, while the other four are interval-scaled variables. One discrete (number of cylinders) and the other three are continuous.

The following code returns the contingency table illustrated in Table 4.16.

Table 4.16 Contingency
table of the number of
cylinders and origin

Num. Cyl.	Origin			
	1	2	3	
4	33	7	26	66
6	13	7	10	30
8	12	0	1	13
Total	58	14	37	109

Table 4.17 Contingency
table of turn diameter versus
miles/gallon city

Turn diameter	Miles/gallon city			
	12–18	19–24	25–	Total
27.0–30.6	2	0	4	6
30.7–34.2	4	12	15	31
34.3–37.8	10	26	6	42
37.9–	15	15	0	30
Total	31	53	25	109

```
car = mistat.load_data('CAR')
count_table = car[['cyl', 'origin']].pivot_table(
            index='cyl', columns='origin', aggfunc=len, fill_value=0)
print(count_table)
```

```
origin   1   2   3
cyl
4       33   7  26
6       13   7  10
8       12   0   1
```

One can prepare a contingency table also from continuous data, by selecting the number and length of intervals for each variable and counting the frequencies of each cell in the table. For example, for the car data, if we wish to construct a contingency table of turn diameter versus miles/gallon, we obtain the contingency table presented in Table 4.17.

4.8.2   Indices of association for contingency tables

In the present section, we construct several indices of association, which reflect the degree of dependence or association between variables. For the sake of simplicity, we consider here indices for two-way tables, i.e., association between two variables.

4.8.2.1   Two Interval-Scaled Variables

If the two variables are continuous ones, measured on an interval scale, or some transformation of it, we can use some of the dependence indices discussed earlier. For example, we can represent each interval by its mid-point and compute the correlation coefficient between these mid-points. As in Sect. 4.2, if variable X is classified into k intervals,

$$(\xi_0, \xi_1), (\xi_1, \xi_2), \cdots, (\xi_{k-1}, \xi_k)$$

and variable Y is classified into m intervals $(\eta_0, \eta_1), \cdots, (\eta_{m-1}, \eta_m)$, let

$$\tilde{\xi}_i = \frac{1}{2}(\xi_{i-1} + \xi_i), \quad i = 1, \cdots, k$$

$$\tilde{\eta}_j = \frac{1}{2}(\eta_{j-1} + \eta_j), \quad j = 1, \cdots, m.$$

Let $p_{ij} = f_{ij}/N$ denote the proportional frequency of the (i, j)th cell, i.e., X values in (ξ_{i-1}, ξ_i) and Y values in (η_{j-1}, η_j). Then, an estimate of the coefficient of correlation obtained from the contingency table is

$$\hat{\rho}_{XY} = \frac{\sum_{i=1}^{k} \sum_{j=1}^{m} p_{ij} (\tilde{\xi}_i - \bar{\xi})(\tilde{\eta}_j - \bar{\eta})}{\left[\sum_{i=1}^{k} p_{i.}(\tilde{\xi}_i - \bar{\xi})^2\right]^{1/2} \left[\sum_{j=1}^{m} p_{.j}(\tilde{\eta}_j - \bar{\eta})^2\right]^{1/2}}, \tag{4.73}$$

where

$$p_{i.} = \sum_{j=1}^{m} p_{ij}, \quad i = 1, \cdots, k,$$

$$p_{.j} = \sum_{i=1}^{k} p_{ij}, \quad j = 1, \cdots, m,$$

$$\bar{\xi} = \sum_{i=1}^{k} p_{i.}\tilde{\xi}_i,$$

and

$$\bar{\eta} = \sum_{j=1}^{m} p_{.j}\tilde{\eta}_j.$$

Notice that the sample correlation r_{XY}, obtained from the sample data, is different from $\hat{\rho}_{XY}$, due to the reduced information that is given by the contingency table. We illustrate this in the following example.

Example 4.20 Consider the data in file **CAR.csv**. The sample correlation between the turn diameter, X, and the gas consumption (Miles/Gal) in a city is $r_{XY} = -0.541$. If we compute this correlation on the basis of the data in Table 4.17, we obtain $\hat{\rho}_{XY} = -.472$. The approximation given by $\hat{\rho}_{XY}$ depends on the number of intervals, k and m, on the length of the intervals and the sample size N.   ■

4.8.2.2   Indices of Association for Categorical Variables

If one of the variables or both are categorical, there is no meaning to the correlation coefficient. We should devise another index of association. Such an index should not depend on the labeling or ordering of the categories. Common indices of association are based on comparison of the observed frequencies f_{ij} of the cells $(i = 1, \cdots, k; j = 1, \cdots, m)$ to the expected ones if the events associated with the categories are independent. The concept of independence, in a probability sense, is defined in Chap. 3. We have seen earlier conditional frequency distributions. If $N_{i\cdot} = \sum_{j=1}^{m} f_{ij}$, the conditional **proportional** frequency of the jth category of Y, given the ith category of X, is

$$p_{j|i} = \frac{f_{ij}}{N_{i\cdot}}, \quad j = 1, \cdots, m.$$

We say that X and Y are **not associated** if

$$p_{j|i} = p_{\cdot j} \quad \text{for all } i = 1, \cdots, k,$$

where

$$p_{\cdot j} = \frac{N_{\cdot j}}{N} \quad j = 1, \cdots, m$$

and

$$N_{\cdot j} = \sum_{i=1}^{k} f_{ij}.$$

Accordingly, the expected frequency of cell (i, j), if there is **no association**, is

$$\tilde{f}_{ij} = \frac{N_{i\cdot} N_{\cdot j}}{N}, \quad i = 1, \cdots, k, j = 1, \cdots, m.$$

A common index of discrepancy between f_{ij} and $\tilde{f}_{ij}$ is

$$X^2 = \sum_{i=1}^{k} \sum_{j=1}^{m} \frac{(f_{ij} - \tilde{f}_{ij})^2}{\tilde{f}_{ij}}. \tag{4.74}$$

Table 4.18 Observed and expected frequencies of turn diameter by miles/gallon, **CAR.csv** ($\tilde{f}_{ij}$ under $\hat{f}_{ij}$)

Turn diameter	Miles/gallon city			Total
	12–18	19–24	25–	
27.0–30.6	2	0	4	6
	1.71	2.92	1.38	
30.7–34.2	4	12	15	31
	8.82	15.07	7.11	
34.3–37.8	10	26	6	42
	11.94	20.42	9.63	
37.9–	15	15	0	30
	8.53	14.59	6.88	
Total	31	53	25	109

This index is called the **chi-squared statistic**. This statistic can be computed with the scipy chi2_contingency command.

Example 4.21 For the **CAR.csv** data, the chi-squared statistic for the association between origin and num cycl is $X^2 = 12.13$. In Chapter 6 Industrial Statistics book, we will study how to assess the statistical significance of such a magnitude of X^2.

The chi2_contingency command returns a tuple of the chi2 statistic, the associated p-value, the degrees of freedom, and the matrix of expected frequencies.

```
chi2 = stats.chi2_contingency(count_table)
print(f'chi2 statistic {chi2[0]:.2f}')
```

```
chi2 statistic 12.13
```

The chi2 statistic can of course also be applied to the frequency Table 4.17. Table 4.18 shows the actual and expected frequencies. The chi-squared statistic is

$$X^2 = \frac{(2 - 1.71)^2}{1.71} + \cdots + \frac{6.88^2}{6.88} = 34.99.$$

■

There are several association indices in the literature, based on the X^2. Three popular indices are:

Mean Squared Contingency

$$\Phi^2 = \frac{X^2}{N} \tag{4.75}$$

Tschuprow's index

$$T = \Phi/\sqrt{(k-1)(m-1)} \tag{4.76}$$

Cramér's index

$$C = \Phi/\sqrt{\min(k-1, m-1)} \qquad (4.77)$$

No association corresponds to $\Phi^2 = T = C = 0$. The larger the index, the stronger the association. For the data of Table 4.18,

$$\Phi^2 = \frac{34.99}{109} = 0.321$$

$$T = 0.283$$

$$C = 0.401.$$

We provide an additional example of contingency tables analysis, using the Cramér index.

Example 4.22 CompuStar, a service company providing technical support and sales of personal computers and printers, decided to investigate the various components of customer satisfaction that are specific to the company. A special questionnaire with 13 questions was designed and, after a pilot run, was mailed to a large sample of customers with a self-addressed stamped envelope and a prize incentive. The prize was to be awarded by lottery among the customers who returned the questionnaire.

The customers were asked to rate, on a 1–6 ranking order, various aspects of the service. The rating of 1 corresponding to very poor and the rating of 6 to very good. These questions include:

Q1: First impression of service representative
Q2: Friendliness of service representative
Q3: Speed in responding to service request
Q4: Technical knowledge of service representative
Q5: Professional level of service provided
Q6: Helpfulness of service representative
Q7: Additional information provided by service representative
Q8: Clarity of questions asked by service representative
Q9: Clarity of answers provided by service representative
Q10: Efficient use of time by service representative
Q11: Overall satisfaction with service
Q12: Overall satisfaction with product
Q13: Overall satisfaction with company

The response ranks are:

1. Very poor
2. Poor
3. Below average
4. Above average

Table 4.19 Two-by-two contingency table of customer responses, for Q3 and Q13

Q3\Q13	1	2	3	4	5	6
1	0	1	0	0	3	1
2	0	2	0	1	0	0
3	0	0	4	2	3	0
4	0	1	1	10	7	5
5	0	0	0	10	71	38
6	0	0	0	1	30	134

Table 4.20 Cramer's indices of Q1–Q10 by Q11–Q13

		Q1	Q2	Q3	Q4	Q5	Q6	Q7	Q8	Q9	Q10
Q11:	Overall satisfaction with service		•	++	•			+	•		
Q12:	Overall satisfaction with product		+	•	•		++	•	•		
Q13:	Overall satisfaction with company		•	++			+	++	•	•	

5. Good

6. Very good

The responses were tallied, and contingency tables were computed linking the questions on overall satisfaction with questions on specific service dimensions. For example, Table 4.19 is a contingency table of responses to Q13 versus Q3.

Cramer's index for Table 4.19 is:

$$C = \frac{1.07}{2.23} = 0.478.$$

There were ten detailed questions (Q1–Q10) and three questions on overall customer satisfaction (Q11–Q13). A table was constructed for every combination of the three overall customer satisfaction questions and the ten specific questions. For each of these 30 tables, Cramer's index was computed, and using a code of graphical symbols, we present these indices in Table 4.20.

The indices are coded according to the following key:

Cramer's index	Code
0–0.2	
0.2–0.3	•
0.3–0.4	+
0.4–0.5	++
0.5–	+++

We can see from Table 4.20 that "Overall satisfaction with company" (Q13) is highly correlated with "Speed in responding to service requests" (Q3). However, the "Efficient use of time" (Q10) was not associated with overall satisfaction.

On the other hand, we also notice that questions Q1, Q5, and Q10 show no correlation with overall satisfaction. Many models have been proposed in the literature for the analysis of customer satisfaction surveys. For a comprehensive review with applications using R, see Kenett and Salini (2012). For more examples of indices of association and graphical analysis of contingency tables, see Kenett (1983). Contingency tables are closely related to the data mining technique of association rules. For more on this, see Kenett and Salini (2008).   ■

4.9   Categorical Data Analysis

If all variables $x_1, \ldots, x_k$ and Y are categorical, we cannot perform the ANOVA without special modifications. In the present section, we discuss the analysis appropriate for such cases.

4.9.1   Comparison of Binomial Experiments

Suppose that we have performed t-independent binomial experiments, each one corresponding to a treatment combination. In the ith experiment, we ran n_i-independent trials. The yield variable, J_i, is the number of successes among the n_i trials ($i = 1, \cdots, t$). We further assume that in each experiment, the n_i trials are independent and have the same, unknown, probability for success, θ_i; i.e., J_i has a binomial distribution $B(n_i, \theta_i)$, $i = 1, \cdots, t$. We wish to compare the probabilities of success, θ_i ($i = 1, \cdots, k$). Accordingly, the null hypothesis is of equal success probabilities, i.e.,

$$H_0 : \theta_1 = \theta_2 = \cdots = \theta_k.$$

We describe here a test, which is good for large samples. Since by the CLT, $\hat{p}_i = J_i/n_i$ has a distribution which is approximately normal for large n_i, with mean θ_i and variance $\frac{\theta_i(1-\theta_i)}{n_i}$, one can show that the large sample distribution of

$$Y_i = 2 \arcsin\left(\sqrt{\frac{J_i + 3/8}{n_i + 3/4}}\right) \tag{4.78}$$

Table 4.20 The arcsin transformation

i	J_i	n_i	Y_i
1	61	108119	0.0476556
2	34	136640	0.0317234
3	10	107338	0.0196631
4	23	105065	0.0298326
5	25	108854	0.0305370
6	9	96873	0.0196752
7	12	107391	0.0214697
8	3	105854	0.0112931
9	13	180630	0.0172102

(in radians) is approximately normal, with mean $\eta_i = 2\arcsin(\sqrt{\theta_i})$ and variance $V\{Y_i\} = 1/n_i, i = 1, \cdots, t$.

Using this result, we obtain that under the assumption of H_0 the sampling distribution of the test statistic

$$Q = \sum_{i=1}^{k} n_i(Y_i - \bar{Y})^2, \tag{4.79}$$

where

$$\bar{Y} = \frac{\sum_{i=1}^{k} n_i Y_i}{\sum_{i=1}^{k} n_i}, \tag{4.80}$$

is approximately chi-squared with $k - 1$ DF, $\chi^2[k-1]$. In this test, we reject H_0, at the level of significance α, if $Q > \chi^2_{1-\alpha}[k-1]$.

Another test statistic for general use in contingency tables will be given in the following section.

Example 4.23 In Table 4.13 we presented the frequency of failures of nine different components in inserting a large number of components automatically. In the present example, we test the hypothesis that the failure probabilities, θ_i, are the same for all components. In Table 4.20 we present the values of J_i (# of failures), n_i, and $Y_i = 2\arcsin\left(\sqrt{\dfrac{J_i + 3/8}{n_i + 3/4}}\right)$, for each component.

We can calculate the Y_i values in Python to obtain Table 4.20.

```
df = pd.DataFrame({
    'i': [1, 2, 3, 4, 5, 6, 7, 8, 9],
    'Ji': [61, 34, 10, 23, 25, 9, 12, 3, 13],
    'ni': [108119,136640,107338,105065,108854,96873,107391,105854,180630],
    })
df['Yi'] = 2*np.arcsin(np.sqrt((df['Ji'] + 3/8)/(df['ni'] + 3/4)))
```

Using this result, we can compute the test statistic Q like this.

```
Ybar = np.sum(df['ni'] * df['Yi']) / np.sum(df['ni'])
Q = np.sum(df['ni'] * (df['Yi'] - Ybar) ** 2)
print(Q)
```

```
105.43139783460373
```

The value of Q is 105.43. The P-value of this statistic is 0. The null hypothesis is rejected. To determine this P-value using Python, since the distribution of Q under H_0 is $\chi^2[8]$, we use the commands:

```
stats.chi2.cdf(105.43, df=8)
```

```
1.0
```

We find that $\Pr\{\chi^2[8] \leq 105.43\} \doteq 1$. This implies that $P = 0$.   ∎

4.10   Chapter Highlights

The main concepts and tools introduced in this chapter include:

- Matrix scatterplots
- 3D-scatterplots
- Multiple boxplots
- Code variables
- Joint, marginal, and conditional frequency distributions
- Sample correlation
- Coefficient of determination
- Simple linear regression
- Multiple regression
- Predicted values, FITS
- Residuals around the regression
- Multiple squared correlation
- Partial regression
- Partial correlation
- Partial-F test
- Sequential SS
- Step-wise regression
- Regression diagnostics
- x-Leverage of a point
- Standard error of predicted value
- Standardized residual
- Cook distance
- Fits distance, DFIT
- Analysis of variance
- Treatment combinations

- Simultaneous confidence intervals
- Multiple comparisons
- Contrasts
- Scheffé's method
- Contingency tables analysis
- Categorical data analysis
- Arcsin transformation
- Chi-squared test for contingency tables

4.11 Exercises

Exercise 4.1 Use file **CAR.csv** to prepare multiple or matrix scatterplots of turn diameter versus horsepower versus miles per gallon. What can you learn from these plots?

Exercise 4.2 Make a multiple (side-by-side) boxplots of the turn diameter by car origin, for the data in file **CAR.csv**. Can you infer that turn diameter depends on the car origin?

Exercise 4.3 Data file **HADPAS.csv** contains the resistance values (ohms) of five resistors placed in 6 hybrids on 32 ceramic substrates. The file contains eight columns. The variables in these columns are:

1. Record number
2. Substrate number
3. Hybrid number
4. Res 3
5. Res 18
6. Res 14
7. Res 7
8. Res 20

 (i) Make a multiple boxplot of the resistance in Res 3, by hybrid.
 (ii) Make a matrix plot of all the Res variables. What can you learn from the plots?

Exercise 4.4 Construct a joint frequency distribution of the variables horsepower and MPG/city for the data in file **CAR.csv**.

Exercise 4.5 Construct a joint frequency distribution for the resistance values of res3 and res14, in data file **HADPAS.csv**. [Code the variables first; see instructions in Example 4.3.]

Exercise 4.6 Construct the conditional frequency distribution of `res3`, given that the resistance values of `res14` is between 1300 and 1500 (ohms).

Exercise 4.7 In the present exercise, we compute the **conditional** means and standard deviations of one variable given another one. Use file **HADPAS.csv**.

We classify the data according to the values of Res 14 (`res14`) to five subgroups. Bin the values for Res 14 using bin edges at [900, 1200, 1500, 1800, 2100, 3000], and use the `groupby` method to split the hadpas data set by these bins. For each group, determine the mean and standard deviation of the Res 3 (`res3`) column. Collect the results and combine into a data frame for presentation.

Exercise 4.8 Given below are four data sets of (X, Y) observations:

(i) Compute the least squares regression coefficients of Y on X, for the four data sets.
(ii) Compute the coefficient of determination, R^2, for each set.

Data set 1		Data set 2		Data set 3		Data set 4	
$X^{(1)}$	$Y^{(1)}$	$X^{(2)}$	$Y^{(2)}$	$X^{(3)}$	$Y^{(3)}$	$X^{(4)}$	$Y^{(4)}$
10.0	8.04	10.0	9.14	10.0	7.46	8.0	6.68
8.0	6.95	8.0	8.14	8.0	6.67	8.0	5.76
13.0	7.58	13.0	8.74	13.0	12.74	8.0	7.71
9.0	8.81	9.0	8.77	9.0	7.11	8.0	8.84
11.0	8.33	11.0	9.26	11.0	7.81	8.0	8.47
14.0	9.96	14.0	8.10	14.0	8.84	8.0	7.04
6.0	7.24	6.0	6.13	6.0	6.08	8.0	5.25
4.0	4.26	4.0	3.10	4.0	5.39	19.0	12.50
12.0	10.84	11.0	9.13	12.0	8.16	8.0	5.56
7.0	4.82	7.0	7.26	7.0	6.42	8.0	7.91
5.0	5.68	5.0	4.74	5.0	5.73	8.0	6.89

Exercise 4.9 Compute the correlation matrix of the variables turn diameter, horsepower, and miles per gallon/city for the data in file **CAR.csv**.

Exercise 4.10

(i) Differentiate partially the quadratic function

$$\text{SSE} = \sum_{i=1}^{n} (Y_i - \beta_0 - \beta_1 X_{i1} - \beta_2 X_{i2})^2$$

with respect to β_0, β_1, and β_2 to obtain the linear equations in the least squares estimates b_0, b_1, and b_2. These linear equations are called **the normal equations**.

(ii) Obtain the formulae for b_0, b_1, and b_2 from the normal equations.

Exercise 4.11 Consider the variables miles per gallon, horsepower, and turn diameter in the data set **CAR.csv**. Find the least squares regression line of MPG (y) on horsepower (x_1) and turn diameter (x_2). For this purpose use first the equations in Sect. 4.4 and then verify your computations by using statsmodels ols method.

Exercise 4.12 Compute the partial correlation between miles per gallon and horsepower, given the number of cylinders, in data file **CAR.csv**.

Exercise 4.13 Compute the partial regression of miles per gallon and turn diameter, given horsepower, in data file **CAR.csv**.

Exercise 4.14 Use the three-stage algorithm of Sect. 4.4.2 to obtain the multiple regression of Exercise 4.11 from the results of 4.13.

Exercise 4.15 Consider Example 4.10. From the calculation output, we see that, when regression CapDiam on Diam1, Diam2, and Diam3, the regression coefficient of Diam2 is not significant (P value $= .925$), and this variable can be omitted. Perform a regression of CapDiam on Diam2 and Diam3. Is the regression coefficient for Diam2 significant? How can you explain the difference between the results of the two regressions?

Exercise 4.16 Regress the yield in **GASOL.csv** on all the four variables x_1, x_2, $astm$, and $endPt$:

(i) What is the regression equation?
(ii) What is the value of R^2?
(iii) Which regression coefficient(s) is (are) non-significant?
(iv) Which factors are important to control the yield?
(v) Are the residuals from the regression distributed normally? Make a graphical test.

Exercise 4.17

(i) Show that the matrix $(H) = (X)(B)$ is idempotent, i.e., $(H)^2 = (H)$.
(ii) Show that the matrix $(Q) = (I - H)$ is idempotent, and therefore $s_e^2 = \mathbf{y}'(Q)\mathbf{y}/(n - k - 1)$.

Exercise 4.18 Show that the vectors of fitted values, $\hat{\mathbf{y}}$, and of the residuals, $\hat{\mathbf{e}}$, are orthogonal, i.e., $\hat{\mathbf{y}}'\hat{\mathbf{e}} = 0$.

Exercise 4.19 Show that the $1 - R^2_{y|(x)}$ is proportional to $||\hat{\mathbf{e}}||^2$, which is the squared Euclidean norm of $\hat{\mathbf{e}}$.

Exercise 4.20 In Sect. 2.5.2 we presented properties of the $\text{cov}(X, Y)$ operator. Prove the following generalization of property (iv). Let $\mathbf{X}' = (X_1, \ldots, X_n)$ be a vector of n random variables. Let $(\boldsymbol{\Sigma})$ be an $n \times n$ matrix whose (i, j)th element is $\Sigma_{ij} = \text{cov}(X_i, X_j)$, $i, j = 1, \ldots, n$. Notice that the diagonal elements of $(\boldsymbol{\Sigma})$ are the variances of the components of $\mathbf{X}$. Let $\boldsymbol{\beta}$ and $\boldsymbol{\gamma}$ be two n-dimensional vectors. Prove that $\text{cov}(\boldsymbol{\beta}'\mathbf{X}, \boldsymbol{\gamma}'\mathbf{X}) = \boldsymbol{\beta}'(\boldsymbol{\Sigma})\boldsymbol{\gamma}$. [The matrix $\boldsymbol{\Sigma}$ is called the variance-covariance matrix of $\mathbf{X}$.]

Exercise 4.21 Let $\mathbf{X}$ be an n-dimensional random vector, having a variance-covariance matrix $(\boldsymbol{\Sigma})$. Let $\mathbf{W} = (\mathbf{B})\mathbf{X}$, where (B) is an $m \times n$ matrix. Show that the variance-covariance matrix of $\mathbf{W}$ is $(\mathbf{B})(\boldsymbol{\Sigma})(\mathbf{B})'$.

Exercise 4.22 Consider the linear regression model $\mathbf{y} = (X)\boldsymbol{\beta} + \mathbf{e}$. $\mathbf{e}$ is a vector of random variables, such that $E\{e_i\} = 0$ for all $i = 1, \ldots, n$ and

$$\text{cov}(e_i, e_j) = \begin{cases} \sigma^2, & \text{if } i = j \\ 0, & \text{if } i \neq j \end{cases}$$

$i, j = 1, \ldots, n$. Show that the variance-covariance matrix of the LSE $\mathbf{b} = (\mathbf{B})\mathbf{y}$ is $\sigma^2[(\mathbf{X})'(\mathbf{X})]^{-1}$.

Exercise 4.23 Consider **SOCELL.csv** data file. Compare the slopes and intercepts of the two simple regressions of ISC at time t_3 on ISC at time t_1 and ISC at t_3 on ISC at t_2.

Exercise 4.24 The following data (see Draper and Smith 1998) gives the amount of heat evolved in hardening of element (in calories per gram of cement) and the percentage of four various chemicals in the cement (relative to the weight of clinkers from which the cement was made). The four regressors are

x_1 : Amount of tricalcium aluminate

x_2 : Amount of tricalcium silicate

x_3 : Amount of tetracalcium alumino ferrite

x_4 : Amount of dicalcium silicate

The regressant Y is the amount of heat evolved. The data are given in the following table and as data set **CEMENT.csv**.

Compute in a sequence the regressions of Y on X_1; of Y on X_1, X_2; of Y on X_1, X_2, X_3; and of Y on X_1, X_2, X_3, X_4. For each regression compute the partial-

X_1	X_2	X_3	X_4	Y
7	26	6	60	78.5
1	29	15	52	74.3
11	56	8	20	104.3
11	31	8	47	87.6
7	52	6	33	95.9
11	55	9	22	109.2
3	71	17	6	102.7
1	31	22	44	72.5
2	54	18	22	93.1
21	47	4	26	115.9
1	40	23	34	83.8
11	66	9	12	113.3
10	68	8	12	109.4

F of the new regression added, the corresponding partial correlation with Y, and the sequential SS.

Exercise 4.25 For the data of Exercise 4.24, construct a linear model of the relationship between Y and $X_1, \ldots, X_4$, by the forward step-wise regression method.

Exercise 4.26 Consider the linear regression of miles per gallon on horsepower for the cars in data file **CAR.csv**, with origin = 3. Compute for each car the residuals, RESI; the standardized residuals, SRES; the leverage HI; and the Cook distance, D.

Exercise 4.27 A simulation of the operation of a piston is available as the piston simulator function *pistonSimulation*. In order to test whether changing the piston weight from 30 to 60 [kg] affects the cycle time significantly, run the simulation program four times at weight 30, 40, 50, and 60 [kg], keeping all other factors at their low level. In each run make $n = 5$ observations. Perform a one-way ANOVA of the results, and state your conclusions.

Exercise 4.28 In experiments performed for studying the effects of some factors on the integrated circuits fabrication process, the following results were obtained, on the pre-etch line width (μ_m)
Perform an ANOVA to find whether the results of the three experiments are significantly different by using Python. Do the two test procedures (normal and bootstrap ANOVA) yield similar results?

Exercise 4.29 In manufacturing film for industrial use, samples from two different batches gave the following film speed:

Exp. 1	Exp. 2	Exp. 3
2.58	2.62	2.22
2.48	2.77	1.73
2.52	2.69	2.00
2.50	2.80	1.86
2.53	2.87	2.04
2.46	2.67	2.15
2.52	2.71	2.18
2.49	2.77	1.86
2.58	2.87	1.84
2.51	2.97	1.86

Batch A: 103, 107, 104, 102, 95, 91, 107, 99, 105, 105

Batch B: 104, 103, 106, 103, 107, 108, 104, 105, 105, 97

Test whether the differences between the two batches are significant, by using (i) a randomization test and (ii) an ANOVA.

Exercise 4.30 Use a randomization test to test the significance of the differences between the results of the three experiments in Exercise 4.28.
   Use this statistic:

$$\delta = \frac{\sum_{k=1}^{3} n_k \bar{x}_k^2 - n\bar{x}^2}{S_x^2}$$

with $n = n_1 + n_2 + n_3$ and x the combined set of all results.

Exercise 4.31 In data file **PLACE.csv**, we have 26 samples, each one of size $n = 16$ and of x-, y-, and θ-deviations of components placements. Make an ANOVA, to test the significance of the sample means in the x-deviation. Classify the samples into homogeneous groups such that the differences between sample means in the same group are not significant and those in different groups are significant. Use the Scheffé coefficient S_α for $\alpha = .05$.

Exercise 4.32 The frequency distribution of cars by origin and number of cylinders is given in the following table.

Num. cylinders	USA	Europe	Asia	Total
4	33	7	26	66
6 or more	25	7	11	43
Total	58	14	37	109

Perform a chi-square test of the dependence of the number of cylinders and the origin of car.

Exercise 4.33 Perform a chi-squared test of the association between turn diameter and miles/gallon based on Table 4.17.

Exercise 4.34 In a customer satisfaction survey, several questions were asked regarding specific services and products provided to customers. The answers were on a 1–5 scale, where 5 means "very satisfied with the service or product" and 1 means "very dissatisfied." Compute the mean squared contingency, Tschuprow's index, and Cramer's index for both contingency tables.

Question 3	Question 1				
	1	2	3	4	5
1	0	0	0	1	0
2	1	0	2	0	0
3	1	2	6	5	1
4	2	1	10	23	13
5	0	1	1	15	100

Question 3	Question 2				
	1	2	3	4	5
1	1	0	0	3	1
2	2	0	1	0	0
3	0	4	2	3	0
4	1	1	10	7	5
5	0	0	1	30	134

Chapter 5
Sampling for Estimation of Finite Population Quantities

Preview Techniques for sampling finite populations and estimating population parameters are presented. Formulas are given for the expected value and variance of the sample mean and sample variance of simple random samples with and without replacement. Stratification is studied as a method to increase the precision of estimators. Formulas for proportional and optimal allocation are provided and demonstrated with case studies. The chapter is concluded with a section on prediction models with known covariates.

5.1 Sampling and the Estimation Problem

5.1.1 Basic Definitions

In the present chapter, we consider the problem of estimating quantities (parameters) of a finite population. The problem of testing hypotheses concerning such quantities, in the context of sampling inspection of product quality, will be studied in Chapter 11 (Industrial Statistics book). Estimation and testing of the parameters of statistical models for infinite populations were discussed in Chap. 3.

Let P designate a finite population of N units. It is assumed that the population size, N, is **known**. Also assume that a list (or a frame) of the population units $L_N = \{u_1, \cdots, u_N\}$ is available.

Let X be a variable of interest and $x_i = X(u_i), i = 1, \cdots, N$ the value ascribed by X to the ith unit, u_i, of P.

The population **mean** and population variance, for the variable X, i.e.,

$$\mu_N = \frac{1}{N} \sum_{i=1}^{N} x_i$$

Supplementary Information The online version contains supplementary material available at https://doi.org/10.1007/978-3-031-07566-7_5.

and (5.1)

$$\sigma_N^2 = \frac{1}{N} \sum_{i=1}^{N} (x_i - \mu_N)^2,$$

are called **population quantities**. In some books (Cochran 1977), these quantities are called "population parameters." We distinguish between population quantities and parameters of distributions, which represent variables in infinite populations. Parameters are not directly observable and can only be estimated, while finite population quantities can be determined exactly if the whole population is observed.

The population quantity μ_N is the expected value of the distribution of X in the population, whose c.d.f. is

$$\hat{F}_N(x) = \frac{1}{N} \sum_{i=1}^{N} I(x; x_i),$$

where (5.2)

$$I(x; x_i) = \begin{cases} 1, & \text{if } x_i \leq x \\ \\ 0, & \text{if } x_i > x. \end{cases}$$

σ_N^2 is the variance of $F_N(x)$.

In this chapter we focus attention on estimating the population mean, μ_N, when a sample of size n, $n < N$ is observed. The problem of estimating the population variance σ_N^2 will be discussed in the context of estimating the standard errors of estimators of μ_N.

Two types of sampling strategies will be considered. One type consists of random samples (with or without replacement) from the whole population. Such samples are called **simple random samples**. The other type of sampling strategy is that of **stratified random sampling**. In stratified random sampling, the population is first partitioned to strata (blocks), and then a simple random sample is drawn from each stratum independently. If the strata are determined so that the variability within strata is smaller relative to the general variability in the population, the precision in estimating the population mean μ_N, using a stratified random sampling will generally be higher than that in simple random sampling. This will be shown in Sect. 5.3.

As an example of a case where stratification could be helpful, consider the following. At the end of each production day, we draw a random sample from the lot of products of that day to estimate the proportion of defective item. Suppose that several machines operate in parallel, and manufacture the same item. Stratification by machine will provide higher precision for the global estimate, as well as information on the level of quality of each machine. Similarly, if we can stratify by shift, by vendor, or by other factors that may contribute to the variability, we may increase the precision of our estimates.

5.1.2 Drawing a Random Sample from a Finite Population

Given a finite population consisting of N distinct elements, we first make a **list** of all the elements of the population, which are all labeled for identification purposes. Suppose we wish to draw a random sample of size n from this population, where $1 \leq n \leq N$. We distinguish between two methods of random sampling: (a) **sampling with replacement** and (b) **sampling without replacement**. A sample drawn with replacement is obtained by returning the selected element, after each choice, to the population before the next item is selected. In this method of sampling, there are altogether N^n possible samples. A sample is called **random sample with replacement** (RSWR) if it is drawn by a method which gives every possible sample the same probability to be drawn. A sample is **without replacement** if an element drawn is not replaced and hence cannot be drawn again. There are $N(N-1)\cdots(N-n+1)$ such possible samples of size n from a population of size N. If each of these has the same probability of being drawn, the sample is called **random sample without replacement** (RSWOR). Bootstrapping discussed in Chap. 3 is an application of RSWR.

Practically speaking, the choice of a particular random sample is accomplished with the aid of **random numbers**. Random numbers can be generated by various methods. For example, an integer has to be drawn at random from the set $0, 1, \cdots, 99$. If we had a ten-faced die, we could label its faces with the numbers $0, \cdots, 9$ and cast it twice. The results of these two drawings would yield a two-digit integer, e.g., 13. Since in general we do not have such a die, we could, instead, use a coin and, flipping it seven times, generate a random number between 0 and 99 in the following way. Let X_j $(j = 1, \cdots, 7)$ be 0 or 1, corresponding to whether a head or tail appeared on the jth flip of the coin. We then compute the integer I, which can assume one of the values $0, 1, \cdots, 127$, according to the formula

$$I = X_1 + 2X_2 + 4X_3 + 8X_4 + 16X_5 + 32X_6 + 64X_7.$$

If we obtain a value greater than 99, we disregard this number and flip the coin again seven times. Adding 1 to the outcome produces random numbers between 1 and 100. In a similar manner, a roulette wheel could also be used in constructing random numbers. A computer algorithm for generating pseudo-random numbers was described in Example 2.5. In actual applications, we use ready-made **tables of random numbers** or computer routines for generating random numbers.

Example 5.1 The following ten numbers were drawn by using a random number generator on a computer: 76, 49, 95, 23, 31, 52, 65, 16, 61, and 24. These numbers form a random sample of size 10 from the set $1, \cdots, 100$. If by chance two or more numbers are the same, the sample would be acceptable if the method is RSWR. If the method is RSWOR, any number that was already selected would be discarded. In Python to draw a RSWOR of 10 integers from the set $\{1, \cdots, 100\}$, use `random.sample`.

```
random.sample(range(1, 101), k=10)
```

Note that we specify the range from 1 to 101 as the Python `range(a,b)` command creates a sequence of numbers from a to b excluding b. ∎

5.1.3 Sample Estimates of Population Quantities and Their Sampling Distribution

So far we have discussed the nature of variable phenomena and presented some methods of exploring and presenting the results of experiments. More specifically, the methods of analysis described in Chap. 1 explore the given data, but do not provide an assessment of what might happen in future experiments.

If we draw from the same population several different random samples, of the same size, we will find generally that statistics of interest assume different values at the different samples.

This can be illustrated in Python; we draw samples, with or without replacement, from a collection of numbers (population) which is stored in a vector. To show it, let us store in X the integers $1,2,\cdots,100$. To sample at random with replacement (RSWR) a sample of size $n = 20$ from X, and put the random sample in a vector $X\,Sample$, we use `random.choices`.

```
# range is a generator and needs to be converted to a list
X = list(range(1, 101))
Xsample = random.choices(X, k=20)
```

This can be repeated four times and collected in a data frame. Finally, we calculate the mean and standard deviation of each sample. Table 5.1 shows the result of such a sampling process.

```
df = pd.DataFrame({f'sample {i}': random.choices(X, k=20)
                   for i in range(1, 5)})
df.agg(['mean', 'std'])
```

```
        sample 1    sample 2    sample 3    sample 4
mean   45.300000   46.500000   53.400000   50.500000
std    27.482243   26.116137   27.058222   34.741906
```

Notice that the "population" mean is 50.5, and its standard deviation is 29.011. The sample means and standard deviations are estimates of these population parameters, and as seen above, they vary around the parameters. The distribution of sample estimates of a parameter is called the **sampling distribution of an estimate**.

Table 5.1 Four random
samples with replacement of
size 20, from $\{1, 2, \cdots, 100\}$

Sample			
1	2	3	4
26	54	4	15
56	59	81	52
63	73	87	46
46	62	85	98
1	57	5	44
4	2	52	1
31	33	6	27
79	54	47	9
21	97	68	28
5	6	50	52
94	62	89	39
52	70	18	34
79	40	4	30
33	70	53	58
6	45	70	18
33	74	7	14
67	29	68	14
33	40	49	32
21	21	70	10
8	43	15	52
Means			
37.9	49.6	46.4	33.6
Stand. Dev.			
28.0	23.7	31.3	22.6

Theoretically (hypothetically) the number of possible different random samples, with replacement, is either infinite, if the population is infinite, or of magnitude N^n, if the population is finite (n is the sample size and N is the population size). This number is practically too large even if the population is finite (100^{20} in the above example). We can, however, approximate this distribution by drawing a large number, M, of such samples. In Fig. 5.1 we present the histogram of the sampling distribution of $\bar{X}_n$, for $M = 1,000$ random samples with replacement of size $n = 20$, from the population $\{1, 2, \cdots, 100\}$ of the previous example.

This can be effectively done in Python with the function `compute_bootci` from the `pingouin` package. `compute_bootci` takes the population and a function to calculate a statistic over a sample of size n. The method returns confidence intervals for the statistics and optionally the all resampled values of th statistics.

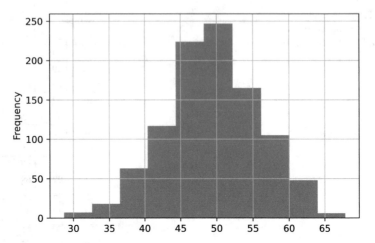

Fig. 5.1 Histogram of 1,000 sample means

```
np.random.seed(1)

X = list(range(100))

# compute_bootci creates samples of the same size as the population
# as we are interested in a smaller sample of size 20, we ignore the
# remaining values when we calculate the mean
def stat_func(sample):
    return np.mean(sample[:20])

B = pg.compute_bootci(X, func=stat_func, n_boot=1000,
                      return_dist=True, seed=1)
print('Mean values of first three mean values: ', B[1][:3])
pd.cut(B[1], bins=12).value_counts()
```

```
Mean values of first three mean values:  [51.8  39.25 45.85]

(28.611, 31.921]      5
(31.921, 35.192]      9
(35.192, 38.462]     30
(38.462, 41.733]     79
(41.733, 45.004]    115
(45.004, 48.275]    191
(48.275, 51.546]    200
(51.546, 54.817]    162
(54.817, 58.088]    108
(58.088, 61.358]     71
(61.358, 64.629]     26
(64.629, 67.9]        4
dtype: int64
```

This frequency distribution is an approximation to the sampling distribution of $\bar{X}_{20}$. It is interesting to notice that this distribution has mean $\bar{\bar{X}} = 50.42$ and standard deviation $\bar{S} = 6.412$. $\bar{\bar{X}}$ is quite close to the population mean 50.5, and $\bar{S}$ is approximately $\sigma/\sqrt{20}$, where σ is the population standard deviation. A proof of this is given in the following section.

Our computer sampling procedure provided a very close estimate of this standard error. Very often we are interested in properties of statistics for which it is difficult to derive formulae for their standard errors. Computer sampling techniques, like bootstrapping discussed in Chap. 3, provide good approximations to the standard errors of sample statistics.

5.2   Estimation with Simple Random Samples

In the present section, we investigate the properties of estimators of the population quantities when sampling is simple random.

The probability structure for simple random samples with or without replacements, RSWR and RSWOR, was studied in Sect. 2.1.4.

Let $X_1, \cdots, X_n$ denote the values of the variable $X(u)$ of the n elements in the random samples. The marginal distributions of X_i ($i = 1, \cdots, n$) if the sample is random, with or without replacement, is the distribution $\hat{F}_N(x)$. If the sample is random **with** replacement, then $X_1, \cdots, X_n$ are **independent**. If the sample is random **without** replacement, then $X_1, \cdots, X_n$ are **correlated** (dependent).

For an estimator of μ_N, we use the sample mean

$$\bar{X}_n = \frac{1}{n} \sum_{j=1}^n X_j.$$

For an estimator of σ_N^2, we use the sample variance

$$S_n^2 = \frac{1}{n-1} \sum_{j=1}^n (X_j - \bar{X}_n)^2.$$

Both estimators are random variables, which may change their values from one sample to another.

An estimator is called **unbiased** if its expected value is equal to the population value of the quantity it estimates. The **precision** of an estimator is the inverse of its sampling variance.

Example 5.2 We illustrate the above with the following numerical example. The population is of size $N = 100$. For simplicity we take $X(u_i) = i$ ($i = 1, \cdots, 100$). For this simple population, $\mu_{100} = 50.5$ and $\sigma_{100}^2 = 833.25$.

Draw from this population 100 independent samples, of size $n = 10$, **with** and **without** replacement.

This can be done in Python as follows.

Table 5.2 Statistics of
sampling distributions

Estimate	RSWR		RSWOR	
	Mean	Std.	Mean	Std.
$\bar{X}_{10}$	51.6	10.6838	50.44	8.3876
S^2_{10}	802.65	243.01	828.04	252.1

```
random.seed(2)
population = list(range(1, 101))

# create samples of size 20 and collect mean and standard deviation
rswr = {'mean': [], 'var': []}
rswor = {'mean': [], 'var': []}
for _ in range(100):
    sample = np.array(random.choices(population, k=10))
    rswr['mean'].append(sample.mean())
    rswr['var'].append(sample.var(ddof=1))

    sample = np.array(random.sample(population, k=10))
    rswor['mean'].append(sample.mean())
    rswor['var'].append(sample.var(ddof=1))

# calculate mean and standard deviation of sample estimates
from collections import namedtuple
SampleStats = namedtuple('SampleStats', 'X10,S2_10')
def calcStats(values):
    return SampleStats(np.mean(values), np.std(values, ddof=1))
rswr['mean'] = calcStats(rswr['mean'])
rswr['var'] = calcStats(rswr['var'])
rswor['mean'] = calcStats(rswor['mean'])
rswor['var'] = calcStats(rswor['var'])
```

The means and standard deviations (Std.) of the 100 sample estimates are summarized in Table 5.2.

As will be shown in the following section, the theoretical expected value of $\bar{X}_{10}$, both in RSWR and RSWOR, is $\mu = 50.5$. We see above that the means of the sample estimates are close to the value of μ. The theoretical standard deviation of $\bar{X}_{10}$ is 9.128 for RSWR and 8.703 for RSWOR. The empirical standard deviations are also close to these values. The empirical means of S^2_{10} are somewhat lower than their expected values of 833.25 and 841.67, for RSWR and RSWOR, respectively. But, as will be shown later, they are not significantly smaller than σ^2. ∎

5.2.1 Properties of $\bar{X}_n$ and S^2_n Under RSWR

If the sampling is RSWR, the random variables $X_1, \cdots, X_n$ are independent, having the same c.d.f. $\hat{F}_N(x)$. The corresponding p.d.f. is

$$p_N(x) = \begin{cases} \dfrac{1}{N}, & \text{if } x = x_j, j = 1, \cdots, N \\[2mm] 0, & \text{otherwise.} \end{cases} \tag{5.3}$$

Accordingly,

$$E\{X_j\} = \frac{1}{N} \sum_{j=1}^{N} x_j = \mu_N, \quad \text{all } j = 1, \cdots, N. \tag{5.4}$$

It follows from the results of Sect. 2.8 that

$$E\{\bar{X}_n\} = \frac{1}{n} \sum_{j=1}^{n} E\{X_j\} \tag{5.5}$$

$$= \mu_N.$$

Thus, **the sample mean is an unbiased estimator of the population mean**.

The variance of X_j is the variance associated with $F_N(x)$, i.e.,

$$V\{X_j\} = \frac{1}{N} \sum_{j=1}^{N} x_j^2 - \mu_N^2$$

$$= \frac{1}{N} \sum_{j=1}^{N} (x_j - \mu_N)^2$$

$$= \sigma_N^2.$$

Moreover, since $X_1, X_2, \cdots, X_n$ are i.i.d,

$$V\{\bar{X}_n\} = \frac{\sigma_N^2}{n}. \tag{5.6}$$

Thus, as explained in Sect. 2.8, the sample mean converges in probability to the population mean, as $n \to \infty$. An estimator having such a property is called **consistent**.

We show now that S_n^2 is an unbiased estimator of σ_N^2.

Indeed, if we write

$$S_n^2 = \frac{1}{n-1} \sum_{j=1}^{n} (X_j - \bar{X}_n)^2$$

$$= \frac{1}{n-1} \left(\sum_{j=1}^{n} X_j^2 - n\bar{X}_n^2 \right).$$

we obtain

$$E\{S_n^2\} = \frac{1}{n-1} \left(\sum_{j=1}^{n} E\{X_j^2\} - nE\{\bar{X}_n^2\} \right).$$

Moreover, since $X_1, \cdots, X_n$ are i.i.d.,

$$E\{X_j^2\} = \sigma_N^2 + \mu_n^2, \quad j = 1, \cdots, n$$

and

$$E\{\bar{X}_n^2\} = \frac{\sigma_N^2}{n} + \mu_N^2.$$

Substituting these in the expression for $E\{S_n^2\}$, we obtain

$$E\{S_n^2\} = \frac{1}{n-1} \left(n(\sigma_N^2 + \mu_N^2) - n \left(\frac{\sigma_N^2}{n} + \mu_N^2 \right) \right) \tag{5.7}$$

$$= \sigma_N^2.$$

An estimator of the standard error of $\bar{X}_n$ is $\dfrac{S_n}{\sqrt{n}}$. This estimator is slightly biased.

In large samples, the distribution of $\bar{X}_n$ is approximately normal, like $N\left(\mu_N, \dfrac{\sigma_N^2}{n}\right)$, as implied by the CLT. Therefore, the interval

$$\left(\bar{X}_n - z_{1-\alpha/2} \frac{S_n}{\sqrt{n}}, \bar{X}_n + z_{1-\alpha/2} \frac{S_n}{\sqrt{n}} \right)$$

has in large samples the property that

$$\Pr\left\{ \bar{X}_n - z_{1-\alpha/2} \frac{S_n}{\sqrt{n}} < \mu_N < \bar{X}_n + z_{1-\alpha/2} \frac{S_n}{\sqrt{n}} \right\} \cong 1 - \alpha.$$

An interval having this property is called a **confidence interval** for μ_N, with an approximate confidence level $(1 - \alpha)$. In the above formula, $z_{1-\alpha/2} = \Phi^{-1}\left(1 - \dfrac{\alpha}{2}\right)$.

It is considerably more complicated to derive the formula for $V\{S_n^2\}$. An approximation for large samples is

$$V\{S_n^2\} \cong \frac{\mu_{4,N} - (\sigma_N^2)^2}{n} + \frac{2(\sigma_N)^2 - \mu_{3,N}}{n^2} + \frac{\mu_{4,N} - 3(\sigma_N^2)^2}{n^3} \tag{5.8}$$

where

$$\mu_{3,N} = \frac{1}{N}\sum_{j=1}^{N}(x_j - \mu_N)^3,   \tag{5.9}$$

and

$$\mu_{4,N} = \frac{1}{N}\sum_{j=1}^{N}(x_j - \mu_N)^4.   \tag{5.10}$$

Example 5.3 In file **PLACE.csv** we have data on x-, y-, and θ-deviations of $N = 416$ placements of components by automatic insertion in 26 PCBs.

Let us consider this record as a finite population. Suppose that we are interested in the population quantities of the variable x-dev. Using Python we find that the population mean, variance, and third and fourth central moments are

$$\mu_N = 0.9124$$

$$\sigma_N^2 = 2.91999$$

$$\mu_{3,N} = -0.98326$$

$$\mu_{4,N} = 14.655.$$

The unit of measurements of the x-dev is 10^{-3} [Inch].

```
place = mistat.load_data('PLACE')
xDev = place['xDev'] / 1e-3
N = len(xDev)
mu_N = xDev.mean()
sigma2_N = xDev.var(ddof=0)
mu_3N = np.sum((xDev - mu_N) ** 3) / N
mu_4N = np.sum((xDev - mu_N) ** 4) / N
print(mu_N.round(4))
print(sigma2_N.round(5))
print(mu_3N.round(5))
print(mu_4N.round(3))
```

```
0.9124
2.91992
-0.98326
14.655
```

Thus, if we draw a simple RSWR, of size $n = 50$, the variance of $\bar{X}_n$ will be $V\{\bar{X}_{50}\} = \dfrac{\sigma_N^2}{50} = 0.0584$. The variance of S_{50}^2 will be

$$V\{S_{50}^2\} \cong \frac{14.655 - (2.9199)^2}{50} + \frac{2(2.9199)^2 + 0.9833}{2500} + \frac{14.655 - 3(2.9199)^2}{125000}$$

$$= 0.1297.$$ ∎

5.2.2 Properties of $\bar{X}_n$ and S_n^2 Under RSWOR

We show first that $\bar{X}_n$ is an unbiased estimator of μ_N, under RSWOR.

Let I_j be an indicator variable, which assumes the value 1 if u_j belongs to the selected sample, s_n, and equal to zero otherwise. Then we can write

$$\bar{X}_n = \frac{1}{n} \sum_{j=1}^{N} I_j x_j. \tag{5.11}$$

Accordingly

$$E\{\bar{X}_n\} = \frac{1}{n} \sum_{j=1}^{N} x_j E\{I_j\}$$

$$= \frac{1}{n} \sum_{j=1}^{N} x_j \Pr\{I_j = 1\}.$$

As shown in Sect. 2.1.4,

$$\Pr\{I_j = 1\} = \frac{n}{N}, \quad \text{all } j = 1, \cdots, N.$$

Substituting this above yields that

$$E\{\bar{X}_n\} = \mu_N. \tag{5.12}$$

It is shown below that

$$V\{\bar{X}_n\} = \frac{\sigma_N^2}{n} \left(1 - \frac{n-1}{N-1} \right). \tag{5.13}$$

To derive the formula for the variance of $\bar{X}_n$, under RSWOR, we use the result of Sect. 2.8 on the variance of linear combinations of random variables. Write first,

$$V\{\bar{X}_n\} = V\left\{ \frac{1}{n} \sum_{i=1}^{N} x_i I_i \right\}$$

$$= \frac{1}{n^2} V\left\{ \sum_{i=1}^{N} x_i I_i \right\}.$$

$\sum_{i=1}^{N} x_i I_i$ is a linear combination of the random variables $I_1, \cdots, I_N$.
First we show that

$$V\{I_i\} = \frac{n}{N}\left(1 - \frac{n}{N}\right), \quad i = 1, \cdots, N.$$

Indeed, since $I_i^2 = I_i$,

$$\begin{aligned}
V\{I_i\} &= E\{I_i^2\} - (E\{I_i\})^2 \\
&= E\{I_i\}(1 - E\{I_i\}) \\
&= \frac{n}{N}\left(1 - \frac{n}{N}\right), \quad i = 1, \cdots, N.
\end{aligned}$$

Moreover, for $i \neq j$,

$$\text{Cov}(I_i, I_j) = E\{I_i I_j\} - E\{I_i\}E\{I_j\}.$$

But,

$$\begin{aligned}
E\{I_i I_j\} &= \Pr\{I_i = 1, I_j = 1\} \\
&= \frac{n(n-1)}{N(N-1)}.
\end{aligned}$$

Hence, for $i \neq j$,

$$\text{Cov}(I_i, I_j) = -\frac{n}{N^2} \cdot \frac{N-n}{N-1}.$$

Finally,

$$V\left\{\sum_{i=1}^{N} x_i I_i\right\} = \sum_{i=1}^{N} x_i^2 V\{I_i\} + \sum\sum_{i\neq j} x_i x_j \text{cov}(X_i, X_j).$$

Substituting these expressions in

$$V\{\bar{X}_n\} = \frac{1}{n^2} V\left\{\sum_{i=1}^{N} x_i I_i\right\},$$

we obtain

$$V\{\bar{X}_n\} = \frac{1}{n^2}\left\{\frac{n}{N}\left(1 - \frac{n}{N}\right)\sum_{i=1}^{N} x_i^2 - \frac{n(N-n)}{N^2(N-1)}\sum\sum_{i\neq j} x_i x_j\right\}.$$

But, $\sum\sum_{i\neq j} x_i x_j = \left(\sum_{i=1}^{N} x_i\right)^2 - \sum_{i=1}^{N} x_i^2$. Hence,

$$V\{\bar{X}_n\} = \frac{N-n}{nN^2}\left\{\frac{N}{N-1}\sum_{i=1}^{N}x_i^2 - \frac{1}{N-1}\left(\sum_{i=1}^{N}x_i\right)^2\right\}$$

$$= \frac{N-n}{n\cdot(N-1)\cdot N}\sum_{i=1}^{N}(x_i-\mu_N)^2$$

$$= \frac{\sigma_N^2}{n}\left(1-\frac{n-1}{N-1}\right).$$

We see that the variance of $\bar{X}_n$ is smaller under RSWOR than under RSWR, by a factor of $\left(1-\dfrac{n-1}{N-1}\right)$. This factor is called the **finite population multiplier**.

The formula we have in Sect. 2.3.2 for the variance of the hypergeometric distribution can be obtained from the above formula. In the hypergeometric model, we have a finite population of size N. M elements have a certain attribute. Let

$$x_i = \begin{cases} 1, & \text{if } w_i \text{ has the attribute} \\ \\ 0, & \text{if } w_i \text{ does not have it.} \end{cases}$$

Since $\sum_{i=1}^{N}x_i = M$ and $x_i^2 = x_i$,

$$\sigma_N^2 = \frac{M}{N}\left(1-\frac{M}{N}\right).$$

If $J_n = \sum_{i=1}^{n}X_i$, we have

$$V\{J_n\} = n^2 V\{\bar{X}_n\}$$

$$= n\frac{M}{N}\left(1-\frac{M}{N}\right)\left(1-\frac{n-1}{N-1}\right). \tag{5.14}$$

To estimate σ_N^2 we can again use the sample variance S_n^2. The sample variance has, however, a slight positive bias. Indeed,

$$E\{S_n^2\} = \frac{1}{n-1}E\left\{\sum_{j=1}^{n}X_j^2 - n\bar{X}_n^2\right\}$$

$$= \frac{1}{n-1}\left(n(\sigma_N^2+\mu_N^2) - n\left(\mu_N^2 + \frac{\sigma^2}{n}\left(1-\frac{n-1}{N-1}\right)\right)\right)$$

$$= \sigma_N^2 \left(1 + \frac{1}{N-1} \right).$$

This bias is negligible if σ_N^2/N is small. Thus, the standard error of $\bar{X}_n$ can be estimated by

$$\text{S.E.}\{\bar{X}_n\} = \frac{S_n}{\sqrt{n}} \left(1 - \frac{n-1}{N-1} \right)^{1/2}. \tag{5.15}$$

When sampling is RSWOR, the random variables $X_1, \cdots, X_n$ are **not independent**, and we cannot justify theoretically the usage of the normal approximation to the sampling distribution of $\bar{X}_n$. However, if n/N is small, the normal approximation is expected to yield good results. Thus, if $\frac{n}{N} < 0.1$, we can approximate the confidence interval, of level $(1 - \alpha)$, for μ_N, by the interval with limits

$$\bar{X}_n \pm z_{1-\alpha/2} \cdot \text{S.E.}\{\bar{X}_n\}.$$

In order to estimate the coverage probability of this interval estimator, when $\frac{n}{N} = 0.3$, we perform the following simulation example.

Example 5.4 We can use Python to select RSWOR of size $n = 30$ from the population $P = \{1, 2, \cdots, 100\}$ of $N = 100$ units, whose values are $x_i = i$.

For this purpose we initialize variable X with the integers $1, \cdots, 100$. Notice that, when $n = 30$, $N = 100$, $\alpha = 0.05$, $z_{1-\alpha/2} = 1.96$, and

$$\frac{1.96}{\sqrt{n}} \left(1 - \frac{n-1}{N-1} \right)^{1/2} = 0.301.$$

```
random.seed(1)
X = list(range(1, 101))

def confInt(x, p, N):
    if p >= 0.5:
        p = 1 - (1 - p) / 2
    else:
        p = 1 - p / 2

    n = len(x)
    z = stats.norm.ppf(p) * np.sqrt(1 - (n-1)/(N-1)) / np.sqrt(n)
    m = np.mean(x)
    s = np.std(x, ddof=1)
    return (m - z * s, m + z * s)

sampled_confInt = []
for _ in range(1000):
    sample = random.sample(X, k=30)
    sampled_confInt.append(confInt(sample, p=0.95, N=100))

# show the first three results
print(sampled_confInt[:3])
```

```
# calculate the ratio of cases where the actual mean of 50.5
# is inside the sample confidence intervals
proportion_coverage = sum(ci[0] < 50.5 < ci[1] for ci in sampled_confInt)
proportion_coverage = proportion_coverage / len(sampled_confInt)
print(proportion_coverage)
```

```
[(36.85179487364511, 55.14820512635489), (44.070802021276585,
60.795864645390076), (43.237002670283765, 59.56299732971623)]
0.943
```

The true population mean is $\mu_N = 50.5$. The estimated coverage probability is the proportion of cases for which $k_1 \leq \mu_N \leq k_2$. In the present simulation, the proportion of coverage is 0.943. The nominal confidence level is $1 - \alpha = 0.95$. The estimated coverage probability is 0.943. Thus, the present example shows that even in cases where $n/N > 0.1$, the approximate confidence limits are quite effective. ∎

5.3 Estimating the Mean with Stratified RSWOR

We consider now the problem of estimating the population mean, μ_N, with stratified RSWOR. Thus, suppose that the population P is partitioned into k strata (subpopulations) $P_1, P_2, \cdots, P_k, k \geq 2$.

Let $N_1, N_2, \cdots, N_k$ denote the sizes; $\mu_{N_1}, \cdots, \mu_{N_k}$ the means; and $\sigma_{N_1}^2, \cdots, \sigma_{N_k}^2$ the variances of these strata, respectively. Notice that the population mean is

$$\mu_N = \frac{1}{N} \sum_{i=1}^{k} N_i \mu_{N_i}, \tag{5.16}$$

and according to the formula of total variance (see Eq. (2.92) in Sect. 2.5.3), the population variance is

$$\sigma_N^2 = \frac{1}{N} \sum_{i=1}^{k} N_i \sigma_{N_i}^2 + \frac{1}{N} \sum_{i=1}^{k} N_i (\mu_{N_i} - \mu_N)^2. \tag{5.17}$$

We see that if the means of the strata are not the same, the population variance is greater than the weighted average of the within strata variances, $\sigma_{N_i}^2$ $(i = 1, \cdots, k)$.

A stratified RSWOR is a sampling procedure in which k-independent random samples without replacement are drawn from the strata. Let n_i, $\bar{X}_{n_i}$, and $S_{n_i}^2$ be the size, mean, and variance of the RSWOR from the ith stratum, P_i $(i = 1, \cdots, k)$.

We have shown in the previous section that $\bar{X}_{n_i}$ is an unbiased estimator of μ_{N_i}. Thus, an unbiased estimator of μ_N is the weighted average

$$\hat{\mu}_N = \sum_{i=1}^{k} W_i \bar{X}_{n_i}, \tag{5.18}$$

where $W_i = \dfrac{N_i}{N}$, $i = 1, \cdots, k$. Indeed,

$$E\{\hat{\mu}_N\} = \sum_{i=1}^{k} W_i E\{\bar{X}_{n_i}\}$$

$$= \sum_{i=1}^{k} W_i \mu_{N_i} \qquad (5.19)$$

$$= \mu_N.$$

Since $\bar{X}_{n_1}, \bar{X}_{n_2}, \cdots, \bar{X}_{n_k}$ are independent random variables, the variance of $\hat{\mu}_N$ is

$$V\{\hat{\mu}_N\} = \sum_{i=1}^{k} W_i^2 V\{\bar{X}_{n_i}\}$$

$$= \sum_{i=1}^{k} W_i^2 \frac{\sigma_{n_i}^2}{n_i} \left(1 - \frac{n_i - 1}{N_i - 1}\right) \qquad (5.20)$$

$$= \sum_{i=1}^{k} W_i^2 \frac{\tilde{\sigma}_{N_i}^2}{n_i} \left(1 - \frac{n_i}{N_i}\right),$$

where

$$\tilde{\sigma}_{N_i}^2 = \frac{N_i}{N_i - 1} \sigma_{N_i}^2.$$

Example 5.5 Returning to the data of Example 5.3, on deviations in the x-direction of automatically inserted components, the units are partitioned to $k = 3$ strata: boards 1–10 in stratum 1, boards 11–13 in stratum 2, and boards 14–26 in stratum 3. The population characteristics of these strata are:

Stratum	Size	Mean	Variance
1	160	-0.966	0.4189
2	48	0.714	1.0161
3	208	2.403	0.3483

The relative sizes of the strata are $W_1 = .385$, $W_2 = .115$, and $W_3 = 0.5$. If we select a stratified RSWOR of sizes $n_1 = 19$, $n_2 = 6$, and $n_3 = 25$, the variance of $\hat{\mu}_N$ will be

$$V\{\hat{\mu}_N\} = (0.385)^2 \frac{0.4189}{19}\left(1 - \frac{18}{159}\right) + (0.115)^2 \frac{1.0161}{6}\left(1 - \frac{5}{47}\right)$$

$$+ (0.5)^2 \frac{0.3483}{25}\left(1 - \frac{24}{207}\right)$$

$$= 0.00798.$$

This variance is considerably smaller than the variance of $\bar{X}_{50}$ in a simple RSWOR, which is

$$V\{\bar{X}_{50}\} = \frac{2.9199}{50}\left(1 - \frac{49}{415}\right)$$

$$= 0.0515.$$

∎

5.4 Proportional and Optimal Allocation

An important question in designing the stratified RSWOR is how to allocate the total number of observations, n, to the different strata, i.e., the determination of $n_i \geq 0$ ($i = 1, \cdots, k$) so that $\sum_{i=1}^{k} n_i = n$, for a given n. This is called the **sample allocation**. One type of sample allocation is the so-called proportional allocation, i.e.,

$$n_i = nW_i, \quad i = 1, \cdots, k. \tag{5.21}$$

The variance of the estimator $\hat{\mu}_N$ under proportional allocation is

$$V_{\text{prop}}\{\hat{\mu}_N\} = \frac{1}{n}\sum_{i=1}^{k} W_i \bar{\sigma}_{N_i}^2 \left(1 - \frac{n}{N}\right)$$

$$= \frac{\bar{\sigma}_N^2}{n}\left(1 - \frac{n}{N}\right), \tag{5.22}$$

where

$$\bar{\sigma}_N^2 = \sum_{i=1}^{k} W_i \bar{\sigma}_{N_i}^2,$$

is the weighted average of the within strata variances.

We have shown in the previous section that if we take a simple RSWOR, the variance of $\bar{X}_n$ is

$$V_{\text{simple}}\{\bar{X}_n\} = \frac{\sigma_N^2}{n}\left(1 - \frac{n-1}{N-1}\right)$$

$$= \frac{\tilde{\sigma}_N^2}{n}\left(1 - \frac{n}{N}\right),$$

where

$$\tilde{\sigma}_N^2 = \frac{N}{N-1}\sigma_N^2.$$

In large-sized populations, σ_N^2 and $\tilde{\sigma}_N^2$ are very close, and we can write

$$V_{\text{simple}}\{\bar{X}_n\} \cong \frac{\sigma_N^2}{N}\left(1 - \frac{n}{N}\right)$$

$$= \frac{1}{n}\left(1 - \frac{n}{N}\right)\left\{\sum_{i=1}^{k} W_i \sigma_{N_i}^2 + \sum_{i=1}^{k} W_i(\mu_{N_i} - \mu_N)^2\right\}$$

$$\cong V_{\text{prop}}\{\hat{\mu}_N\} + \frac{1}{n}\left(1 - \frac{n}{N}\right)\sum_{i=1}^{k} W_i(\mu_{N_i} - \mu_N)^2.$$

This shows that $V_{\text{simple}}\{\bar{X}_n\} > V_{\text{prop}}\{\hat{\mu}_N\}$; i.e., the estimator of the population mean, μ_N, under stratified RSWOR, with proportional allocation, generally has smaller variance (more precise) than the estimator under a simple RSWOR. The difference grows with the variance between the strata means, $\sum_{i=1}^{k} W_i(\mu_{N_i} - \mu_N)^2$. Thus effective stratification is one which partitions the population to strata which are homogeneous within (small values of $\sigma_{N_i}^2$) and heterogeneous between (large value of $\sum_{i=1}^{k} W_i(\mu_{N_i} - \mu_N)^2$). If sampling is stratified RSWR, then the variance $\hat{\mu}_N$, under proportional allocation, is

$$V_{\text{prop}}\{\hat{\mu}_N\} = \frac{1}{n}\sum_{i=1}^{k} W_i \sigma_{N_i}^2. \tag{5.23}$$

This is strictly smaller than the variance of $\bar{X}_n$ in a simple RSWR. Indeed

$$V_{\text{simple}}\{\bar{X}_n\} = \frac{\sigma_N^2}{n}$$

$$= V_{\text{prop}}\{\hat{\mu}_N\} + \frac{1}{n}\sum_{i=1}^{k} W_i(\mu_{N_i} - \mu_N)^2.$$

Example 5.6 Defective circuit breakers are a serious hazard since their function is to protect electronic systems from power surges or power drops. Variability in power supply voltage levels can cause major damage to electronic systems. Circuit breakers are used to shield electronic systems from such events. The proportion of potentially defective circuit breakers is a key parameter in designing redundancy levels of protection devices and preventive maintenance programs. A lot of $N = 10,000$ circuit breakers was put together by purchasing the products from $k = 3$ different vendors. We want to estimate the proportion of defective breakers, by sampling and testing $n = 500$ breakers. Stratifying the lot by vendor, we have three strata of sizes $N_1 = 3,000$; $N_2 = 5,000$; and $N_3 = 2,000$. Before installing the circuit breakers, we draw from the lot a stratified RSWOR, with proportional allocation, i.e., $n_1 = 150$, $n_2 = 250$, and $n_3 = 100$. After testing we find in the first sample $J_1 = 3$ defective circuit breakers, in the second sample $J_2 = 10$, and in the third sample $J_3 = 2$ defectives. Testing is done with a special purpose device, simulating intensive usage of the product.

In the present case, we set $X = 1$ if the item is defective and $X = 0$ otherwise. Then μ_N is the proportion of defective items in the lot. μ_{N_i} ($i = 1, 2, 3$) is the proportion defectives in the ith stratum.

The unbiased estimator of μ_N is

$$\hat{\mu}_N = 0.3 \times \frac{J_1}{150} + 0.5 \times \frac{J_2}{250} + 0.2 \times \frac{J_3}{100}$$

$$= 0.03.$$

The variance within each stratum is $\sigma_{N_i}^2 = P_{N_i}(1 - P_{N_i})$, $i = 1, 2, 3$, where P_{N_i} is the proportion in the ith stratum. Thus, the variance of $\hat{\mu}_N$ is

$$V_{\text{prop}}\{\hat{\mu}_N\} = \frac{1}{500}\bar{\sigma}_N^2 \left(1 - \frac{500}{10,000}\right)$$

where

$$\bar{\sigma}_N^2 = 0.3\tilde{\sigma}_{N_1}^2 + 0.5\tilde{\sigma}_{N_2}^2 + 0.2\tilde{\sigma}_{N_3}^2,$$

or

$$\bar{\sigma}_N^2 = 0.3 \times \frac{3000}{2999} P_{N_1}(1 - P_{N_1}) + 0.5\frac{5000}{4999} P_{N_2}(1 - P_{N_2}) + 0.2\frac{2000}{1999} P_{N_3}(1 - P_{N_3}).$$

Substituting $\frac{3}{150}$ for an estimate of P_{N_1}, $\frac{10}{250}$ for that of P_{N_2}, and $\frac{2}{100}$ for P_{N_3}, we obtain the estimate of $\bar{\sigma}_N^2$,

$$\bar{\sigma}_N^2 = 0.029008.$$

Finally, an estimate of $V_{\text{prop}}\{\hat{\mu}_N\}$ is

$$\hat{V}_{\text{prop}}\{\hat{\mu}_N\} = \frac{0.029008}{500}\left(1 - \frac{500}{10,000}\right)$$

$$= 0.00005511.$$

The standard error of the estimator is 0.00742.
  Confidence limits for μ_N, at level $1 - \alpha = .95$, are given by

$$\hat{\mu}_N \pm 1.96 \times \text{S.E.}\{\hat{\mu}_N\} = \begin{cases} 0.0446 \\ \\ 0.0154. \end{cases}$$

These limits can be used for spare parts policy.   ∎

  When the variances $\tilde{\sigma}_N^2$ within strata are known, we can further reduce the variance of μ_N by an allocation, which is called **optimal allocation**.
  We wish to minimize

$$\sum_{i=1}^{k} W_i^2 \frac{\tilde{\sigma}_{N_i}^2}{n_i}\left(1 - \frac{n_i}{N_i}\right)$$

subject to the constraint:

$$n_1 + n_2 + \cdots + n_k = n.$$

This can be done by minimizing

$$L(n_1, \cdots, n_k, \lambda) = \sum_{i=1}^{k} W_i^2 \frac{\tilde{\sigma}_{N_i}^2}{n_i} - \lambda\left(n - \sum_{i=1}^{k} n_i\right),$$

with respect to $n_1, \cdots, n_k$ and λ. This function is called the **Lagrangian**, and λ is called the **Lagrange multiplier**.
  The result is

$$n_i^0 = n\frac{W_i \tilde{\sigma}_{N_i}}{\sum_{j=1}^{k} W_j \tilde{\sigma}_j}, \quad i = 1, \cdots, k. \tag{5.24}$$

We see that the proportional allocation is optimal when all $\tilde{\sigma}_{N_i}^2$ are equal.
  The variance of $\hat{\mu}_N$, corresponding to the optimal allocation, is

$$V_{\text{opt}}\{\hat{\mu}_N\} = \frac{1}{N}\left(\sum_{i=1}^{k} W_i \tilde{\sigma}_{N_i}\right)^2 - \frac{1}{N}\sum_{i=1}^{k} W_i \tilde{\sigma}_{N_i}^2. \qquad (5.25)$$

5.5   Prediction Models with Known Covariates

In some problems of estimating the mean μ_N of a variable Y in a finite population, we may have information on variables $X_1, X_2, \cdots, X_k$ which are related to Y. The variables $X_1, \cdots, X_k$ are called **covariates**. The model relating Y to $X_1, \cdots, X_k$ is called a **prediction model**. If the values of Y are known only for the units in the sample, while the values of the covariates are known for all the units of the population, we can utilize the prediction model to improve the precision of the estimator. The method can be useful, for example, when the measurements of Y are destructive, while the covariates can be measured without destroying the units. There are many such examples, like the case of measuring the compressive strength of a concrete cube. The measurement is destructive. The compressive strength Y is related to the ratio of cement to water in the mix, which is a covariate that can be known for all units. We will develop the ideas with a simple prediction model.

Let $\{u_1, u_2, \cdots, u_N\}$ be a finite population, P. The values of $x_i = X(u_i)$, $i = 1, \cdots, N$ are known for all the units of P. Suppose that $Y(u_i)$ is related linearly to $X(u_i)$ according to the prediction model

$$y_i = \beta x_i + e_i, \quad i = 1, \cdots, N, \qquad (5.26)$$

where β is an unknown regression coefficient and $e_1, \cdots, e_N$ are i.i.d. random variables such that

$$E\{e_i\} = 0, \quad i = 1, \cdots, N$$

$$V\{e_i\} = \sigma^2, \quad i = 1, \cdots, N.$$

The random variable e_i in the prediction model is due to the fact that the linear relationship between Y and X is not perfect, but subject to random deviations.

We are interested in the population quantity $\bar{y}_N = \frac{1}{N}\sum_{i=1}^{N} y_i$. We cannot however measure all the Y values. Even if we know the regression coefficient β, we can only predict $\bar{y}_N$ by $\beta\bar{x}_N$, where $\bar{x}_N = \frac{1}{N}\sum_{j=1}^{N} x_j$. Indeed, according to the prediction model, $\bar{y}_N = \beta\bar{x}_N + \bar{e}_N$, and $\bar{e}_N$ is a random variable with

$$E\{\bar{e}_N\} = 0, \quad V\{\bar{e}_N\} = \frac{\sigma^2}{N}. \qquad (5.27)$$

Thus, since $\bar{y}_N$ has a random component, and since $E\{\bar{y}_N\} = \beta\bar{x}_N$, we say that a predictor of $\bar{y}_N$, say $\hat{y}_N$, is **unbiased**, if $E\{\hat{Y}_N\} = \beta\bar{x}_N$. Generally, β is unknown. Thus, we draw a sample of units from P and measure their Y values, in order to estimate β. For estimating β we draw a simple RSWOR from P of size n, $1 < n < N$.

Let $(X_1, Y_1), \cdots, (X_n, Y_n)$ be the values of X and Y in the random sample. A predictor of $\bar{y}_N$ is some function of the observed sample values. Notice that after drawing a random sample we have **two** sources of variability. One due to the random error components $e_1, \cdots, e_n$, associated with the sample values, and the other one is due to the random sampling of the n units of P. Notice that the error variables $e_1, \cdots, e_n$ are independent of the X values and thus independent of $X_1, X_2, \cdots, X_n$, randomly chosen to the sample. In the following, expectation and variances are taken with respect to the errors model and with respect to the sampling procedure. We will examine now a few alternative predictors of $\bar{y}_N$:

(i) **The sample mean**, $\bar{Y}_n$.

Since

$$\bar{Y}_n = \beta\bar{X}_n + \bar{e}_n,$$

we obtain that

$$E\{\bar{Y}_n\} = \beta E\{\bar{X}_n\} + E\{\bar{e}_n\}$$

$E\{\bar{e}_n\} = 0$ and since the sampling is RSWOR, $E\{\bar{X}_n\} = \bar{x}_N$. Thus $E\{\bar{Y}_n\} = \beta\bar{x}_N$, and the predictor is **unbiased**. The variance of the predictor is, since $\bar{e}_n$ is independent of $\bar{X}_n$,

$$V\{\bar{Y}_n\} = \beta^2 V\{\bar{X}_n\} + \frac{\sigma^2}{n}$$

$$= \frac{\sigma^2}{n} + \frac{\beta^2\sigma_x^2}{n}\left(1 - \frac{n-1}{N-1}\right). \tag{5.28}$$

where

$$\sigma_x^2 = \frac{1}{N}\sum_{j=1}^{N}(x_j - \bar{x}_N)^2.$$

(ii) **The ratio predictor**,

$$\hat{Y}_R = \bar{x}_N\frac{\bar{Y}_n}{\bar{X}_n}. \tag{5.29}$$

The ratio predictor will be used when all $x_i > 0$. In this case $\bar{X}_n > 0$ in every possible sample. Substituting $\bar{Y}_n = \beta \bar{X}_n + \bar{e}_n$, we obtain

$$E\{\hat{Y}_R\} = \beta \bar{x}_N + \bar{x}_N E \left\{ \frac{\bar{e}_n}{\bar{X}_n} \right\}.$$

Again, since $\bar{e}_n$ and $\bar{X}_n$ are independent, $E \left\{ \dfrac{\bar{e}_n}{\bar{X}_n} \right\} = 0$, and $\hat{Y}_R$ is an **unbiased** predictor. The variance of $\hat{Y}_R$ is

$$V\{\hat{Y}_R\} = (\bar{x}_N)^2 V \left\{ \frac{\bar{e}_n}{\bar{X}_n} \right\}.$$

Since $\bar{e}_n$ and $\bar{X}_n$ are independent, and $E\{\bar{e}_n\} = 0$, the law of the total variance implies that

$$
\begin{aligned}
V\{\hat{Y}_R\} &= \frac{\sigma^2}{n} \bar{x}_N^2 E \left\{ \frac{1}{\bar{X}_n^2} \right\} \\
&= \frac{\sigma^2}{n} E \left\{ \left(1 + \frac{(\bar{X}_n - \bar{x}_N)}{\bar{x}_N} \right)^{-2} \right\} \\
&= \frac{\sigma^2}{n} E \left\{ 1 - \frac{2}{\bar{x}_N}(\bar{X}_n - \bar{x}_N) + \frac{3}{\bar{x}_N^2}(\bar{X}_n - \bar{x}_N)^2 + \cdots \right\} \\
&\cong \frac{\sigma^2}{n} \left(1 + \frac{3\gamma_x^2}{n} \left(1 - \frac{n-1}{N-1} \right) \right)
\end{aligned}
$$
(5.30)

where $\gamma_x = \sigma_x / \bar{x}_N$ is the coefficient of variation of X. The above approximation is effective in large samples.

Using the large sample approximation, we see that the ratio predictor $\hat{Y}_R$ has a smaller variance than $\bar{Y}_n$ if

$$\frac{3\sigma^2 \gamma_x^2}{n^2} \left(1 - \frac{n-1}{N-1} \right) < \frac{\beta^2 \sigma_x^2}{n} \left(1 - \frac{n-1}{N-1} \right)$$

or if

$$n > \frac{3\sigma^2}{(\beta \bar{x}_N)^2}.$$

Other possible predictors for this model are

$$\hat{Y}_{RA} = \bar{x}_N \cdot \frac{1}{N} \sum_{i=1}^{n} \frac{Y_i}{X_i} \tag{5.31}$$

and

$$\hat{Y}_{RG} = \bar{x}_N \cdot \frac{\sum_{i=1}^{n} Y_i X_i}{\sum_{i=1}^{N} X_i^2}. \tag{5.32}$$

We leave it as an exercise to prove that both $\hat{Y}_{RA}$ and $\hat{Y}_{RG}$ are unbiased predictors and to derive their variances.

What happens, under the above prediction model, if the sample drawn is not random, but the units are chosen to the sample by some non-random fashion?

Suppose that a non-random sample $(x_1, y_1), \cdots, (x_n, y_n)$ is chosen. Then

$$E\{\bar{y}_n\} = \beta \bar{x}_n$$

and

$$V\{\bar{y}_n\} = \frac{\sigma^2}{n}.$$

The predictor $\bar{y}_n$ is biased, unless $\bar{x}_n = \bar{x}_N$. A sample which satisfies this property is called a **balanced sample** with respect to X. Generally, the **mean squared error** (MSE) of $\bar{y}_n$, under non-random sampling, is

$$\text{MSE}\{\bar{y}_n\} = E\{(\bar{y}_n - \beta \bar{x}_N)^2\}$$
$$= \frac{\sigma^2}{n} + \beta^2 (\bar{x}_n - \bar{x}_N)^2. \tag{5.33}$$

Thus, if the sample is balanced with respect to X, then $\bar{y}_n$ is a more precise predictor than all the above, which are based on simple random samples.

Example 5.7 Electronic systems such as television sets, radios, or computers contain printed circuit boards with electronic components positioned in patterns determined by design engineers. After assembly (either by automatic insertion machines or manually) the components are soldered to the board. In the relatively new surface-mount technology, minute components are simultaneously positioned and soldered to the boards. The occurrence of defective soldering points impacts the assembly plant productivity and is therefore closely monitored. In file **PRED.csv** we find 1,000 records on variable X and Y. X is the number of soldering points on a board, and Y is the number of defective soldering points. The mean of Y is $\bar{y}_{1000} = 7.495$ and that of X is $\bar{x}_{1000} = 148.58$. Moreover, $\sigma_x^2 = 824.562$ and the coefficient of variation is $\gamma_x = .19326$. The relationship between X and Y is $y_i = \beta x_i + e_i$, where $E\{e_i\} = 0$ and $V\{e_i\} = 7.5$, $\beta = 0.05$. Thus, if we have to

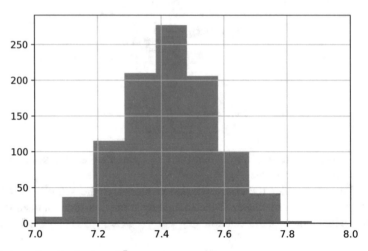

Fig. 5.2 Sampling distribution of $\bar{Y}$

predict $\bar{y}_{1000}$ by a predictor based on a RSWR, of size $n = 100$, the variances of $\bar{Y}_{100}$ and $\hat{Y}_R = \bar{x}_{1000}\dfrac{\bar{Y}_{100}}{\bar{X}_{100}}$ are

$$V\{\bar{Y}_{100}\} = \frac{7.5}{100} + \frac{0.0025 \times 824.562}{100} = 0.0956.$$

On the other hand, the large sample approximation yields

$$V\{\hat{Y}_R\} = \frac{7.5}{100}\left(1 + \frac{3 \times 0.037351}{100}\right)$$

$$= 0.07508.$$

We see that, if we have to predict $\bar{y}_{1000}$ on the basis of an RSWR of size $n = 100$, the ratio predictor, $\hat{Y}_R$, is more precise.   ∎

In Figures 5.2 and 5.3 we present the histograms of 500 predictors $\bar{Y}_{100}$ and 500 $\hat{Y}_R$ based on RSWR of size 100 from this population.

5.6   Chapter Highlights

The main concepts and definitions introduced in this chapter include:

- Population quantiles
- Simple random samples
- Stratified random samples

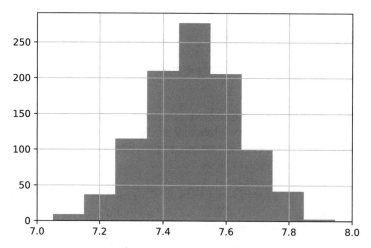

Fig. 5.3 Sampling distribution of $\hat{Y}_R$

- Unbiased estimators
- Precision of an estimator
- Finite population multiplier
- Sample allocation
- Proportional allocation
- Optimal allocation
- Prediction models
- Covariates
- Ratio predictor
- Prediction unbiasedness
- Prediction MSE

5.7 Exercises

Exercise 5.1 Consider a finite population of size N, whose elements have values $x_1, \cdots, x_N$. Let $\hat{F}_N(x)$ be the c.d.f., i.e.,

$$\hat{F}_N(x) = \frac{1}{N} \sum_{i=1}^{N} I\{x_i \leq x\}.$$

Let $X_1, \cdots, X_n$ be the values of a RSWR. Show that $X_1, \cdots, X_n$ are independent having a common distribution $\hat{F}_N(x)$.

Exercise 5.2 Show that if $\bar{X}_n$ is the mean of a RSWR, then $\bar{X}_n \to \mu_N$ as $n \to \infty$ in probability (WLLN).

Exercise 5.3 What is the large sample approximation to $\Pr\{\sqrt{n} \mid \bar{X}_n - \mu_N \mid < \delta\}$ in RSWR?

Exercise 5.4 Use Python to draw random samples with or without replacement from data file **PLACE.csv**. Write a function which computes the sample correlation between the x-dev and y-dev in the sample values. Execute this function 100 times, and make a histogram of the sample correlations.

Exercise 5.5 Use file **CAR.csv** and Python. Construct samples of 50 records at random, without replacement (RSWOR). For each sample, calculate the median of the variables turn diameter, horsepower, and mpg. Repeat this 200 times, and present the histograms of the sampling distributions of the medians.

Exercise 5.6 In continuation of Example 5.5, how large should the sample be from the three strata, so that the SE $\{\bar{X}_i\}$ $(i = 1, \ldots, 3)$ will be smaller than $\delta = 0.05$?

Exercise 5.7 The proportion of defective chips in a lot of $N = 10,000$ chips is $P = 5 \times 10^{-4}$. How large should a RSWOR be so that the width of the confidence interval for P, with coverage probability $1 - \alpha = .95$, will be 0.002?

Exercise 5.8 Use Python to perform stratified random samples from the three strata of the data file **PLACE.csv** (see Example 5.5). Allocate 500 observations to the three samples proportionally. Estimate the population mean (of x-dev). Repeat this 100 times, and estimate the standard error or your estimates. Compare the estimated standard error to the exact one.

Exercise 5.9 Derive the formula for n_i^0 $(i = 1, \cdots, k)$ in the optimal allocation, by differentiating $L(n_1, \cdots, n_k, \lambda)$ and solving the equations.

Exercise 5.10 Consider the prediction model

$$y_i = \beta + e_i, \quad i = 1, \ldots, N$$

where $E\{e_i\} = 0$, $V\{e_i\} = \sigma^2$ and $\mathrm{COV}(e_i, e_j) = 0$ for $i \neq j$. We wish to predict the population mean $\mu_N = \frac{1}{N} \sum_{i=1}^{N} y_i$. Show that the sample mean $\bar{Y}_n$ is prediction unbiased. What is the prediction MSE of $\bar{Y}_n$?

Exercise 5.11 Consider the prediction model

$$y_i = \beta_0 + \beta_1 x_i + e_i, \quad i = 1, \ldots, N,$$

where $e_1, \ldots, e_N$ are independent r.v.s with $E\{e_i\} = 0$, $V\{e_i\} = \sigma^2 x_i$ ($i = 1, \ldots, n$). We wish to predict $\mu_N = \dfrac{1}{N} \sum_{i=1}^{N} y_i$. What should be a good predictor for μ_N?

Exercise 5.12 Prove that $\hat{Y}_{RA}$ and $\hat{Y}_{RG}$ are unbiased predictors and derive their prediction variances.

Chapter 6
Time Series Analysis and Prediction

Preview In this chapter, we present essential parts of time series analysis, with the objective of predicting or forecasting its future development. Predicting future behavior is generally more successful for stationary series, which do not change their stochastic characteristics as time proceeds. We develop and illustrate time series which are of both types, namely, covariance stationary and non-stationary.

We started by fitting a smooth function, showing the trend to a complex time series consisting of the Dow Jones Industrial Average index in the 302 trading days of 1941. We defined then the notions of covariance stationarity of the deviations from the trend function and their lag-correlation and partial correlation. Starting with the simple white noise, we studied the properties of moving averages of white noise, which are covariance stationary. After this we introduced the auto-regressive time series, in which the value of an observed variable at a given time is a linear function of several past values in the series plus a white noise error. A criterion for covariance stationarity of auto regressive series was given in terms of the roots of its characteristic polynomial. We showed also how to express these stationary series as an infinite linear combination of white noise variables. More complex covariance stationary time series, which are combinations of auto-regressive and moving averages, called ARMA series, and integrated ARMA series, called ARIMA, were discussed too.

The second part of the chapter deals with prediction of future values. We started with the optimal linear predictor of covariance stationary time series. These optimal predictors are based on moving windows of the last n observations in the series. We demonstrated that even for the **DOW1941.csv** series, if we apply the optimal linear predictor on windows of size $n = 20$, of the deviations from the trend, and then added the forecasts to the trend values, we obtain very good predictors for the next day index.

Supplementary Information The online version contains supplementary material available at https://doi.org/10.1007/978-3-031-07566-7_6.

329

For cases where it cannot be assumed that the deviations are covariance stationary, we developed a prediction algorithm which is based on the values of the original series. Again, based on moving windows of size n, we fit by least squares a quadratic polynomial to the last n values and extrapolate s time units forward. As expected, this predictor is less accurate than the optimal linear for covariance stationary series but can be useful for predicting one unit ahead, $s = 1$.

The third part of the chapter deals with dynamic linear models, which can incorporate Bayesian analysis and vector-valued observations.

6.1 The Components of a Time Series

A time series $\{X_t, t = 1, 2, \ldots\}$ is a sequence of random variables ordered according to the observation time. The analysis of the fluctuation of a time series assists us in analyzing the current behavior and forecasting the future behavior of the series. In the following sections, we introduce elementary concepts. There are three important components of a time series: the trend, the correlation structure among the observations, and the stochastic nature of the random deviations around the trend (the noise). If these three components are known, a reasonably good prediction or forecasting can be made. However, there are many types of time series in which the future behavior of the series is not necessarily following the past behavior. In the present chapter, we discuss these two types of forecasting situations. For more details see Box et al. (2015), Zacks (2009), and Shumway and Stoffer (2010).

6.1.1 The Trend and Covariances

The function $f(t) : t \mapsto E\{X_t\}$ is called the **trend** of the time series. A smooth trend can often be fitted to the time series data. Such a trend could be locally described as a polynomial of certain order plus a trigonometric Fourier sequence of orthogonal periodic functions.

Example 6.1 In Fig. 6.1 we present the time series **DOW1941.csv**, with the Dow Jones index in 302 working days of year 1941. The smooth curve traced among the points is the fitted local trend given by the function

$$
\begin{aligned}
f(t) =\,& 123.34 + 27.73\frac{t-151}{302} - 15.83\left(\frac{t-151}{302}\right)^2 - 237.00\left(\frac{t-151}{302}\right)^3 \\
& + 0.1512\cos\frac{4\pi t}{302} + 1.738\sin\frac{4\pi t}{302} + 1.770\cos\frac{8\pi t}{302} - 0.208\sin\frac{8\pi t}{302} \\
& - 0.729\cos\frac{12\pi t}{302} + 0.748\sin\frac{12\pi t}{302}.
\end{aligned}
$$

Fig. 6.1 Dow Jones values in 1941

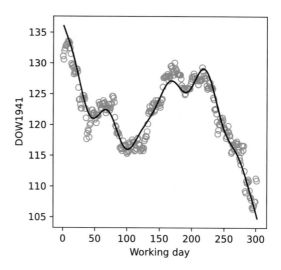

```
dow1941 = mistat.load_data('DOW1941')
t = np.arange(1, len(dow1941) + 1)
x = (t - 151) / 302
omega = 4 * np.pi * t / 302
ft = (123.34 + 27.73 * x - 15.83* x ** 2 - 237.00 * x**3
      + 0.1512 * np.cos(omega) + 1.738 * np.sin(omega)
      + 1.770 * np.cos(2 * omega) - 0.208 * np.sin(2 * omega)
      - 0.729 * np.cos(3 * omega) + 0.748 * np.sin(3 * omega))

fig, ax = plt.subplots(figsize=[4, 4])
ax.scatter(dow1941.index, dow1941, facecolors='none', edgecolors='grey')
ax.plot(t, ft, color='black')
ax.set_xlabel('Working day')
ax.set_ylabel('DOW1941')
plt.show()
```

To fit such a trend by the method of least squares, we use the multiple linear regression technique, in which the dependent variable Y is the vector of the time series, while the X vectors are the corresponding polynomial and trigonometric variables $((t - 151)/302)^j, j = 0, 1, 2, 3$ and $\cos(j4\pi t/302), \sin(j4\pi t/302)$, $j = 1, 2, 3$. ■

6.1.2 Analyzing Time Series with Python

In this subsection, we show how Python can be used to analyze time series using functionality available in `pandas` and `statsmodels`.

As a first step, we load the **DOW1941_DATE.csv** dataset and convert it to a time series.

```
dow1941 = mistat.load_data('DOW1941_DATE')

# convert Date column to Python datetime
dates = pd.to_datetime(dow1941['Date'], format='%Y-%m-%d')
dow1941_ts = pd.Series(dow1941['Open'], name='Dow_Jones_Index')
dow1941_ts.index = pd.DatetimeIndex(dates)

dow1941_ts.head()
```

```
Date
1941-01-02    131.1
1941-01-03    130.6
1941-01-04    132.0
1941-01-06    132.4
1941-01-07    132.8
Name: Dow_Jones_Index, dtype: float64
```

We next fit a series of additive models. The initial one uses just a linear trend. Instead of using the date as x, we create a data frame that contains a column with a sequence of 1, 2, …for each time step. The statsmodels package has the function tsatools.add_trend that add this trend column.

```
from statsmodels.tsa import tsatools
dow1941_df = tsatools.add_trend(dow1941_ts, trend='ct')
dow1941_df.head()
```

```
            Dow_Jones_Index   const   trend
Date
1941-01-02            131.1     1.0     1.0
1941-01-03            130.6     1.0     2.0
1941-01-04            132.0     1.0     3.0
1941-01-06            132.4     1.0     4.0
1941-01-07            132.8     1.0     5.0
```

It is now straightforward to fit a linear regression model of Dow_Jones_Index. Figure 6.2 shows the result we get using statsmodels.

```
from statsmodels.tsa import tsatools
dow1941_df = tsatools.add_trend(dow1941_ts, trend='ct')
model_1 = smf.ols(formula='Dow_Jones_Index ~ trend + 1', data=dow1941_df).fit()
print(model_1.params)
print(f'r2-adj: {model_1.rsquared_adj:.3f}')

ax = dow1941_ts.plot(color='grey')
ax.set_xlabel('Time')
ax.set_ylabel('Dow Jones index')
model_1.predict(dow1941_df).plot(ax=ax, color='black')
plt.show()
```

```
Intercept    125.929262
trend         -0.026070
dtype: float64
r2-adj: 0.151
```

It is clear that a linear model doesn't adequately describe the change over time. We can extend the model by adding quadratic and cubic terms. The resulting model is shown in Fig. 6.3 and is clearly an improvement. However be careful, a polynomial fit will not be very reliable for extrapolating beyond the actual data range.

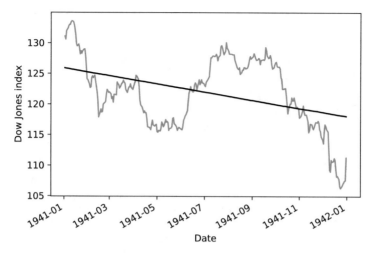

Fig. 6.2 Decomposition of the Dow1941 time series using an additive model with a linear trend

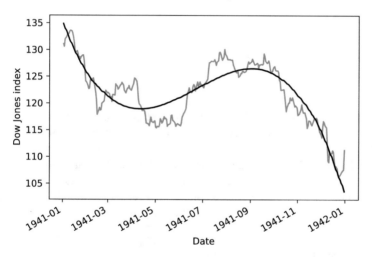

Fig. 6.3 Decomposition of the Dow1941 time series using an additive model with a linear, quadratic, and cubic trend

```
dow1941_df = tsatools.add_trend(dow1941_ts, trend='ct')
formula = 'Dow_Jones_Index ~ I(trend**3) + I(trend**2) + trend + 1'
model_2 = smf.ols(formula=formula, data=dow1941_df).fit()
print(model_2.params)
print(f'r2-adj: {model_2.rsquared_adj:.3f}')

ax = dow1941_ts.plot(color='grey')
ax.set_xlabel('Time')
ax.set_ylabel('Dow Jones index')
model_2.predict(dow1941_df).plot(ax=ax, color='black')
plt.show()
```

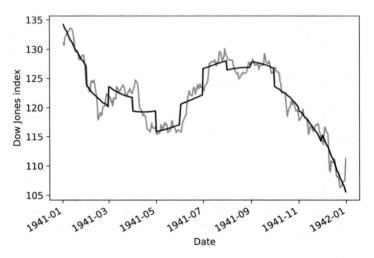

Fig. 6.4 The Dow1941 time series with a fit to a cubic model with monthly seasonal effects

```
Intercept               135.306337
np.power(trend, 3)       -0.000009
np.power(trend, 2)        0.003788
trend                    -0.450896
dtype: float64
r2-adj: 0.809
```

We can see that there are still deviations from the actual data that are unexplained by the model. These may be due to seasonality. Seasonality in this case means a periodic factor. This could be hour of day, day of week, week of year, season, and so on. Here we consider a monthly seasonality. Note that we are only considering a year of data, so this is more like a monthly adjustment of the data.

We extend the data frame with a column that labels the month of the year and include it in the regression. The resulting model is shown in Fig. 6.4.

```
dow1941_df = tsatools.add_trend(dow1941_ts, trend='ct')
dow1941_df['month'] = dow1941_df.index.month
formula = 'Dow_Jones_Index ~ C(month) + I(trend**3) + I(trend**2) + trend + 1'
model_3 = smf.ols(formula=poly_formula, data=dow1941_df).fit()
print(model_3.params)
print(f'r2-adj: {model_3.rsquared_adj:.3f}')

ax = dow1941_ts.plot(color='grey')
ax.set_xlabel('Time')
ax.set_ylabel('Dow Jones index')
model_3.predict(dow1941_df).plot(ax=ax, color='black')
plt.show()
```

```
Intercept               134.580338
C(month)[T.2]            -3.247325
C(month)[T.3]             0.244691
C(month)[T.4]            -2.046448
C(month)[T.5]            -5.612228
C(month)[T.6]            -2.137326
C(month)[T.7]             2.308240
```

```
C(month)[T.8]              0.667104
C(month)[T.9]              1.662265
C(month)[T.10]            -1.316258
C(month)[T.11]            -1.758233
C(month)[T.12]            -0.419919
np.power(trend, 3)        -0.000007
np.power(trend, 2)         0.002990
trend                     -0.359448
dtype: float64
r2-adj: 0.909
```

The adjusted R square for this model is 90.9%. The jumps of the fitted curve in Fig. 6.4 are due to the monthly effects. The model, with the coefficient estimates and non-centered polynomial terms, is

$$134.58 - 0.35944 Day + 0.002990 Day^2 - 0.000007 Day^3 + Match(Month)$$

We see that the effects of May and July, beyond the cubic trend in the data, are substantial (-5.61 and $+2.31$, respectively).

```
fig, axes = plt.subplots(figsize=[4, 5], nrows=3)
def residual_plot(model, ax, title):
  model.resid.plot(color='grey', ax=ax)
  ax.set_xlabel('')
  ax.set_ylabel(f'Residuals\n{title}')
  ax.axhline(0, color='black')
residual_plot(model_1, axes[0], 'Model 1')
residual_plot(model_2, axes[1], 'Model 2')
residual_plot(model_3, axes[2], 'Model 3')
axes[2].set_xlabel('Time')
plt.tight_layout()
plt.show()
```

Figure 6.5 presents the residuals from the three models used to fit the data. The residual plots of the first two models show that these models are not producing white noise and, therefore, there is more structure in the data to account for. The residuals from the third model look more like white noise, compared to the residuals of the first and second model. A formal assessment using a normal probability plot confirms this.

```
def plotLag(ts, lag, ax, limits):
  ax.scatter(ts[:-lag], ts[lag:], facecolors='none', edgecolors='black')
  ax.set_title(f'Lag {lag}')
  ax.set_xlim(*limits)
  ax.set_ylim(*limits)

fig, axes = plt.subplots(figsize=[6, 6], nrows=2, ncols=2)
limits = [dow1941_ts.min(), dow1941_ts.max()]
plotLag(dow1941_ts, 1, axes[0][0], limits)
plotLag(dow1941_ts, 5, axes[0][1], limits)
plotLag(dow1941_ts, 15, axes[1][0], limits)
plotLag(dow1941_ts, 60, axes[1][1], limits)

plt.tight_layout()
plt.show()
```

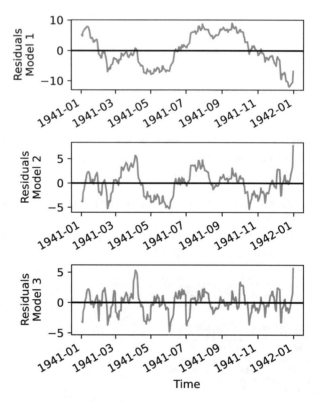

Fig. 6.5 Residuals from the model used in Figs. 6.2, 6.3, and 6.4 to fit the Dow1941 data

A simple analysis used to identify the lag-correlation mentioned above is to draw scatterplots of the data versus lagged data. Figure 6.6 shows such scatterplots for lags of 60 days, 15 days, 5 days, and 1 day. We observe a high correlation between the current and yesterday's Dow index and barely no relationship with values from 2 months earlier.

This correlation structure is not accounted for by the least squares regression models used in Fig. 6.4 to model the DOW1941 time series. This lack of independence between successive observations is affecting our ability to properly predict future observations. In the next sections we show how to account for such autocorrelations.

6.2   Covariance Stationary Time Series

Let $X_t = f(t) + U_t$. It is assumed that $E\{U_t\} = 0$ for all $t = 1, 2, \ldots$. Furthermore, the sequence of residuals $\{U_t, t = 1, 2, \ldots\}$ is called covariance stationary if $K(h) = cov(U_t, U_{t+h})$ is independent of t for all $h = 1, 2, \ldots$. Notice that

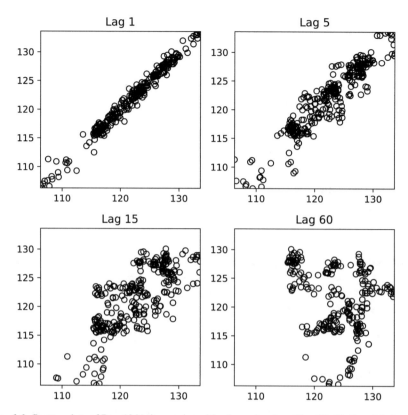

Fig. 6.6 Scatter plots of Dow1941 time series with a lagged series of lag 60, 15, 5, and 1 day

in this case the variance of each U_t is $K(0)$. The lag-correlation of order h is $\rho(h) = K(h)/K(0), h = 1, 2, \ldots$. The simplest covariance stationary time series is $\{e_t, t = 0, \pm 1, \pm 2, \ldots\}$, where $E\{e_t\} = 0$, the variance $V\{e_t\} = \sigma^2$ for all t, and $\rho(h) = 0$, for all $|h| > 0$. Such a time series is called a **white noise**. We denote it by $WN(0, \sigma^2)$.

6.2.1 Moving Averages

A linear combination of WN random variables is called a **moving average**. A moving average of order q, $MA(q)$, is the linear combination of q WN variables, i.e.,

$$X_t = \sum_{j=0}^{q} \beta_j e_{t-j}. \tag{6.1}$$

where the coefficients β_j are the same for all t. The covariance function of an $MA(q)$ is stationary and is given by

$$K(h) = \begin{cases} \sigma^2 \sum_{j=0}^{q-|h|} \beta_j \beta_{j+|h|} & |h| = 0, \dots, q \\ 0 & |h| > q. \end{cases} \tag{6.2}$$

Notice that $K(h) = K(-h)$, and a moving average of infinite order exists, if $\sum_{j=-\infty}^{\infty} |\beta_j| < \infty$.

Example 6.2 Consider an $MA(3)$ in which $X = 3e_t + 2.5e_{t-1} - 1.5e_{t-2} + e_{t-3}$ and $\sigma^2 = 1$.

This covariance stationary time series has $K(0) = 9 + 6.25 + 2.25 + 1 = 18.50$, $K(1) = 7.5 - 3.75 - 1.5 = 2.25$, $K(2) = -4.5 + 2.5 = -2$, and $K(3) = 3$.

The lag-correlations are $\rho(0) = 1$, $\rho(1) = 0.1216$, $\rho(2) = -0.1081$, and $\rho(3) = 0.1622$. All lag-correlations for $|h| > 3$ are zero.   ∎

6.2.2   Auto-Regressive Time Series

Another important class of time series is the **auto-regressive** model. A time series is called auto-regressive of order p, $AR(p)$, if $E\{X_t\} = \mu$, for all t, and $X_t = \sum_{j=1}^{p} \gamma_j X_{t-j} + e_t$, for all t, where e_t is a $WN(0, \sigma^2)$. Equivalently, we can specify an $AR(p)$ time series as

$$X_t + a_1 X_{t-1} + \cdots + a_p X_{t-p} = e_t. \tag{6.3}$$

This time series can be converted to a moving average time series by applying the **Z-transform**

$$A_p(z) = 1 + a_1 z^{-1} + \cdots + a_p z^{-p},$$

p is an integer, $A_0(z) = 1$ and $z^{-j} X_t = X_{t-j}$. Accordingly, $A_p(z) X_t = X_t + a_1 X_{t-1} + \cdots + a_p X_{t-p} = e_t$. From this we obtain that

$$X_t = (A_p(z))^{-1} e_t = \sum_{j=0}^{\infty} \beta_j e_{t-j}, \tag{6.4}$$

where

$$(A_p(z))^{-1} = 1/(1 + a_1 z^{-1} + \cdots + a_p z^{-p}) = \sum_{j=0}^{\infty} \beta_j z^{-j}. \qquad (6.5)$$

This inverse transform can be computed by the algebra of power series. The inverse power series always exists since $\beta_0 \neq 0$. We can obtain the coefficients β_j as illustrated in the following example. Notice that an infinite power series obtained in this way might not converge. If it does not converge, the inversion is not useful. The transform $(A_p(z))^{-1}$ is called a **transfer function**.

The polynomial $A_p^*(z) = z^p A_p(z)$ is called the **characteristic polynomial** of the $AR(p)$. The auto-regressive time series $AR(p)$ is covariance stationary only if all its characteristic roots belong to the interior of the unit circle, or the roots of $A_p(z)$ are all outside the unit circle.

The covariance function $K(h)$ can be determined by the following equations, called the **Yule-Walker equations**

$$K(0) + a_1 K(1) + \cdots + a_p K(p) = \sigma^2,$$
$$K(h) + a_1 K(h-1) + \cdots + a_p K(h-p) = 0, \qquad \text{for } h > 0. \qquad (6.6)$$

Example 6.3   Consider the $AR(2)$

$$X_t - X_{t-1} + 0.89 X_{t-2} = e_t, \qquad t = 0, \pm 1, \pm 2, \ldots$$

In this case, the characteristic polynomial is $A_2^*(z) = 0.89 - z + z^2$. The two characteristic roots are the complex numbers $\zeta_1 = 0.5 + 0.8i$ and $\zeta_2 = 0.5 - 0.8i$. These two roots are inside the unit circle, and thus this AR(2) is covariance stationary.

Using series expansion we obtain

$$\frac{1}{1 - z^{-1} + 0.89 z^{-2}} = 1 + z^{-1} + 0.11 z^{-2} - 0.78 z^{-3} - 0.8779 z^{-4} - 0.1837 z^{-5} + \ldots$$

The corresponding $MA(5)$ which is

$$X_t^* = e_t + e_{t-1} + 0.11 e_{t-2} - 0.78 e_{t-3} - 0.8779 e_{t-4} - 0.1837 e_{t-5}$$

is a finite approximation to the infinite order MA representing X_t. This approximation is not necessarily good. To obtain a good approximation to the variance and covariances, we need a longer moving average.

As for the above AR(2), the $K(h)$ for $h = 0, 1, 2$ are determined by solving the Yule-Walker linear equations

$$\begin{pmatrix} 1 & a_1 & a_2 \\ a_1 & 1+a_2 & 0 \\ a_2 & a_1 & 1. \end{pmatrix} \begin{pmatrix} K(0) \\ K(1) \\ K(2) \end{pmatrix} = \begin{pmatrix} \sigma^2 \\ 0 \\ 0 \end{pmatrix}$$

We obtain for $\sigma = 1$, $K(0) = 6.6801$, $K(1) = 3.5344$, and $K(2) = -2.4108$. Correspondingly, the lag-correlations are $\rho(0) = 1$, $\rho(1) = 0.5291$, and $\rho(2) = -0.3609$. For $h \geq 3$ we use the recursive equation

$$K(h) = -a_1 K(h-1) - a_2 K(h-2).$$

Accordingly, $K(3) = 2.4108 - 0.89 * 3.5344 = -0.7348$ and so on.   ∎

An important tool for determining the order p of an auto-regressive time series is the **partial lag-correlation**, denoted as $\rho^*(h)$. This index is based on the lag-correlations in the following manner.

Let R_k denote a symmetric $(k \times k)$ matrix called the **Toeplitz matrix** which is:

$$\begin{pmatrix} 1 & \rho(1) & \rho(2) & \rho(3) & \dots & \rho(k-1) \\ \rho(1) & 1 & \rho(1) & \rho(2) & \dots & \rho(k-2) \\ \rho(2) & \rho(1) & 1 & \rho(1) & \dots & \rho(k-3) \\ \rho(3) & \rho(2) & \rho(1) & 1 & \dots & \rho(k-4) \\ \vdots & \vdots & \vdots & \vdots & \ddots & \vdots \\ \rho(k-1) & \rho(k-2) & \rho(k-3) & \rho(k-4) & \dots & 1 \end{pmatrix}$$

The solution $\boldsymbol{\phi}^{(k)}$ of the normal equations

$$R_k \boldsymbol{\phi}^{(k)} = \boldsymbol{\rho}_k \tag{6.7}$$

yields least squares estimators of X_t and of X_{t+k+1}, based on the values of $X_{t+1}, \dots, X_{t+k}$. These are

$$\hat{X}_t = \sum_{j=1}^{k} \phi_j^{(k)} X_{t+j} \tag{6.8}$$

and

$$\hat{X}_{t+k+1} = \sum_{j=1}^{k} \phi_j^{(k)} X_{t+k+1-j}. \tag{6.9}$$

One obtains the following formula for the partial correlation of lag $k+1$,

Table 6.1 Lag-correlations
and partial lag-correlations
for the DOW1941 data

k	$\rho(k)$	$\rho^*(k)$
0	1.0000	1.0000
1	0.9805	0.9838
2	0.9521	−0.2949
3	0.9222	0.0220
4	0.8907	−0.0737
5	0.8592	0.0117
6	0.8290	0.0165
7	0.8009	0.0360
8	0.7738	−0.0203
9	0.7451	−0.0859
10	0.7162	0.0048
11	0.6885	0.0179
12	0.6607	−0.0346
13	0.6309	−0.0839
14	0.6008	0.0053
15	0.5708	−0.0349

$$\rho^*(k+1) = \frac{\rho(k+1) - \boldsymbol{\rho}'_k R_k^{-1} \boldsymbol{\rho}^*_k}{1 - \boldsymbol{\rho}'_k R_k^{-1} \boldsymbol{\rho}_k} \tag{6.10}$$

where $\boldsymbol{\rho}_k = (\rho(1), \ldots, \rho(k))'$ and $\boldsymbol{\rho}^*_k = (\rho(k), \ldots, \rho(1))'$.
In package mistat the function toeplitz forms the above matrix.

Example 6.4 We can use the statsmodels functions acf and pacf to calculate lag-correlations and partial lag-correlations for the DOW1941   data set. The package also has convenience functions to visualize the results. The calculated lag-correlations are shown in Table 6.1 and visualized in Fig. 6.7.

```
from statsmodels.tsa.stattools import acf, pacf
from statsmodels.graphics.tsaplots import plot_acf, plot_pacf

dow_acf = acf(dow1941_ts, nlags=15, fft=True)
dow_pacf = pacf(dow1941_ts, nlags=15)

fig, axes = plt.subplots(ncols=2, figsize=[8, 3.2])
plot_acf(dow1941_ts, lags=15, ax=axes[0])
plot_pacf(dow1941_ts, lags=15, method='ywm', ax=axes[1])
plt.tight_layout()
plt.show()
```

The 15 lag-correlations in Table 6.1 are all significant since the DOW1941 series is not stationary. Only the first two partial lag-correlations are significantly different from zero. On the other hand, if we consider the residuals around the trend curve f(t), we get the following the lag-correlations in Table 6.2 and Fig. 6.8. We see that in this case the first four lag-correlations are significant.

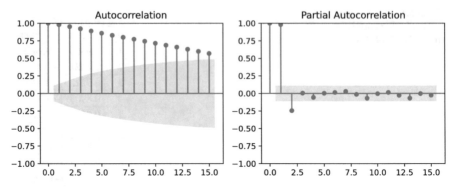

Fig. 6.7 Visualization of lag-correlations and partial lag-correlations for the DOW1941 data set

```
dow_acf = acf(model_3.resid, nlags=15, fft=True)
fig, axes = plt.subplots(ncols=2, figsize=[8, 3.2])
plot_acf(model_3.resid, lags=15, ax=axes[0])
plot_pacf(model_3.resid, lags=15, method='ywm', ax=axes[1])
plt.show()
```

∎

Table 6.2 Lag-correlations for the residuals of the cubic model with monthly seasonal effects of the DOW1941 data

k	$\rho(k)$
0	1.0000
1	0.8172
2	0.6041
3	0.4213
4	0.2554
5	0.1142
6	0.0137
7	−0.0366
8	−0.0646
9	−0.0895
10	−0.0905
11	−0.0682
12	−0.0790
13	−0.1214
14	−0.1641
15	−0.2087

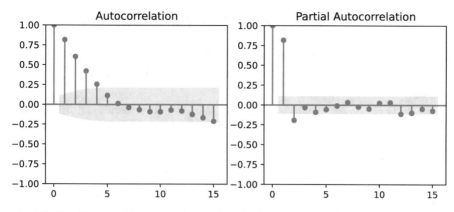

Fig. 6.8 Visualization of lag-correlations and partial lag-correlations for the cubic model with monthly seasonal effects of the DOW1941 data set

6.2.3   *Auto-Regressive Moving Average Time Series*

A time series of the form

$$X_t + a_1 X_{t-1} + \cdots + a_p X_{t-p} = e_t + b_1 e_{t-1} + \cdots + b_q e_{t-q} \tag{6.11}$$

is called an $ARMA(p, q)$ time series. If the characteristic roots of $A_p^*(z)$ are within the unit disk, then this time series is covariance stationary. We could write

$$X_t = \sum_{j=0}^{\infty} \beta_j z^{-j} (1 + b_1 z^{-1} + \cdots + b_q z^{-q}) e_t$$

$$= \sum_{k=0}^{\infty} \upsilon_k z^{-k} e_t. \tag{6.12}$$

Here $\upsilon_k = \sum_{l=0}^{k} \beta_l b_{k-l}$, where $b_{k-l} = 0$. When $k - l > q$, one can obtain this by series expansion of $B_q(x)/A_p(x)$.

Example 6.5 We consider here the ARMA(2,2), where, as in Example 6.2, $A_2(z) = 1 - z^{-1} + 0.89z^{-2}$ and $B_2(z) = 1 + z^{-1} - 1.5z^{-2}$. In this case

$$(1+x-1.5x^2)/(1-x+0.89x^2) = 1+2x-0.39x^2-2.17x^3-1.8229x^4+0.1084x^5+\ldots$$

Thus, we can approximate the ARMA(2,2) by an MA(5), namely,

$$X_t = e_t + 2e_{t-1} - 0.39e_{t-2} - 2.17e_{t-3} - 1.8229e_{t-4} + 0.1084e_{t-5}.$$

∎

6.2.4 Integrated Auto-Regressive Moving Average Time Series

A first-order difference of a time series is $\Delta\{X_t\} = (1 - z^{-1})\{X_t\} = \{X_t - X_{t-1}\}$.
Similarly, a kth-order difference of $\{X_t\}$ is $\Delta^k\{X_t\} = (1 - z^{-1})^k\{X_t\} = \sum_{j=0}^{k}\binom{k}{j}(-1)^j z^{-j}\{X_t\}$.

If $\Delta^k\{X_t\}$ is an $ARMA(p, q)$, we call the time series an **integrated** $ARMA(p, q)$ of order k or in short $ARIMA(p, k, q)$. This time series has the structure

$$A_p(z)(1 - z^{-1})^k X_t = B_q(z)e_t \tag{6.13}$$

where $e_t \sim i.i.d.(0, \sigma^2)$.

Accordingly, we can express

$$[A_p(z)/B_q(z)](1 - z^{-1})^k X_t = e_t. \tag{6.14}$$

Furthermore,

$$[A_p(z)/B_q(z)](1 - z^{-1})^k X_t = \sum_{j=0}^{\infty}\varphi_j z^{-j} X_t. \tag{6.15}$$

It follows that

$$X_t = \sum_{j=1}^{\infty}\pi_j X_{t-j} + e_t, \tag{6.16}$$

where $\pi_j = -\varphi_j$, for all $j = 1, 2, \ldots$. This shows that the $ARIMA(p, k, q)$ time series can be approximated by a linear combination of its previous values. This can be utilized, if the coefficients of $A_p(z^1)$ and those of $B_q(z)$ are known, for prediction of future values. We illustrate it in the following example.

Example 6.6 As in Example 6.5, consider the time series $ARIMA(2, 2, 2)$, where $A_2(z) = 1 - z^{-1} + 0.89z^{-2}$ and $B_2(z^1) = 1 + z^{-1} - 1.5z^{-2}$.
We have

$$\frac{1 - x + 0.89x^2}{1 + x - 1.5x^2} = 1 - 2x + 4.39x^2 - 7.39x^3 + 13.975x^4 - 25.060x^5 \ldots.$$

Multiplying by $(1 - x)^2$ we obtain that

$$X_t = 4X_{t-1} - 9.39X_{t-2} + 18.17X_{t-3} - 33.24X_{t-4}$$
$$+ 60.30X_{t-5} - 73.625X_{t-6} + 25.06X_{t-7} - + \cdots + e_t.$$

The predictors of X_{t+m} are obtained recursively from the above, as follows:

$$\hat{X}_{t+1} = 4X_t - 9.39X_{t-1} + 18.17X_{t-2} - 33.24X_{t-3} + - \dots$$

$$\hat{X}_{t+2} = 4\hat{X}_{t+1} - 9.39X_t + 18.17X_{t-1} - 33.24X_{t-2} + - \dots$$

$$\hat{X}_{t+3} = 4\hat{X}_{t+2} - 9.39\hat{X}_{t+1} + 18.17X_t - 33.24X_{t-1} + - \dots$$

∎

6.2.5   Applications with Python

To conclude this section, we revisit the Dow1941 data using an ARMA model
with `statsmodels`. This additional modeling effort aims at picking up the
autocorrelation in the data. We fit an ARMA model to the residuals from the cubic
and monthly effect model used in Fig. 6.4. Note that even though the method is
called ARIMA, we fit an ARMA model by setting `differences` in the order
argument to 0. The auto-regressive and moving average order systematically varies
between 0 and 4. The best fitting model is ARMA(2,2). The estimates of the model
parameters are shown in Fig. 6.9.

```
from statsmodels.tsa.arima.model import ARIMA
from statsmodels.graphics.tsaplots import plot_predict

# Identify optimal ARMA options using the AIC score
bestModel = None
bestAIC = None
for ar in range(0, 5):
```

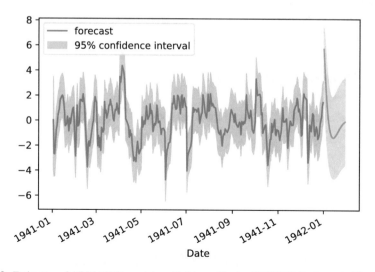

Fig. 6.9 Estimates of ARMA(2,2) model applied to residuals of DOW1941 shown in Fig. 6.4

```
for ma in range(0, 5):
    model = ARIMA(model_3.resid, order=(ar, 0, ma)).fit()
    if bestAIC is None or bestAIC > model.aic:
        bestAIC = model.aic
        bestModel = (ar, 0, ma)
print(f'Best model: {bestModel}')

model = ARIMA(model_3.resid, order=bestModel).fit()

prediction = model.get_forecast(30).summary_frame()
prediction['date'] = [max(dow1941_ts.index) + datetime.timedelta(days=i)
                      for i in range(1, len(prediction) + 1)]

plot_predict(model)
ax = plt.gca()

ax.plot(prediction['date'], prediction['mean'])
ax.fill_between(prediction['date'],
                prediction['mean_ci_lower'], prediction['mean_ci_upper'],
                color='lightgrey')
plt.show()
```

```
Best model: (2, 0, 2)
```

The ARMA(2,2) model, with 95% confidence intervals, and 30 predictions are shown in Fig. 6.9.

To generate predictions for January 1942 values of the DOW, we use the prediction of residuals shown in Fig. 6.9 and add these to the cubic and month effect model shown in Fig. 6.4.

6.3   Linear Predictors for Covariance Stationary Time Series

Let $\{X_t\}$ be a covariance stationary time series, such that $X_t = \sum_{j=0}^{\infty} w_j e_{t-j}$ and $e_t \sim WN(0, \sigma^2)$. Notice that $E\{X_t\} = 0$ for all t. The covariance function is $K(h)$.

6.3.1   Optimal Linear Predictors

A linear predictor of X_{t+s} based on the n data points $\mathbf{X}_t^{(n)} = (X_t, X_{t-1}, \dots, X_{t-n+1})'$ is

$$\hat{X}_{t+s}^{(n)} = (\boldsymbol{b}_s^{(n)})' \mathbf{X}_t^{(n)}, \tag{6.17}$$

where

$$\mathbf{b}_s^{(n)} = (b_{0,s}^{(n)}, b_{1,s}^{(n)}, \dots, b_{n-1,s}^{(n)})'.$$

The prediction mean squared error (PMSE) of this linear predictor is

$$E\{(\hat{X}_{t+s}^{(n)} - X_{t+s})^2\} = V\{(\boldsymbol{b}_s^{(n)})'\mathbf{X}_t^{(n)}\}$$

$$- 2COV(\hat{X}_{t+s}, (\boldsymbol{b}_s^{(n)})'\mathbf{X}_t^{(n)}) + V\{X_{t+s}\}. \qquad (6.18)$$

The covariance matrix of $\mathbf{X}_t^{(n)}$ is the Toeplitz matrix $K(0)R_n$, and the covariance of X_{t+s} and $\mathbf{X}_t^{(n)}$ is

$$\gamma_s^{(n)} = K(0) \begin{pmatrix} \rho(s) \\ \rho(s+1) \\ \rho(s+2) \\ \cdots \\ \rho(n+s-1) \end{pmatrix} = K(0)\boldsymbol{\rho}_s^{(n)}. \qquad (6.19)$$

Hence, we can write the prediction PMSE as

$$PMSE(\hat{X}_{t+s}^{(n)}) = K(0)\left((\boldsymbol{b}_s^{(n)})'R_n\boldsymbol{b}_s^{(n)} - 2(\boldsymbol{b}_s^{(n)})'\boldsymbol{\rho}_s^{(n)} + 1\right). \qquad (6.20)$$

It follows that the best linear predictor based on $\mathbf{X}_t^{(n)}$ is

$$BLP(X_{t+s}|\mathbf{X}_t^{(n)}) = (\boldsymbol{\rho}_s^{(n)})'R_n^{-1}\mathbf{X}_t^{(n)}. \qquad (6.21)$$

The minimal PMSE of this BLP is

$$PMSE^* = K(0)(1 - (\boldsymbol{\rho}_s^{(n)})'R_n^{-1}\boldsymbol{\rho}_s^{(n)}). \qquad (6.22)$$

Notice that the minimal PMSE is an increasing function of s.

Example 6.7 Let

$$X_t = a_{-1}e_{t-1} + a_0 e_t + a_1 e_{t+1}, \qquad t = 0, \pm 1, \ldots$$

where $e_t \sim WN(0, \sigma^2)$. This series is a special kind of MA(2), called **moving smoother**, with

$$K(0) = \sigma^2(a_{-1}^2 + a_o^2 + a_1^2) = B_0\sigma^2,$$
$$K(1) = \sigma^2(a_{-1}a_0 + a_0a_1) = B_1\sigma^2,$$
$$K(2) = \sigma^2 a_{-1}a_1 = \sigma^2 B_2,$$

and $K(h) = 0$, for all $|h| > 2$. Moreover, $\gamma_n^{(s)} = 0$ if $s > 2$. This implies that $\hat{X}_{t+s}^{(n)} = 0$, for all $s > 2$. If $s = 2$, then

$$\gamma_2^{(n)} = \sigma^2(B_2, 0'_{n-1})$$

and

$$\hat{\mathbf{b}}_2^{(n)} = \sigma^2 R_n^{-1} (B_2, 0'_{n-1})'.$$

In the special case of $a_{-1} = 0.25$, $a_0 = .5$, $a_1 = 0.25$, $\sigma = 1$, the Toeplitz matrix is

$$R_3 = \frac{1}{16} \begin{pmatrix} 6 & 4 & 1 \\ 4 & 6 & 4 \\ 1 & 4 & 6 \end{pmatrix}.$$

Finally, the best linear predictors are

$$\hat{X}_{t+1}^{(3)} = 1.2X_t - 0.9X_{t-1} + 0.4X_{t-2},$$

$$\hat{X}_{t+2}^{(3)} = 0.4X_t - 0.4X_{t-1} + 0.2X_{t-2}.$$

The PMSE are, correspondingly, 0.1492 and 0.3410.   ■

Example 6.8 In the previous example, we applied the optimal linear predictor to a simple case of an MA(2) series. Now we examine the optimal linear predictor in a more elaborate case, where the deviations around the trend are covariance stationary. We use the **DOW1941.csv** time series stationary ARMA series.

```
predictedError = mistat.optimalLinearPredictor(model_2.resid,11,nlags=10)
predictedTrend = model_2.predict(dow1941_df)
correctedTrend = predictedTrend + predictedError

fig, ax = plt.subplots()
ax.scatter(dow1941_ts.index, dow1941_ts,
           facecolors='none', edgecolors='grey')
predictedTrend.plot(ax=ax, color='grey')
correctedTrend.plot(ax=ax, color='black')
ax.set_xlabel('Time')
ax.set_ylabel('Dow Jones index')
plt.show()

print(f'PMSE(trend) = {np.mean((predictedTrend - dow1941_ts)**2):.4f}')
print(f'PMSE(corrected) = {np.mean((correctedTrend-dow1941_ts)**2):.4f}')
```

```
PMSE(trend) = 6.3315
PMSE(corrected) = 0.5580
```

In Fig. 6.10 we present the one-day ahead predictors for the DOW1941 data. This was obtained by applying the function `optimalLinearPredictor` from the `mistat` package to the series of the deviations around the trend function f(t) for the DOW1941 data, with a window of size 10, and adding the results to the predicted trend data. The corresponding total prediction risk is 0.558.   ■

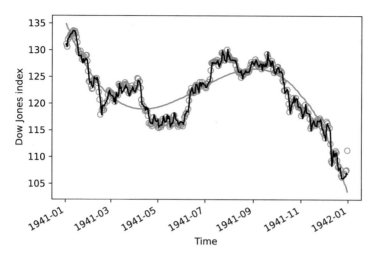

Fig. 6.10 One-step ahead predictors for the DOW1941 data using an optimal linear predictors with window of size 10

6.4   Predictors for Non-stationary Time Series

The linear predictor discussed in the previous section was based on the covariance stationarity of the deviations from the trend function. Such predictors are valid if we can assume that the future behavior of the time series is similar to the observed part of the past. This however is seldom the case. In the present section, we develop an adaptive procedure, which extrapolates the observed trend in a small window of the past. Such predictors would generally be good ones only for small values of time units in the future.

6.4.1   Quadratic LSE Predictors

For a specified window size n, $n > 5$, we fit a polynomial of degree $p = 2$ (quadratic) to the last n observations. We then extrapolate to estimate $f(t + s)$, $s \geq 1$. This approach is based on the assumption that in a close neighborhood of t, say $(t - n, t + n)$, $f(t)$ can be approximated by a quadratic whose parameters may change with t, i.e.,

$$f_2(t) = \beta_0(t) + \beta_1(t)t + \beta_2(t)t^2. \tag{6.23}$$

The quadratic moving LSE algorithm applies the method of ordinary least squares to estimate $\beta_j(t)$, $j = 0, 1, 2$, based on the data in the moving window $\{X_{t-n+1}, \ldots X_t\}$. With these estimates it predicts X_{t+s} with $f_2(t + s)$.

We provide here some technical details. In order to avoid the problem of unbalanced matrices when t is large, we shift in each cycle of the algorithm the origin to t. Thus, let $\mathbf{X}_t^{(n)} = (X_t, X_{t-1}, \ldots X_{t-n+1})'$ be a vector consisting of the values in the window. Define the matrix

$$A_{(n)} = \begin{pmatrix} 1 & 0 & 0 \\ 1 & -1 & 1 \\ \vdots & \vdots & \vdots \\ 1 & -(n-1) & (n-1)^2 \end{pmatrix}.$$

Then the LSE of $\beta_j(t)$, $j = 0, 1, 2$ is given in the vector

$$\hat{\boldsymbol{\beta}}^{(n)}(t) = (A'_{(n)} A_{(n)})^{-1} A'_{(n)} \mathbf{X}_t^{(n)}. \tag{6.24}$$

With these LSEs the predictor of X_{t+s} is

$$\hat{X}_{t+s}^{(n)}(t) = \hat{\beta}_0^{(n)}(t) + \hat{\beta}_1^{(n)}(t)s + \hat{\beta}_2^{(n)}(t)s^2. \tag{6.25}$$

Example 6.9 In the present example, we illustrate the quadratic predictor on a non-stationary time series. We use the quadraticPredictor function in the mistat package. With this function, we can predict the outcomes of an autoregressive series, step by step, after the first n observations. In Fig. 6.11 we see the one-step ahead prediction, $s = 1$, with a window of size $n = 20$ for the data of DOW1941.

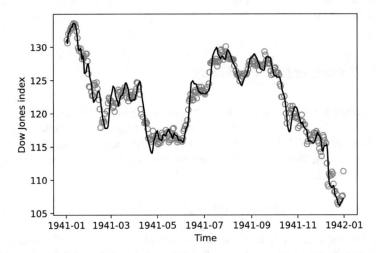

Fig. 6.11 One-step ahead prediction, $s = 1$, for the DOW1941 data using quadratic LSE predictors with window of size 20

```
quadPrediction = mistat.quadraticPredictor(dow1941_ts, 20, 1)

fig, ax = plt.subplots()
ax.scatter(dow1941_ts.index, dow1941_ts,
           facecolors='none', edgecolors='grey')
ax.plot(dow1941_ts.index, quadPrediction, color='black')
ax.set_xlabel('Time')
ax.set_ylabel('Dow Jones index')
plt.show()

print(f'PMSE(quadratic) = {np.mean((quadPrediction-dow1941_ts)**2):.4f}')
```

```
PMSE(quadratic) = 1.9483
```

We see in this figure that the quadratic LSE predictor is quite satisfactory. The results depend strongly on the size of the window, n. In the present case, the PMSE is 1.9483. ∎

6.4.2   Moving Average Smoothing Predictors

A moving average smoother, $MAS(m)$, is a sequence which replaces X_t by the a fitted polynomial based on the window of size $n = 2m + 1$, around X_t. The simplest smoother is the linear one. That is, we fit by LSE a linear function to a given window. Let $S_{(m)} = (A_{(m)})'(A_{(m)})$. In the linear case,

$$S_{(m)} = \begin{pmatrix} 2m + 1 & 0 \\ 0 & m(m + 1)(2m + 1)/3 \end{pmatrix}. \tag{6.26}$$

Then, the vector of coefficients is

$$\hat{\beta}^{(m)}(t) = S_{(m)}(A_{(m)})'X_t^{(m)}. \tag{6.27}$$

The components of this vector are

$$\beta_0^{(m)}(t) = \frac{1}{2m + 1} \sum_{j=-m}^{m} X_{t+j}$$

$$\beta_1^{(m)}(t) = \frac{3}{m(m + 1)(2m + 1)} \sum_{j=1}^{m} j(X_{t+j} - X_{t-j}). \tag{6.28}$$

Example 6.10 In Fig. 6.12 we present this linear smoother predictor for the DOW1941 data, with $m = 3$ and $s = 1$. The calculation uses the function masPrediction from the mistat package. The observed PMSE is 1.4917.

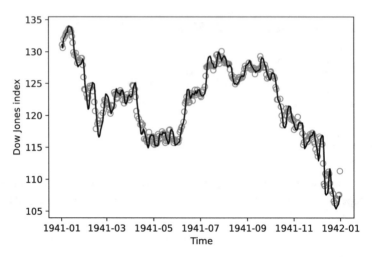

Fig. 6.12 Linear smoother predictor for the DOW1941, with $m = 3$ and $s = 1$

```
masPrediction = mistat.masPredictor(dow1941_ts, 3, 1)

fig, ax = plt.subplots()
ax.scatter(dow1941_ts.index, dow1941_ts,
           facecolors='none', edgecolors='grey')
ax.plot(dow1941_ts.index, masPrediction, color='black')
ax.set_xlabel('Time')
ax.set_ylabel('Dow Jones index')
plt.show()

print(f'PMSE(MAS) = {np.mean((masPrediction - dow1941_ts)**2):.4f}')
```

```
PMSE(MAS) = 1.4917
```

■

6.5   Dynamic Linear Models

The dynamic linear model (DLM) relates recursively the current observation, possibly vector of several dimensions, to a linear function of parameters, possibly random. Formally, we consider the random linear model

$$\mathbf{X_t} = \mathbf{A_t}\boldsymbol{\theta_t} + \boldsymbol{\epsilon_t}, \tag{6.29}$$

where

$$\boldsymbol{\theta_t} = \mathbf{G_t}\boldsymbol{\theta_{t-1}} + \boldsymbol{\eta_t}. \tag{6.30}$$

In this model, $\mathbf{X_t}$ and $\boldsymbol{\epsilon_t}$ are q-dimensional vectors. $\boldsymbol{\theta_t}$ and $\boldsymbol{\eta_t}$ are p-dimensional vectors. $\mathbf{A}$ is a $q \times p$ matrix of known constants, and $\mathbf{G}$ is a $p \times p$ matrix of known constants. $\{\boldsymbol{\epsilon_t}\}$ and $\{\boldsymbol{\eta_t}\}$ are mutually independent vectors. Furthermore, for each t, $\epsilon_t \sim N(0, \mathbf{V}_t)$ and $\boldsymbol{\eta_t} \sim N(0, \mathbf{W}_t)$.

6.5.1   Some Special Cases

Different kinds of time series can be formulated as dynamic linear models. For example, a time series with a polynomial trend can be expressed as Eq. (6.29) with $q = 1$, $\mathbf{A}_t = (1, t, t^2, \ldots, t^{p-1})$, and $\boldsymbol{\theta}_t$ is the random vector of the coefficients of the polynomial trend, which might change with t.

6.5.1.1   The Normal Random Walk

The case of $p = 1$, $q = 1$, $A = 1 \times 1$, and $G = 1 \times 1$ is called a Normal Random Walk. For this model, $V_0 = v$ and $W_0 = w$ are prior variances, and the prior distribution of θ is $N(m, c)$. The posterior distribution of θ given X_t is $N(m_t, c_t)$, where

$$c_t = (c_{t-1} + w)v/(c_{t-1} + v + w), \tag{6.31}$$

and

$$m_t = (1 - c_{t-1}/v)m_{t-1} + (c_{t-1}/v)X_t. \tag{6.32}$$

Example 6.11 Figure 6.13 represents a Normal Random Walk with a Bayesian prediction. The prediction line in Fig. 6.13 is the posterior expectation m_t. This figure was computed using the function `normRandomWalk` in the `mistat` package. Each application of this function yields another random graph.

```
res = mistat.normRandomWalk(100, 3, 1, 1, seed=2)

fig, ax = plt.subplots()
ax.scatter(res.t, res.X, facecolors='none', edgecolors='grey')
ax.plot(res.t, res.predicted, color='black')
ax.set_xlabel('Time')
ax.set_ylabel('TS')

plt.show()
```

■

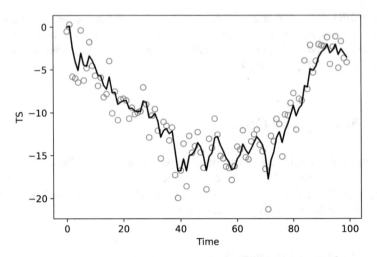

Fig. 6.13 Normal Random Walk (open circle) with a Bayesian prediction (black line)

6.5.1.2 Dynamic Linear Model With Linear Growth

Another application is the dynamic linear model with linear growth. In this case, $X_t = \theta_0 + \theta_1 t + \epsilon_t$, with random coefficient θ. As a special case, consider the following coefficients, $A_t = (1, t)'$ and $G_t = I_2$. Here m_t and C_t are the posterior mean and covariance matrix, given recursively by m

$$\mathbf{m}_t = \mathbf{m}_{t-1} + (1/r_t)(X_t - A_t'\mathbf{m}_{t-1})(C_{t-1} + \mathbf{W})\mathbf{A}_t, \qquad (6.33)$$

$$r_t = v + \mathbf{A_t}'(\mathbf{C_{t-1}} + \mathbf{W})\mathbf{A_t} \qquad (6.34)$$

and

$$\mathbf{C_t} = \mathbf{C}_{t-1} + \mathbf{W} - (1/r_t)(\mathbf{C}_{t-1} + \mathbf{W})\mathbf{a_t}\mathbf{a_t'}(\mathbf{C}_{t-1} + \mathbf{W}). \qquad (6.35)$$

The predictor of X_{t+1} at time t is $\hat{X}_{t+1}(t) = \mathbf{A'_{t+1}}\mathbf{m_t}$.

Example 6.12 In Fig. 6.14, we present the one-day ahead prediction ($s = 1$) for the DOW1941 data. We applied the function `dlmLinearGrowth` from the `mistat` package with parameters X, C_0, v, W, and M_0. For M_0 and C_0 we used the LSE of a regression line fitted to the first 50 data points. These are $M_0 = (134.234, -0.3115)'$, with covariance matrix

$$C_0 = \begin{pmatrix} 0.22325 & -0.00668 \\ -0.00668 & 0.00032 \end{pmatrix},$$

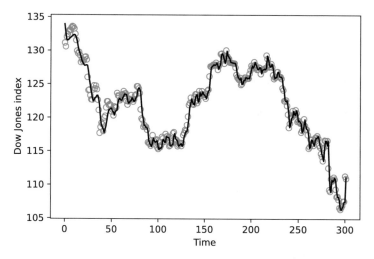

Fig. 6.14 One-step ahead prediction (s = 1) for the DOW1941 data using a dynamic linear model with linear growth

and the random vector η has the covariance matrix

$$W = \begin{pmatrix} 0.3191 & -0.0095 \\ -0.0095 & 0.0004 \end{pmatrix}.$$

The value for v is set to 1. As seen in Fig. 6.14, the prediction using this method is very good.

```
C0 = np.array([[0.22325, -0.00668], [-0.00668, 0.00032]])
M0 = np.array([134.234, -0.3115])
W = np.array([[0.3191, -0.0095], [-0.0095, 0.0004]])
v = 1

dow1941 = mistat.load_data('DOW1941.csv')
predicted = mistat.dlmLinearGrowth(dow1941, C0, v, W, M0)

fig, ax = plt.subplots()
ax.scatter(dow1941.index, dow1941, facecolors='none', edgecolors='grey')
ax.plot(dow1941.index, predicted, color='black')
ax.set_xlabel('Time')
ax.set_ylabel('Dow Jones index')
plt.show()
```

∎

6.5.1.3 Dynamic Linear Model for ARMA(p,q)

In this model, for a time series stationary around a mean zero, we can write

$$X_t = \sum_{j=1}^{p} a_j X_{t-j} + \sum_{j=0}^{q} b_j \epsilon_{t-j}, \qquad (6.36)$$

where $b_0 = 1, a_p \neq 0, b_q \neq 0$. Let $n = \max(1 + q, p)$. If $p < n$, we insert the extra coefficients $a_{p+1} = \cdots = a_n = 0$, and if $q < n - 1$, we insert $b_{q+1} = \cdots = b_{n-1} = 0$. We let $\mathbf{A} = (1, 0, \ldots, 0)$, and

$$\mathbf{G} = \begin{pmatrix} a_1 & a_2 & \ldots & a_n \\ 1 & 0 & \ldots & 0 \\ \vdots & 1 & \ldots & 0 \\ \vdots & \vdots & 1 & 0 \end{pmatrix}.$$

Furthermore, let $\mathbf{h}' = (1, b_1, \ldots, b_{n-1})$, $\boldsymbol{\theta}'_t = (X_t, \ldots, X_{t-n+1})$, and $\boldsymbol{\eta}'_t = (\mathbf{h}'\epsilon_t, 0, \ldots, 0)$. Then, we can write the $ARMA(p,q)$ time series as $\{Y_t\}$, where

$$Y_t = \mathbf{A}\boldsymbol{\theta}_t, \qquad (6.37)$$

$$\boldsymbol{\theta}_t = \mathbf{G}\boldsymbol{\theta}_{t-1} + \boldsymbol{\eta}_t, \qquad (6.38)$$

with $V = 0$ and $\mathbf{W} = (W_{i,j} : i, j = 1, \ldots, n)$, in which $W_{i,j} = I\{i = j = 1\}\sigma^2(1 + \sum_{j=1}^{n-1} b_j^2)$. The recursive formulas for $\mathbf{m}_t$ and $\mathbf{C}_t$ are

$$\mathbf{m}_t = (\sum_{i=1}^{n} a_i m_{t-1,i}, m_{t-1,1}, \ldots, m_{t-1,n-1})'$$

$$+ (1/r_t)(Y_t - \sum_{i=1}^{n} a_i m_{t-1,i})(\mathbf{G}\mathbf{C_{t-1}}\mathbf{G}' + \mathbf{W})(1, 0')', \qquad (6.39)$$

where

$$r_t = W_{11} + \mathbf{a}'\mathbf{C_{t-1}}\mathbf{a}, \quad \text{and} \quad \mathbf{a}' = (a_1, \ldots, a_n). \qquad (6.40)$$

We start the recursion with $\mathbf{m}'_0 = (X_p, \ldots, X_1)$ and

$$\mathbf{C}_0 = K_X(0) \, \text{Toeplitz}(1, \rho_X(1), \ldots, \rho_X(n - 1)).$$

The predictor of X_{t+1} at time t is

$$\hat{X}_{t+1} = \mathbf{A}\mathbf{G}\mathbf{m}_t = \mathbf{a}'\mathbf{X}_t. \qquad (6.41)$$

We illustrate this method in the following example.

Example 6.13 Consider the stationary ARMA(3,2) given by

$$X_t = 0.5X_{t-1} + 0.3X_{t-2} + 0.1X_{t-3} + \epsilon_t + 0.3\epsilon_{t-1} + 0.5\epsilon_{t-2},$$

in which $\{\epsilon_t\}$ is an i.i.d. sequence of N(0,1) random variables. The initial random variables are $X_1 \sim N(0, 1)$, $X_2 \sim X_1 + (0, 1)$ and $X_3 \sim X_1 + X_2 + N(0, 1)$. Here $a = (0.5, 0.3, 0.1)$, and $b_1 = 0.3, b_2 = 0.5$. The matrix $\mathbf{G}$ is

$$\mathbf{G} = \begin{pmatrix} 0.5 & 0.3 & 0.1 \\ 1 & 0 & 0 \\ 0 & 1 & 0 \end{pmatrix},$$

$V = 0$, and

$$\mathbf{W} = \begin{pmatrix} 1.34 & 0 & 0 \\ 0 & 0 & 0 \\ 0 & 0 & 0 \end{pmatrix}.$$

We start with $\mathbf{m}_0 = (X_3, X_2, X_1)$. All the characteristic roots are in the unit circle. Thus this time series is covariance stationary. The Yule-Walker equations yield the covariances: $K_X(0) = 7.69, K_X(1) = 7.1495, K_X(2) = 7.0967, and\, K_X(3) = 6.4622$. Thus we start the recursion with

$$\mathbf{C}_0 = \begin{pmatrix} 7.6900 & 7.1495 & 7.0967 \\ 7.1495 & 7.6900 & 7.1495 \\ 7.0967 & 7.1495 & 7.6900 \end{pmatrix}.$$

In Fig. 6.15 we see a random realization of this ARMA(3,2) and the corresponding prediction line for $s = 1$. The time series was generated by the function `simulateARMA`. Its random realization is given by the dots in the figure. The one-day ahead prediction was computed by the function `predictARMA` and is given by the solid line. Both functions are from the `mistat` package. The empirical PMSE is 0.9411.

```
a = [0.5, 0.3, 0.1]
b = [0.3, 0.5]
ts = pd.Series(mistat.simulateARMA(100, a, b, seed=1))
predicted = mistat.predictARMA(ts, a)

fig, ax = plt.subplots()
ax.scatter(ts.index, ts, facecolors='none', edgecolors='grey')
ax.plot(ts.index, predicted, color='black')
ax.set_xlabel('Time')
ax.set_ylabel('TS')
plt.show()

print(f'PMSE(ARMA) = {np.mean((predicted - ts)**2):.4f}')
```

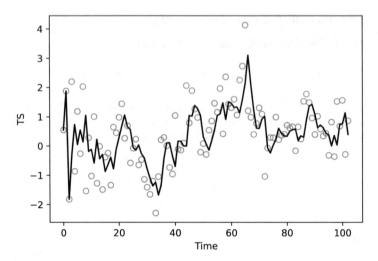

Fig. 6.15 Random realization of ARMA(3,2) and the corresponding prediction line for $s = 1$

```
PMSE(ARMA) = 0.8505
```

■

6.6   Chapter Highlights

The main concepts and definitions introduced in this chapter include:

- Trend function
- Covariance stationary
- White noise
- Lag-correlation
- Partial lag-correlation
- Moving averages
- Auto-regressive
- z-Transform
- Characteristic polynomials
- Yule-Walker equations
- Toeplitz matrix
- MA model
- ARMA model
- ARIMA model
- Linear predictors
- Polynomial predictors
- Moving average smoother
- Dynamic linear models

6.7 Exercises

Exercise 6.1 Evaluate trends and peaks in the data on COVID-19-related mortality available in https://www.euromomo.eu/graphs-and-maps/. Evaluate the impact of the time window on the line chart pattern. Identify periods with changes in mortality and periods with stability in mortality.

Exercise 6.2 The data set "SeasCom" provides the monthly demand for a seasonal commodity during 102 months:

(i) Plot the data to see the general growth of the demand.
(ii) Fit to the data the trend function:

$$f(t) = \beta_1 + \beta_2((t - 51)/102) + \beta_3 \cos(\pi t/6) + \beta_4 \sin(\pi t/6).$$

(iii) Plot the deviations of the data from the fitted trend, i.e., $\hat{U}_t = X_t - \hat{f}(t)$.
(iv) Compute the correlations between $(\hat{U}_t, \hat{U}_{t+1})$ and $(\hat{U}_t, \hat{U}_{t+2})$.
(v) What can you infer from these results?

Exercise 6.3 Write the formula for the lag-correlation $\rho(h)$ for the case of stationary $MA(q)$.

Exercise 6.4 For a stationary $MA(5)$, with coefficients $\beta' = (1, 1.05, .76, -.35, .45, .55)$, make a table of the covariances $K(h)$ and lag-correlations $\rho(h)$.

Exercise 6.5 Consider the infinite moving average $X_t = \sum_{j=0}^{\infty} q^j e_{t-j}$ where $e_t \sim WN(0, \sigma^2)$, where $0 < q < 1$. Compute

(i) $E\{X_t\}$
(ii) $V\{X_t\}$

Exercise 6.6 Consider the AR(1) given by $X_t = 0.75X_{t-1} + e_t$, where $e_t \sim WN(0, \sigma^2)$, and $\sigma^2 = 5$. Answer the following:

(i) Is this sequence covariance stationary?
(ii) Find $E\{X_t\}$,
(iii) Determine $K(0)$ and $K(1)$.

Exercise 6.7 Consider the auto-regressive series AR(2), namely,

$$X_t = 0.5X_{t-1} - 0.3X_{t-2} + e_t,$$

where $e_t \sim WN(0, \sigma^2)$.

(i) Is this series covariance stationary?

(ii) Express this AR(2) in the form $(1 - \phi_1 z^{-1})(1 - \phi_2 z^{-1})X_t = e_t$, and find the values of ϕ_1 and ϕ_2.

(iii) Write this AR(2) as an $MA(\infty)$. (Hint: Write $(1 - \phi z^{-1})^{-1} = \sum_{j=0}^{\infty} \phi^j z^{-j}$.)

Exercise 6.8 Consider the AR(3) given by $X_t - 0.5X_{t-1} + 0.3X_{t-2} - 0.2X_{t-3} = e_t$, where $e_t \sim WN(0, 1)$. Use the Yule-Walker equations to determine $K(h)$, $|h| = 0, 1, 2, 3$.

Exercise 6.9 Write the Toeplitz matrix R_4 corresponding to the series in Exercise 6.7.

Exercise 6.10 Consider the series $X_t - X_{t-1} + 0.25X_{t-2} = e_t + .4e_{t-1} - .45e_{t-2}$:

(i) Is this an ARMA(2,2) series?

(ii) Write the process as an MA(∞) series.

Exercise 6.11 Consider the second-order difference, $\Delta^2 X_t$, of the DOW1941 series:

(i) Plot the acf and the pacf of these differences.

(ii) Can we infer that the DOW1941 series is an integrated ARIMA(1,2,2)?

Exercise 6.12 Consider again the data set SeasCom and the trend function $f(t)$, which was determined in Exercise 6.2. Apply the function optimalLinearPredicto to the deviations of the data from its trend function. Add the results to the trend function to obtain a one-day ahead prediction of the demand.

Chapter 7
Modern Analytic Methods: Part I

Preview This chapter is a door opener to computer age statistics. It covers a range of supervised and unsupervised learning methods and demonstrates their use in various applications.

7.1 Introduction to Computer Age Statistics

Big data and data science applications have been facilitated by hardware developments in computer science. As data storage began to increase, more advanced software was required to process it. This led to the development of cloud computing and distributed computing. Parallel machine processing was enhanced by the development of Hadoop, based on off-the-shelf Google File System (GFS) and Google MapReduce, for performing distributed computing.

New analytic methods were developed to handle very large data sets that are being processed through distributed computing. These methods are typically referred to as machine learning, statistical learning, data mining, big data analytics, data science, or AI (artificial intelligence). Breiman (2001b) noted that models used in big data analytics are developed with a different purpose than traditional statistical models. Computer age models do not assume a probability-based structure for the data, such as $y = X\beta + \epsilon$, where $\epsilon \sim NID(0, \sigma^2)$. In general, they make no assumptions as to a "true" model producing the data. The advantage of these computer age methods is that no assumptions are being made about model form or error, so that standard goodness of fit assessment is not necessary. However, there are still assumptions being made, and overfitting is assessed with holdout sets and cross-validation. Moreover, conditions of non-stationarity and strong data stratification of the data pose complex challenges in assessing predictive capabilities of such models (Efron and Hastie 2016).

Supplementary Information The online version contains supplementary material available at https://doi.org/10.1007/978-3-031-07566-7_7.

Without making any assumptions about the "true" form of the relationship between the x and the y, there is no need to estimate population parameters. Rather, the emphasis of predictive analytics, and its ultimate measure of success, is prediction accuracy. This is computed by first fitting a training set and then calculating measures such as root-mean-square error or mean absolute deviation. The next step is moving on to such computations on holdout data sets, on out-of-sample data, or with cross-validation. Ultimately, the prediction error is assessed on new data collected under new circumstances. In contrast to classical statistical methods, this approach leads to a totally different mindset in developing models. Traditional statistical research is focused on understanding the process that generated the observed data and modeling the process using methods such as least squares or maximum likelihood. Computer age statistics is modeling the data per se and focuses on the algorithmic properties of the proposed methods. This chapter is about such algorithms.

7.2 Data Preparation

Following problem elicitation and data collection, a data preparation step is needed. This involves assessing missing data, duplicated records, missing values, outliers, typos, and many other issues that weaken the quality of the data and hinder advanced analysis.

Lawrence (2017) proposed a classification of the status of data into quality bands labeled C, B, A, AA, and AAA. These represent the level of usability of data sets.

Band C (conceive) refers to the stage that the data is still being ingested. If there is information about the data set, it comes from the data collection phase and how the data was collected. The data has not yet been introduced to a programming environment or tool in a way that allows operations to be performed on the data set. The possible analyses to be performed on the data set in order to gain value from the data possibly haven't been conceived yet, as this can often only be determined after inspecting the data itself.

Band B (believe) refers to the stage in which the data is loaded into an environment that allows cleaning operations. However, the correctness of the data is not fully assessed yet, and there may be errors or deficiencies that invalidate further analysis. Therefore, analyses performed on data at this level are often more cursory and exploratory with visualization methods to ascertain the correctness of the data.

In **band A** (analyze), the data is ready for deeper analysis. However, even if there are no more factual errors in the data, the quality of an analysis or machine learning model is greatly influenced by how the data is represented. For instance, operations such as feature selection and normalization can greatly increase the accuracy of machine learning models. Hence, these operations need to be performed before arriving at accurate and adequate machine learning models or analyses.

In **band AA** (allow analysis), we consider the context in which the data set is allowed to be used. Operations in this band detect, quantify, and potentially address any legal, moral, or social issues with the data set, since the consequences of using illegal, immoral, or biased data sets can be enormous. Hence, this band is about verifying whether analysis can be applied without (legal) penalties or negative social impact. One may argue that legal and moral implications are not part of data cleaning, but rather distinct parts of the data process. However, we argue that readiness is about learning the ins and outs of your data set and detecting and solving any potential problems that may occur when analyzing and using a data set.

Band AAA is reached when you determine that the data set is clean. The data is self-contained, and no further input is needed from the people that collected or created the data.

In Chap. 8 in the Industrial Statistics book, in a section on analytic pipelines, we introduce a Python application providing scores to data sets based on these bands. The application is available in https://github.com/pywash/pywash. Instructions on how to use the `pywash` application on our GitHub site are available at https://gedeck.github.io/mistat-code-solutions.

7.3 The Information Quality Framework

Breiman (2001b) depicts two cultures in the use of statistical modeling to reach conclusions from data, data modeling, and algorithmic analysis. The information quality framework (InfoQ) presented in this subsection addresses outputs from both approaches, in the context of business, academic, services, and industrial data analysis applications.

The InfoQ framework provides a structured approach for evaluating analytic work. InfoQ is defined as the utility, U, derived by conducting a certain analysis, f, on a given data set X, with respect to a given goal g. For the mathematically inclined

$$\text{InfoQ}(U, f, X, g) = U(f(X|g)).$$

As an example, consider a cellular operator that wants to reduce churn by launching a customer retention campaign. His goal, g, is to identify customers with high potential for churn—the logical target of the campaign. The data, X, consists of customer usage, lists of customers who changed operators, traffic patterns, and problems reported to the call center. The data scientist plans to use a decision tree, f, which will help define business rules that identify groups of customers with high churn probabilities. The utility, U, is increased profits by targeting this campaign only on customers with a high churn potential.

InfoQ is determined by eight dimensions that are assessed individually in the context of the specific problem and goal. These dimensions are:

1. Data resolution: Is the measurement scale, measurement uncertainty, and level of data aggregation appropriate relative to the goal?

2. Data structure: Are the available data sources (including both structured and unstructured data) comprehensive with respect to goal?
3. Data integration: Are the possibly disparate data sources properly integrated together? Note: This step may involve resolving poor and confusing data definitions, different units of measure, and varying time stamps.
4. Temporal relevance: Is the time-frame in which the data were collected relevant to the goal?
5. Generalizability: Are results relevant in a wider context? In particular, is the inference from the sample population to target population appropriate (statistically generalizable, Chap. 3)? Can other considerations be used to generalize the findings?
6. Chronology of data and goal: Are the analyses and needs of the decision-maker synched up in time?
7. Operationalization: Are results presented in terms that can drive action?
8. Communication: Are results presented to decision-makers at the right time and in the right way?

Importantly, InfoQ helps structure discussions about trade-offs, strengths, and weaknesses in data analysis projects (Kenett and Redman 2019). Consider the cellular operator noted above, and consider a second potential data set X^*. X^* includes everything X has, plus data on credit card churn, but that additional data won't be available for 2 months. With such data, resolution (the first dimension) goes up, while temporal resolution (the fourth) goes down. In another scenario, suppose a new machine learning analysis, f^*, has been conducted in parallel, but results from f and f^* don't quite line up. "What to do?" These are the examples of discussions between decision-makers, data scientists, and CAOs. Further, the InfoQ framework can be used in a variety of settings, not just helping decision-makers become more sophisticated. It can be used to assist in the design of a data science project, as a mid-project assessment and as a post mortem to sort out lessons learned. See Kenett and Shmueli (2016) for a comprehensive discussion of InfoQ and its applications in risk management, healthcare, customer surveys, education, and official statistics.

7.4 Determining Model Performance

The performance of a model can be measured in various ways. The Python package `scikit-learn` contains a wide variety of different metrics. A few of them are listed in Table 7.1.

In order to avoid overfitting, one needs to compare results derived from fitting the model with a training set to results with a validation set not involved in fitting the model. There are basically two approaches to achieve this.

A first approach is applicable with large data sets. In this context one can randomly select a subset, through uniform or stratified sampling. This results in

Table 7.1 Model performance metrics

Classification	
Accuracy	Accuracy is defined as the number of correct predictions made by the model on a data set
Balanced accuracy	Modification of accuracy suitable for imbalanced data sets
ROC	
Regression	
R^2	Coefficient of determination (see Sect. 4.3.2.1)
R^2_{adj}	Adjusted coefficient of determination (see Sect. 4.3.2.1)
MSE	Mean squared error is defined as the mean squared difference between actual and predicted y
MAE	Mean absolute error is defined as the mean absolute difference between actual and predicted y
AIC	Akaike information criterion
BIC	Bayesian information criterion

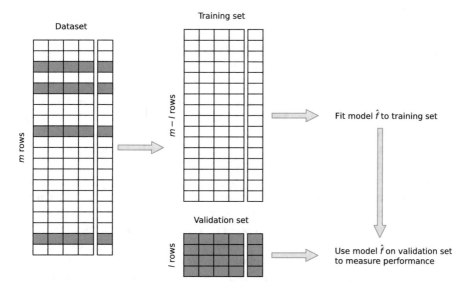

Fig. 7.1 Splitting data set into training and validation set for model validation

two distinct data sets. One is used for fitting a model, the training set, and the other for evaluating its performance, the validation set (see Fig. 7.1). A variation on this approach is to split the data into three parts, a training set, a tuning set, and a validation set. The tuning set is being used in fine-tuning the model, such as determining the number of optimal splits in a decision tree introduced in Sect. 7.5 and the actual splits being determined by the training set. In most applications only one has a training set and a validation set.

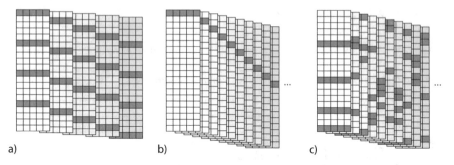

Fig. 7.2 (**a**) k-Fold cross-validation, (**b**) leave-one-out cross-validation, and (**c**) bootstrapping

When the number of records is not large, data partitioning might not be advisable as each partition will contain too few records for model building and performance evaluation. An alternative to data partitioning is cross-validation, which is especially useful with small samples. Cross-validation is a procedure that starts with partitioning the data into "folds," or non-overlapping subsamples. Often, we choose $k = 5$ folds, meaning that the data are randomly partitioned into five equal parts, where each fold has 20% of the observations (see Fig. 7.2a). A model is then fit k times. Each time, one of the folds is used as the validation set, and the remaining $k - 1$ folds serve as the training set. The result is that each fold is used once as the validation set, thereby producing predictions for every observation in the data set. We can then combine the model's predictions on each of the k validation sets in order to evaluate the overall performance of the model. If the number of folds is equal to the number of data points, this variant of k-fold cross-validation is also known as leave-one-out cross-validation (see Fig. 7.2b).

Cross-validation is often used for choosing the algorithm's parameters, i.e., tuning the model. An example is the number of splits in a decision tree. Cross-validation estimates out-of-sample prediction error and enables the comparison of statistical models. It is applied in supervised settings, such as regression and decision trees, but does not easily extend to unsupervised methods, such as dimensionality reduction methods or clustering. By fitting the model on the training data set and then evaluating it on the testing set, the over-optimism of using data twice is avoided. Craven and Wahba (1978) and Seeger (2008) use cross-validated objective functions for statistical inference by integrating out-of-sample prediction error estimation and model selection, into one step.

Consider a data set $\mathbf{A}$ where rows of $\mathbf{A}$ correspond to m observations of n independent/predictor variables and the $m \times 1$ vector b corresponding to one dependent variables. A multiple linear regression model, where $\mathbf{A}$ is an $m \times n$ data matrix, x is vector with n parameters, and b is a vector containing m responses.

Cross-validation can also be performed by randomly picking l observations (in Fig. 7.2c, $l = 4$) and fitting a model using $m - l$ data points. The predictions on the l singled-out observations are then compared to their actual values and the results of repeated sampling aggregated to assess the predictive performance.

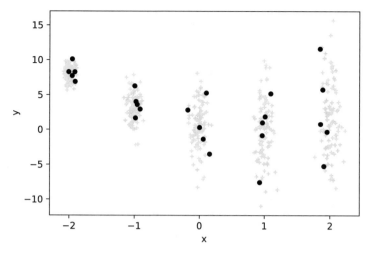

Fig. 7.3 Artificial data set for Example 7.1. The black points represent the data set which is sampled from an underlying distribution shown by the grey crosses

If the data is collected with an underlying structure such as scores from football games, players, or time windows, the cross-validation can account for the structure by singling out specific players, games, or game periods. In these situations, all observations for the selected unit of observation are excluded from the training set.

For such structured multilevel data, the use of cross-validation for estimating out-of-sample prediction error and model selection deserves close attention. Specifically, in order to test a model accounting for such a structure, the holdout set cannot be a simple random sample of the data, but, as mentioned above, it needs to have some multilevel structure where groups as well as individual observations are held out. An example in the context of the analysis of a survey is provided in Price et al. (1996). In general, there are no specific guidelines for conducting cross-validation in multilevel structured data (Wang and Gelman 2015).

Example 7.1 We use an artificial data set to demonstrate the importance of stratified sampling. The data set (see Fig. 7.3) has five sets of five data points. Such a data set is typical for DOE results (design of experiments; see Chap. 5 in the Industrial Statistics book). As the underlying distribution shows, the error is small on the left and increases to the right.

In the following Python code, we first create 100 stratified samples by RSWR of four data points from each set of five, create a quadratic model, and record the model performance using r^2. In the second block, we create a non-stratified sample by RSWR of 20 data points from the full set of 25.

```
formula = 'y ~ 1 + x + np.power(x, 2)'
def sample80(df):
  """ Sample 80% of the dataset using RSWR """
  return df.sample(int(0.8 * len(df)), replace=True)
stratR2 = []
```

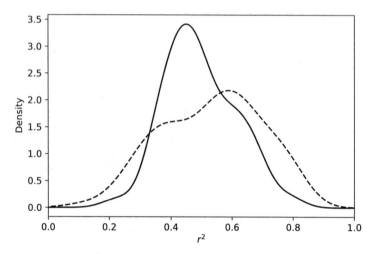

Fig. 7.4 Distribution of r^2 for model using stratified (solid line) and fully random (dashed line) RSWR sampling

```
for _ in range(100):
    stratSample = df.groupby('group').apply(lambda x: sample80(x))
    model = smf.ols(formula=formula, data=stratSample).fit()
    stratR2.append(model.rsquared)
sampleR2 = []
for _ in range(100):
    sample = sample80(df)
    model = smf.ols(formula=formula, data=sample).fit()
    sampleR2.append(model.rsquared)
```

Figure 7.4 shows the distribution of the r^2 values from the two sampling approaches. The distribution for stratified sampling is tighter. An assessment of model performance based on stratified sampling results will therefore be more reliable.

∎

7.5 Decision Trees

Partition models, also called decision trees, are non-parametric tools used in supervised learning in the context of classification and regression. In supervised learning you observe multiple covariate and one or more target variables. The goal is to predict or classify the target using the values of covariates. Decision trees are based on splits in covariates or predictors that create separate but homogeneous groups. Splits are not sensitive to outliers but are based on a "greedy" one -step look ahead, without accounting for overall performance. Breiman et al. (1984) implement a decision tree procedure called CART (Classification And Regression Trees). Other procedures are C4.5 and CHAID (Chi-square Automatic Interaction

Detector). Trees can handle missing data without the need to perform imputation. Moreover they produce rules on the predators that can be effectively communicated and implemented. Single trees, sometimes called exploratory decision trees, are however poor predictors. This can be improved with random forests, bootstrap forests, and boosted trees that we discuss later.

To evaluate the performance of a decision tree, it is important to understand the notion of class confusion and the confusion matrix. A confusion matrix for a target variable we want to predict involving n classes is an $n \times n$ matrix with the columns labeled with predicted classes and the rows labeled with actual classes.[1] Each data point in a training or validation set has an actual class label as well as the class predicted by the decision tree (the predicted class). This combination determines the confusion matrix.

For example, consider a two-class problem with a target response being "Pass" or "Fail." This will produce a 2×2 confusion matrix, with actual Pass predicted as Pass and actual Fail predicted as Fail, these are on the diagonal. The lower left and top right values are actual Pass predicted as Fail and actual Fail predicted as Pass. These off-diagonal values correspond to misclassifications. When one class is rare, for example, in large data sets with most data points being "Pass" and a relatively small number of "Fail," the confusion matrix can be misleading. Because the "Fail" class is rare among the general population, the distribution of the target variable is highly unbalanced. As the target distribution becomes more skewed, evaluation based on misclassification (off diagonal in the confusion matrix) breaks down. For example, consider a domain where the unusual class appears in 0.01% of the cases. A simple rule that would work is to always choose the most prevalent class; this gives 99.9% accuracy but is useless. In quality control it would never detect a Fail item, or, in fraud detection, it would never detect rare cases of fraud, and misclassification can be greatly misleading. With this background, let us see a decision tree in action.

Example 7.2 Data set **SENSORS.csv** consists of 174 measurements from 63 sensors tracking performance of a system under test. Each test generates values for these 63 sensors and a status determined by the automatic test equipment. The distribution of the test results is presented in Fig. 7.5. Our goal is to predict the outcome recorded by the testing equipment, using sensor data. The test results are coded as Pass (corresponding to "Good," 47% of the observations) and Fail (all other categories, marked in grey). The column **Status** is therefore a dichotomized version of the column **Test result**.

```
sensors = mistat.load_data('SENSORS.csv')
dist = sensors['testResult'].value_counts()
dist = dist.sort_index()
ax = dist.plot.bar(color='lightgrey')
ax.patches[dist.index.get_loc('Good')].set_facecolor('black')
plt.show()
```

---

[1] Note that this is the convention we use in this book. Some texts label the rows with predicted classes and columns with actual classes.

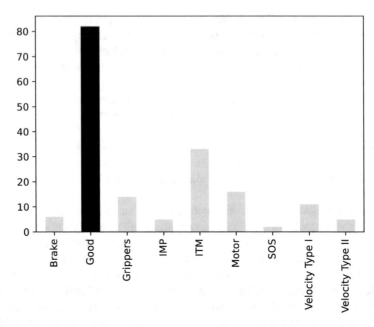

Fig. 7.5 Distribution of test results. The test result "Good" coded as Pass is highlighted in dark grey

The goal is to predict the outcome recorded by the testing equipment, using sensor data. We can use `scikit-learn` for this. It has decision tree implementations for classification and regression. Here, we create a classification model for Pass-Fail using the 67 sensors.

```
from sklearn.tree import DecisionTreeClassifier, plot_tree, export_text

predictors = [c for c in sensors.columns if c.startswith('sensor')]
outcome = 'status'
X = sensors[predictors]
y = sensors[outcome]

# Train the model
clf = DecisionTreeClassifier(ccp_alpha=0.012, random_state=0)
clf.fit(X, y)

# Visualization of tree
plot_tree(clf, feature_names=list(X.columns))
plt.show()
```

`Scikit-learn` provides two convenience functions to create text and graph representations (see Fig. 7.6) of the resulting tree (`plot_tree`, `export_text`).

```
# Text representation of tree
print(export_text(clf, feature_names=list(X.columns)))
```

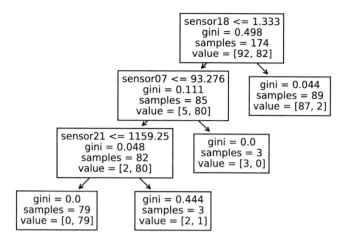

Fig. 7.6 Decision tree of sensor data with three splits

```
|--- sensor18 <= 1.33
|   |--- sensor07 <= 93.28
|   |   |--- sensor21 <= 1159.25
|   |   |   |--- class: Pass
|   |   |--- sensor21 >  1159.25
|   |   |   |--- class: Fail
|   |--- sensor07 >  93.28
|   |   |--- class: Fail
|--- sensor18 >  1.33
|   |--- class: Fail
```

Figure 7.7 shows an alternative representation using the `dtreeviz` package with additional information for each of the nodes. A similar view is available for a regression tree.

The first split is on sensor 18 with cutoff point 1.33. Eighty-seven of the 89 observations with sensor $18 > 1.33$ are classified as Fail. Most observations with sensor $18 \leq 1.33$ are classified as Pass (80 out of 85). By splitting this subset on sensor $7 > 93.276$, we find three Fails. The remaining 82 observations are split a third time on sensor $21 \leq 1159.25$ giving a subset of 79 Pass and a second subset with two Fail and one Pass. As we controlled the complexity of the tree using the argument `ccp_alpha`, no further splits are found.

The decision tree provides us an effective way to classify tests as Pass-Fail with just three sensors (out of 67). These splits are based on vertical splits by determining cutoff values for the predictor variables, the 67 sensors. We present next how such splits are determined in the **SENSORS.csv** data and how the performance of a decision tree is evaluated.   ∎

The node splitting in the `scikit-learn` implementation of the decision tree can be based on two different criteria for classification, Gini impurity, and entropy. These criteria measure the classification outcome of a given node. We first determine the probability of finding each class k in the node m, $p_{mk} = N_{mk}/N_m$.

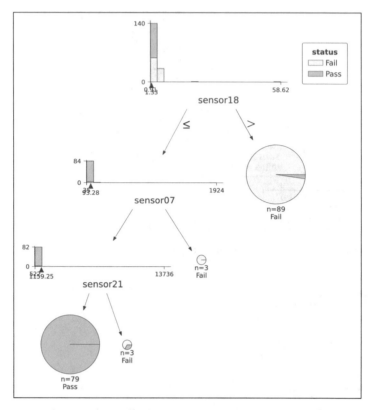

Fig. 7.7 Decision tree visualization of classification tree

Using these probabilities we get

$$\text{Gini impurity} \qquad H(\mathbf{p}_m) = \sum_{i \in \text{classes}} p_{mk}(1 - p_{mk}) \qquad (7.1)$$

$$\text{Entropy} \qquad H(\mathbf{p}_m) = - \sum_{i \in \text{classes}} p_{mk} \log_2(p_{mk}). \qquad (7.2)$$

The quality of a split is now defined as a weighted sum of the criteria of the two child nodes:

$$G_m = \frac{N_m^{\text{left}}}{N_m} H(\mathbf{p}_m^{\text{left}}) + \frac{N_m^{\text{right}}}{N_m} H(\mathbf{p}_m^{\text{right}}). \qquad (7.3)$$

The split that minimizes this sum is used in the decision tree. Figure 7.8 shows the change of the G_m measure with changing sensor 18 value. We see that the Gini criterion gives a split at 1.333. The split value for entropy is slightly higher at 1.425.

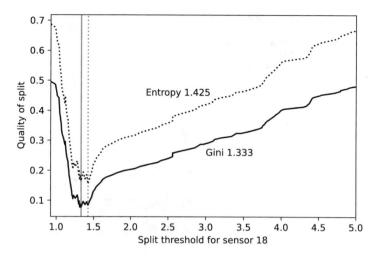

Fig. 7.8 Determination of initial optimal split of sensor 18 using the Gini (solid) and entropy (dotted) criteria

For continuous responses, `scikit-learn` implements the following criteria:

MSE
$$H(\mathbf{y}_m) = \frac{1}{N_m} \sum_{y \in y_m} (y - \bar{y}_m)^2 \qquad (7.4)$$

MAE
$$H(\mathbf{y}_m) = \frac{1}{N_m} \sum_{y \in y_m} |y - \text{median}(\bar{y}_m)| \qquad (7.5)$$

Poisson
$$H(\mathbf{y}_m) = \frac{1}{N_m} \sum_{y \in y_m} \left(y - \log \frac{y}{\bar{y}_m} - y + \bar{y}_m \right). \qquad (7.6)$$

Mean squared error (MSE) and mean absolute error (MAE) are the usual metrics for regression. The Poisson criterion can be used if the outcome is a count or a frequency.

   Assessing the performance of a decision tree is based on an evaluation of its predictive ability. The observations in each leaf are classified, as a group, according to the leaf probability and a cut of threshold. The default cutoff is typically 50% implying that all types of misclassification carry the same cost. The terminal node that is reached by the path through the decision tree that always goes left classifies all samples as Pass (see Fig. 7.6). In all other terminal nodes, the majority of samples are classified as Fail (87 out of 89, 3 out of 3, and 2 out of 3). The ratio of classified to total is used in `scikit-learn` to determine a prediction probability for a sample. If, for example, a sample ends up in the node where two out of three training samples are "Fail," `predict_proba` will return a probability of 0.66 for Fail and 0.33 for Pass.

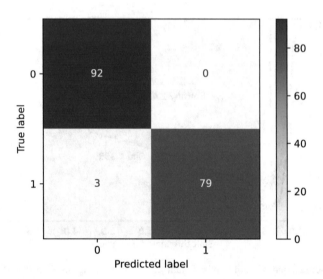

Fig. 7.9 Confusion matrix for the decision tree in Fig. 7.6

Based on these probabilities, a classification of the observations in the leaves is conducted using the recorded values and the predicted values. In the case of the Pass-Fail data of sensor data, with four slits, this generates a 2×2 confusion matrix displayed in Fig. 7.9. Only three observations are misclassified. They correspond to observations which were a Pass in the test equipment and were classified as Fail by the decision tree. The default cutoff for assigning the prediction is 50%. If we lower the probability for Pass to 30%, we can reduce the number of false positives by 1 and however increase the number of false negatives at the same time by 2. This might improve the number of working products we falsely discard; however it would also increase the risk of sending out defective products.

```
# missclassification probabilities
print('Probabilities of missclassified data points')
print(clf.predict_proba(X.loc[clf.predict(X) != y,:]))

# actual in rows / predicted in columns
print('Confusion matrix')
cm = confusion_matrix(y, clf.predict(X))
print(cm)

disp = ConfusionMatrixDisplay(confusion_matrix=cm, display_labels=clf.classes_)
disp.plot(cmap=plt.cm.Blues)
plt.show()
```

```
Probabilities of missclassified data points
[[0.97752809 0.02247191]
 [0.97752809 0.02247191]
 [0.66666667 0.33333333]]
Confusion matrix
[[92  0]
 [ 3 79]]
```

The decision tree analysis can be conducted on the original data set with a target consisting of nine values. When the target is a continuous variable, the same approach produces a regression tree where leaves are characterized not by counts but by average and standard deviation values. We do not expand here on such cases.

Two main properties of decision trees are:

1. Decision trees use decision boundaries that are perpendicular to the data set space axes. This is a direct consequence of the fact that trees select a single attribute at a time.
2. Decision trees are "piecewise" classifiers that segment the data set space recursively using a divide-and-conquer approach. In principle, a classification tree can cut up the data set space arbitrarily finely into very small regions.

It is difficult to determine, in advance, if these properties are a good match to a given data set. A decision tree is understandable to someone without a statistics or mathematics background. If the data set does not have two instances with exactly the same covariates, but different target values, and we continue to split the data, we are left with a single observation at each leaf node. This essentially corresponds to a lookup table. The accuracy of this tree is perfect, predicting correctly the class for every training instance. However, such a tree does not generalize and will not work as well on a validation set or new data. When providing a lookup table, unseen instances do not get classified. On the other hand, a decision tree will give a nontrivial classification even for data not seen before.

Tree-structured models are very flexible in what they can represent and, if allowed to grow without bound, can fit up to an arbitrary precision. But the trees may need to include a large number of splits in order to do so. The complexity of the tree lies in the number of splits. Using a training set and a validation set, we can balance accuracy and complexity.

Some strategies for obtaining a proper balance are (i) to stop growing the tree before it gets too complex and (ii) to grow the tree until it is too large and then "prune" it back, reducing its size (and thereby its complexity). There are various methods for accomplishing both. The simplest method to limit tree size is to specify a minimum number of observations that must be present in a leaf (argument `min_samples_leafint`, default 1!). Cost-complexity pruning combines the misclassification rate $R(T)$ of the tree T with the number of terminal nodes $|N_{\text{terminal}}|$:

$$R_\alpha(T) = R(T) + \alpha|N_{\text{terminal}}|. \tag{7.7}$$

This expression can be used to define an effective α_{eff} for each node. By comparing this effective α_{eff} with a given α, we can successively remove nodes and prune the tree. Figure 7.10 shows the effect of increasing `cpp_alpha` on tree size for the **SENSORS.csv** data set. A suitable value can be derived using cross-validation.

The data at the leaf is used to derive statistical estimates of the value of the target variable for future cases that would fall to that leaf. If we make predictions of the target based on a very small subset of data, they can be inaccurate. A further option

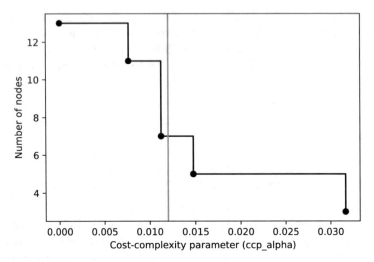

Fig. 7.10 Controlling tree depth using cost-complexity pruning. The value `ccp_alpha` was used to get the tree in Fig. 7.6

is to derive a training set, a validation set, and a testing set. The training set is used to build a tree, the validation set is used to prune the tree to get a balance between accuracy and complexity, and the testing set is used to evaluate the performance of the tree with fresh data.

7.6 Ensemble Models

Following this overview of decision trees, we move on to present a variation based on computer-intensive methods that enhances the stability of the decision tree predictions. A well-known approach to reduce variability in predictions is to generate several predictions and compute a prediction based on majority votes or averages. This "ensemble" method will work best if the combined estimates are independent. In this case, the ensemble-based estimate will have a smaller variability by a factor of square root of n, the number of combined estimates.

Bootstrap forests and boosted trees are ensembles derived from several decision trees fit to versions of the original data set. Both are available as alternatives to the decision tree method in Python. Popular implementations can be found in the packages `scikit-learn` and `xgboost`. In these algorithms we fit a number of trees, say 100 trees, to the data and use "majority vote" among the 100 trees to classify a new observation (Breiman 2001a). For prediction, we average the 100 trees. An ensemble of 100 trees makes up a "forest." However, fitting 100 trees to the same data produces redundantly the same tree, 100 times. To address this redundancy, we randomly pick subsets of the data. Technically we use bootstrapping to create a new data set for each of the 100 trees by sampling the original data, with replacement (see Sect. 3.10). This creates new data sets that are based on the original data but are not identical to it. These 100 alternative data sets produce 100 different trees. The integration of multiple models, created through bootstrapping, is known as "bootstrapped aggregation" or "bagging." This is the approach in random forests. In bootstrap forests, besides picking the data at random, also the set of predictors used in the tree is picked at random. When determining the variable upon which to split, bootstrap forests consider a randomly selected subset of the original independent variables. Typically, the subset is of size around the square root of the number of predictors. This gets the algorithm to consider variables not considered in a standard decision tree. Looking at the ensemble of 100 trees produces more robust and unbiased predictions. Reporting the number of bootstrap forest splits, on each predictor variable, provides useful information on the relative importance of the predictor variables.

Creating a random forest model in Python is straightforward with `scikit-learn`.

```
predictors = [c for c in sensors.columns if c.startswith('sensor')]
outcome = 'status'
X = sensors[predictors]
y = sensors[outcome]

# Train the model
clf = RandomForestClassifier(ccp_alpha=0.012, random_state=0)
clf.fit(X, y)

# actual in rows / predicted in columns
print('Confusion matrix')
print(confusion_matrix(y, clf.predict(X)))
```

```
Confusion matrix
[[92  0]
 [ 0 82]]
```

We see that random forest classifier classifies all data correctly.

Another highly popular implementation of ensemble models is `xgboost`. It provides classes that can be used in the same way as `scikit-learn` classifiers and regressors.

```
from xgboost import XGBClassifier

predictors = [c for c in sensors.columns if c.startswith('sensor')]
outcome = 'status'
X = sensors[predictors]
# Encode outcome as 0 (Fail) and 1 (Pass)
y = np.array([1 if s == 'Pass' else 0 for s in sensors[outcome]])

# Train the model
xgb = XGBClassifier(objective='binary:logistic', subsample=.63,
                    eval_metric='logloss', use_label_encoder=False)
xgb.fit(X, y)

# actual in rows / predicted in columns
print('Confusion matrix')
print(confusion_matrix(y, xgb.predict(X)))
```

```
Confusion matrix
[[92  0]
 [ 0 82]]
```

The next section is about a competing method to decision trees, the Naïve Bayes classifier.

7.7 Naïve Bayes Classifier

The basic idea of the Naïve Bayes classifier is a simple algorithm. For a given new record to be classified, $x_1, x_2, \ldots, x_n$, find other records like it (i.e., y with same values for the predictors $x_1, x_2, \ldots, x_n$). Following that, identify the prevalent class among those records (the ys), and assign that class to the new record. This is applied to categorical variables, and continuous variables must be discretized, binned, and converted to categorical variables. The approach can be used efficiently with very large data sets and relies on finding other records that share same predictor values as the record to be classified. We want to find the "probability of y belonging to class A, given specified values of predictors, $x_1, x_2, \ldots, x_n$." However, even with large data sets, it may be hard to find other records that exactly match the record to be classified, in terms of predictor values. The Naïve Bayes classifier algorithm assumes independence of predictor variables (within each class), and using the multiplication rule computes the probability that the record to be classified belongs to class A, given predictor values $x_1, x_2, \ldots x_n$, without limiting calculation only to records that exactly share these same values. From Bayes' theorem (Chap. 2), we know that

$$P(y|x_1, \ldots, x_n) = \frac{P(y)P(x_1, \ldots, x_n|y)}{P(x_1, \ldots, x_n)} \tag{7.8}$$

where y is the value to be classified and $x_1, \ldots, x_n$ are the predictors.

In the Naïve Bayes classifier, we move from conditioning the predictors $x_1, x_2, \ldots, x_n$ on the target y to conditioning the target y on the predictors $x_1, x_2, \ldots, x_n$.

To calculate the expression on the right, we rely on the marginal distribution of the predictors and assume their independence, hence (7.9)

$$P(y|x_1, \ldots, x_n) = \frac{P(y) \prod P(x_i|y)}{P(x_1, \ldots, x_n)}. \tag{7.9}$$

This is the basis of the Naïve Bayes classifier. It classifies a new observation by estimating the probability that the observation belongs to each class and reports the class with the highest probability. The Naïve Bayes classifier is very efficient in terms of storage space and computation time. Training consists only of storing counts of classes and feature occurrences, as each observation is recorded. However, in spite of its simplicity and the independence assumption, the Naïve Bayes classifier performs surprisingly well. This is because the violation of the independence assumption tends not to hurt classification performance. Consider the following intuitive reasoning. Assume that two observations are strongly dependent so that when one sees one we are also likely to see the other. If we treat them as independent, observing one enhances the evidence for the observed class, and seeing the other also enhances the evidence for its class. To some extent, this double-counts the evidence. As long as the evidence is pointing in the right direction, classification with this double-counting will not be harmful. In fact, the probability estimates are expanded in the correct direction. The class probabilities will be therefore overestimated for the correct class and underestimated for the incorrect classes. Since for classification we pick the class with the highest estimated probability, making these probabilities more extreme in the correct direction is not a problem. It can however become a problem if we use the probability estimates themselves. Naïve Bayes is therefore safely used for ranking where only the relative values in the different classes are relevant. Another advantage of the Naïve Bayes classifier is an "incremental learner." An incremental learner is an induction technique that updates its model, one observation at a time, and does not require to reprocess all past training data when new training data becomes available. Incremental learning is especially advantageous in applications where training labels are revealed in the course of the application, and the classifier needs to reflect this new information as quickly as possible. The Naïve Bayes classifier is included in nearly every machine learning toolkit and serves as a common baseline classifier against which more sophisticated methods are compared.

Example 7.3   To demonstrate the application of a Naïve Bayes classifier, we invoke the results of a customer satisfaction survey, **ABC.csv**. The data consists of 266 responses to a questionnaire with a question on overall satisfaction (q1) and responses to 125 other questions. Figure 7.11 shows the distribution of q1 and five other questions.

```
abc = mistat.load_data('ABC.csv')
all_questions = [c for c in abc.columns if c.startswith('q')]
abc[all_questions] = abc[all_questions].astype('category')

questions = ['q1', 'q4', 'q5', 'q6', 'q7']
q1_5 = (abc['q1'] == 5)

fig, axes = plt.subplots(ncols=len(questions))
for ax, question in zip(axes, questions):
  response = abc[question]
  df = pd.DataFrame([
    {satisfaction: counts for satisfaction, counts
      in response.value_counts().iteritems()},
    {satisfaction: counts for satisfaction, counts
      in response[q1_5].value_counts().iteritems()},
  ])
  df = df.transpose()  # flip columns and rows
  # add rows of 0 for missing satisfaction
  for s in range(6):
    if s not in df.index:
      df.loc[s] = [0, 0]
  df = df.fillna(0)  # change missing values to 0
```

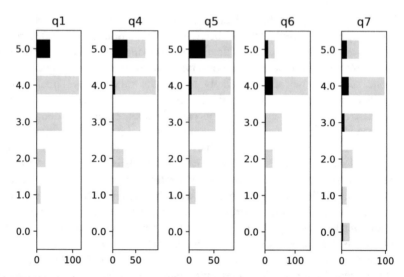

Fig. 7.11 Distribution of q1 and five other questions with response "5" in q1 highlighted

```
    df = df.sort_index()   # the index contains the satisfaction
    df.columns = ['counts', 'counts_q1_5']   # rename the columns
    df['counts'].plot.barh(y='index', ax=ax, color='lightgrey')
    df['counts_q1_5'].plot.barh(y='index', ax=ax, color='black')
    ax.set_ylim(-0.5, 5.5)
    ax.set_title(question)
plt.tight_layout()
plt.show()
```

We can see that the response "5" in q1 corresponds to top-level responses in q4, q5, q6, and q7. Based on such responses in q4–q7, we can therefore confidently predict a response "5" to q1.

The `scikit-learn` package provides two implementations of Naïve Bayes, `MultinomialNB` for count data and `BernoulliNB` for binary data. However, both implementations have the limitation that they are not suitable to handle missing data. In order to derive a model for the survey data, we need to deal with this problem by removing some of the questions and responses. There are various ways of "imputing" missing values. Here we replace the missing values for a question with the most frequent response for this question using the `SimpleImputer` method.

```
predictors = list(all_questions)
predictors.remove('q1')
target = 'q1'
# q1 has missing values - remove rows from dataset
q1_missing = abc[target].isna()
X = abc.loc[~q1_missing, predictors]
y = abc.loc[~q1_missing, target]

imp = SimpleImputer(missing_values=np.nan, strategy='most_frequent')
X = imp.fit_transform(X)
```

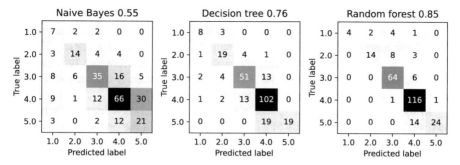

Fig. 7.12 Comparison of performance of the Naïve Bayes, decision tree, and random forest classifier with q1 as target and remaining questions as predictors

The Naïve Bayes classifier can be applied to all 125 responses to the questionnaire. The outputs from this analysis are presented in Fig. 7.12. We observe 119 misclassifications, mostly to respondents who answered "4" to q1.

```
nb_model = MultinomialNB()
nb_model.fit(X, y)
print(confusion_matrix(y, nb_model.predict(X)))
print(accuracy_score(y, nb_model.predict(X)))
```

```
[[ 7  2  2  0  0]
 [ 3 14  4  4  0]
 [ 8  6 35 16  5]
 [ 9  1 12 66 30]
 [ 3  0  2 12 21]]
0.5458015267175572
```

The Naïve Bayes classifier's misclassification rate of 45% was obtained with an easy-to-compute and incremental learning algorithm. A decision tree with 27 splits (`ccp_alpha=0.014`) has a misclassification rate of 24%. The bootstrap forest was much better with a 15% misclassification rate (see Fig. 7.12). ∎

7.8 Neural Networks

A neural network is composed of a set of computational units, called neurons, connected together through weighted connections. Neurons are organized in layers so that every neuron in a layer is exclusively connected to the neurons of the preceding layer and the subsequent layer. Every neuron, also called a node, represents an autonomous computational unit and receives inputs as a series of signals that dictate its activation. Following activation, every neuron produces an output signal. All the input signals reach the neuron simultaneously, so the neuron receives more than one input signal, but it produces only one output signal. Every input signal is associated with a connection weight. The weight determines the relative importance the input signal can have in producing the final impulse transmitted by the neuron. The connections can be exciting, inhibiting, or null according to whether the corresponding weights are, respectively, positive, negative, or null. The weights are adaptive coefficients that, in analogy with the biological model, are modified in response to the various signals that travel on the network according to a suitable learning algorithm. A threshold value θ, called bias, is usually introduced. Bias is similar to an intercept in a regression model.

In formal terms, a neuron j, with a threshold θ_j, receives n input signals $x = [x_1, x_2, \ldots, x_n]$ from the units to which it is connected in the previous layer. Each signal is attached with an importance weight $w_j = [w_{1j}, w_{2j}, \ldots, w_{nj}]$.

The input signals, their importance weights, and the threshold value determine a combination function. The combination function produces a potential or net input. An activation function transforms the potential into an output signal. The combination function is usually linear, so that the potential is a weighted sum of the input values multiplied by the weights of the respective connections. This sum is compared to the threshold. The potential of neuron j, P_j is defined as

$$P_j = \sum_{i=1}^{n} w_{ij} x_i - \theta_j.$$

The bias term in the potential can be simplified by setting $x_0 = 1$ and $w_{0j} = -\theta_j$:

$$P_j = \sum_{i=0}^{n} w_{ij} x_i.$$

The output of the jth neuron, y_j, is derived from applying the activation function to potential P_j:

$$y_j = f(x, w_j) = f\langle(\sum_{i=0}^{n} w_{ij} x_i)\rangle.$$

Neural network are organized in layers: input, output, or hidden. The input layer receives information only from the external environment where each neuron usually corresponds to an explanatory variable. The input layer does not perform any calculation; it transmits information to the next level. The output layer produces the final results, which are sent by the network to the outside of the system. Each of its neurons corresponds to a response variable. In a neural network, there are generally two or more response variables. Between the output layer and the input layer, there can be one or more intermediate layers, called hidden layers, because they are not directly in contact with the external environment. These layers are exclusively for analysis; their function is to take the relationship between the input variables and the output variables and adapt it more closely to the data.

The hidden layers are characterized by the used activation function. Commonly used functions are:

Linear   The identity function. The linear combination of the n input variables is not transformed:

$$f(x) = x.$$

The linear activation function can be used in conjunction with one of the nonlinear activation functions. In this case, the linear activation function is placed in the second layer, and the nonlinear activation functions are placed in the first layer. This is useful if you want to first reduce the dimensionality of the n input variables to m and then have a nonlinear model for the m variables.

If only linear activation functions are used, the model for a continuous output variable y reduces to a linear combination of the X variables and therefore corresponds to linear regression. For a nominal or ordinal y output variable, the model reduces to a logit, logistic, or multinomial logistic regression.

ReLU   The rectified linear activation function. The ReLU function is zero for negative values and linear for positive values:

$$f(x) = \max(x, 0).$$

The ReLU activation function as a replacement of the sigmoid activation function is one of the key changes that made deep learning neural networks possible.

TanH   The hyperbolic tangent or sigmoid function. TanH transforms values to be between -1 and 1 and is the centered and scaled version of the logistic function. The hyperbolic tangent function is

$$f(x) = \frac{e^{2x} - 1}{e^{2x} + 1}.$$

Gaussian   The Gaussian function. This option is used for radial basis function behavior or when the response surface is Gaussian (normal) in shape. The Gaussian function is

$$f(x) = e^{-x^2}.$$

The "architecture" of a neural network refers to the network's organization: the number of layers, the number of units (neurons) belonging to each layer, and the manner in which the units are connected. Four main characteristics are used to classify network topology:

- Degree of differentiation of the input and output layer
- Number of layers
- Direction of flow for the computation

- Type of connections

The variables used in a neural network can be classified by type (qualitative or quantitative) and by their role in the network (input or output). Input and output in neural networks correspond to explanatory and response in statistical methods. In a neural network, quantitative variables are represented by one neuron. The qualitative variables, both explanatory and responses, are represented in a binary way using several neurons for every variable; the number of neurons equals the number of levels of the variable. The number of neurons to represent a variable need not be equal to the number of its levels. Since the value of that neuron will be completely determined by the others, one often eliminates one level or one neuron. Once the variables are coded, a preliminary descriptive analysis may indicate the need of a data transformation or to standardize the input variables. If a network has been trained with transformed input or output, when it is used for prediction, the outputs must be mapped on to the original scale.

The objective of training a neural network with data, to determine its weights on the basis of the available data set, is not to find an exact representation of the training data but to build a model that can be generalized or that allows us to obtain valid classifications and predictions when fed with new data. Similar to tree models, the performance of a supervised neural network can be evaluated with reference to a training data set or validation data set. If the network is very complex and the training is carried out for a large number of iterations, the network can perfectly classify or predict the data in the training set. This could be desirable when the training sample represents a "perfect" image of the population from which it has been drawn, but it is counterproductive in real applications since it implies reduced predictive capacities on a new data set. This phenomenon is known as overfitting. To illustrate the problem, consider only two observations for an input variable and an output variable. A straight line adapts perfectly to the data but poorly predicts a third observation, especially if it is radically different from the previous two. A simpler model, the arithmetic average of the two output observations, will fit the two points worse but may be a reasonable predictor of a third point. To limit the overfitting problem, it is important to control the degree of complexity of the model. A model with few parameters will involve a modest generalization. A model that is too complex may even adapt to noise in the data set, perhaps caused by measurement errors or anomalous observations; this will lead to inaccurate generalizations. Regularization is used to control a neural network's complexity. It consists of the addition of a penalty term to the error function. Some typical penalty functions are summarized in Table 7.2. Regularization is not only important for neural networks. It also forms the basis of many other approaches, e.g., ridge regression or Lasso regression.

To find weights that yield the best predictions, one applies a process that is repeated for all records. At each record, one compares prediction to actual. The difference is the error for the output node. The error is propagated back and distributed to all the hidden nodes and used to update their weights. Weights are updated after each record is run through the network. Completion of all records

Table 7.2 Regularization methods

Method	Penalty function	Description
Squared (L2)	$\sum \beta_i^2$	Use this method if you think that most of your X variables are contributing to the predictive ability of the model
Absolute (L1)	$\sum \lvert \beta_i \rvert$	Use either of these methods if you have a large number of X variables and you think that a few of them contribute more than others to the predictive ability of the model
Weight decay	$\sum \frac{\beta_i^2}{1+\beta_i^2}$	
No penalty	–	Does not use a penalty. You can use this option if you have a large amount of data and you want the fitting process to go quickly. However, this option can lead to models with lower predictive performance than models that use a penalty

through the network is one epoch (also called sweep or iteration). After one epoch is completed, one returns to the first record and repeats the process.

The updating stops: (1) when weights change very little from one iteration to the next, (2) when the misclassification rate reaches a required threshold, or (3) when a limit on runs is reached.

Neural networks are used for classification and prediction. They can capture complicated relationship between the outcome and a set of predictors. The network "learns" and updates its model iteratively as more data are fed into it. A major danger in neural networks is overfitting. It requires large amounts of data and has good predictive performance but is "black box" in nature.

Example 7.4 With the emergence of deep learning, TensorFlow, Keras, or PyTorch are the go-to options for implementing neural networks. It would go beyond the scope of this book to cover these packages in detail. In addition, the rapid development in this field would make examples quickly outdated. Instead we will demonstrate neural networks using the scikit-learn implementation.

After loading, we preprocess the data and impute missing data as described in Example 7.3. Following this, the data need to be range normalized onto the interval 0 to 1. This helps the training process.

```
# scale predictor variables to interval (0, 1)
X = MinMaxScaler().fit_transform(X)

clf = MLPClassifier(hidden_layer_sizes=(4, ), activation='logistic',
                    solver='lbfgs', max_iter=1000,
                    random_state=1)
clf.fit(X, y)
# clf.predict(X)
fig, ax = plt.subplots()
ConfusionMatrixDisplay.from_estimator(clf, X, y, ax=ax,
  cmap=plt.cm.Greys, colorbar=False)

ax.set_title(f'Neural network {accuracy_score(y, clf.predict(X)):.2f}')
plt.tight_layout()
```

Fig. 7.13 Confusion matrix
of neural network classifier
for predicting q1 from the
remaining responses

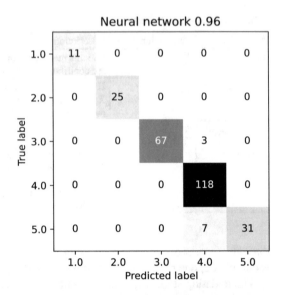

The resulting model has an accuracy of 96% (see Fig. 7.13). However, care must
be taken that this performance is not due to overfitting. This will require monitoring
the training process using validation data.   ∎

7.9   Clustering Methods

Clustering methods are unsupervised methods where the data has no variable labeled
as a target. Our goal is to group similar items together, in clusters. In the previous
sections, we discussed supervised methods where one of the variables is labeled
as a target response and the other variables as predictors that are used to predict
the target. In clustering methods all variables have an equal role. We differentiate
between hierarchical and non-hierarchical clustering methods.

7.9.1   Hierarchical Clustering

Hierarchical clustering is generated by starting with each observation as its own
cluster. Then, clusters are merged iteratively until only a single cluster remains.
The clusters are merged in function of a distance function. The closest clusters are
merged into a new cluster. The end result of an hierarchical clustering method is a
dendrogram, where the j-cluster set is obtained by merging clusters from the $(j+1)$
cluster set.

Example 7.5 To demonstrate clustering methods, we use the **ALMPIN.csv**
data set that consists of six measurements on 70 aluminum pins introduced in

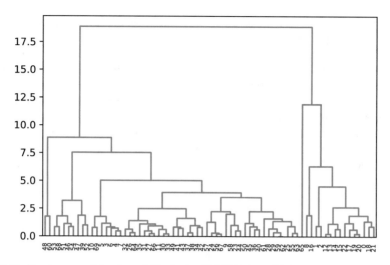

Fig. 7.14 Dendrogram of the six variables in the ALMPIN data with 70 observations

Chap. 4 as Example 4.3. To conduct this analysis in Python, we can cluster the data using implementations in `scipy` or `scikit-learn`. Here, we use the `AgglomerativeClustering` method from `scikit-learn`. The default settings use the Ward algorithm with the Euclidean distance. We also first standardize the data. The dendrogram in Fig. 7.14 is produced by `scipy`. However, as the preparation of the input data for `scipy`'s `dendrogram` method is not straightforward, we provide a function for this with the `mistat` package. The dendrogram starts at the bottom with 70 clusters of individual observations and ends up on the top as one cluster.

```
from sklearn.cluster import AgglomerativeClustering
from sklearn.preprocessing import StandardScaler
from mistat import plot_dendrogram

almpin = mistat.load_data('ALMPIN.csv')

scaler = StandardScaler()
model = AgglomerativeClustering(distance_threshold=0, n_clusters=None)

X = scaler.fit_transform(almpin)
model = model.fit(X)

fig, ax = plt.subplots()
plot_dendrogram(model, ax=ax)
ax.set_title('Dendrogram')
plt.show()
```

∎

The diagram in Fig. 7.14 can be cut across at any level to give any desired number of clusters. Moreover, once two clusters are joined, they remain joined in all higher levels of the hierarchy. The merging of clusters is based on computation of a distance between clusters with a merge on the closest one. There are several possible distance measures described next.

Ward's minimum variance method minimizes the total within-cluster variance. With this method, the distance between two clusters is the ANOVA sum of squares between the two clusters summed over all the variables. At each clustering step, the within-cluster sum of squares is minimized over all partitions obtainable by merging two clusters from the previous generation. Ward's method tends to join clusters with a small number of observations and is strongly biased toward producing clusters with approximately the same number of observations. It is also sensitive to outliers. The distance for Ward's method is

$$D_{KL} = \frac{\left\| \overline{x_K} - \overline{x_L} \right\|^2}{\frac{1}{N_K} + \frac{1}{N_L}} \tag{7.10}$$

where:

C_K is the Kth cluster, subset of 1, 2, ..., n.
N_K is the number of observations in C_K.
$\overline{x_K}$ is the mean vector for cluster C_K .
$\|x\|$ is the square root of the sum of the squares of the elements of x (the Euclidean length of the vector x).

Other methods include single linkage, complete linkage, and average linkage.

Single Linkage
The distance for the single-linkage cluster method is

$$D_{KL} = \min_{i \in C_K} \min_{j \in C_L} d(x_i, x_j) \tag{7.11}$$

with $d(x_i, x_j) = \left\| x_i - x_j \right\|^2$ where x_i is the ith observation.
Complete Linkage
The distance for the complete-linkage cluster method is

$$D_{KL} = \max_{i \in C_K} \max_{j \in C_L} d(x_i, x_j). \tag{7.12}$$

Average Linkage
For average linkage, the distance between two clusters is found by computing the average dissimilarity of each item in the first cluster to each item in the second cluster. The distance for the average linkage cluster method is

$$D_{KL} = \sum_{i \in C_K} \sum_{j \in C_L} \frac{d(x_i, x_j)}{N_K N_L}. \tag{7.13}$$

These distances perform differently on different clustering problems. The dendrograms from single-linkage and complete-linkage methods are invariant under monotone transformations of the pairwise distances. This does not hold for the average-linkage method. Single linkage often leads to long "chains" of clusters, joined by individual points located near each other. Complete linkage tends to produce many small, compact clusters. Average linkage is dependent upon the size of the clusters. Single linkage and complete linkage depend only on the smallest or largest distance, respectively, and not on the size of the clusters.

7.9.2   *K-Means Clustering*

Another clustering methods is K-means. The K-means clustering is formed by an iterative fitting process. The K-means algorithm first selects a set of K points, called cluster seeds, as an initial setup for the means of the clusters. Each observation is assigned to the nearest cluster seed, to form a set of temporary clusters. The seeds are then replaced by the actual cluster means, and the points are reassigned. The process continues until no further changes occur in the clusters.

The K-means algorithm is a special case of the EM algorithm, where E stands for expectation and M stands for maximization. In the case of the K-means algorithm, the calculation of temporary cluster means represents the expectation step, and the assignment of points to the closest clusters represents the maximization step. K-Means clustering supports only numeric columns. K-Means clustering ignores nominal and ordinal data characteristics and treats all variables as continuous.

In K-means you must specify in advance the number of clusters, K. However, you can compare the results of different values of K in order to select an optimal number of clusters for your data. For background on K-means clustering, see Hastie et al. (2009).

Figure 7.15 is a graphical representation of a K-means analysis of the ALMPIN data. To derive this analysis in Python, use the KMeans method from scikit-learn. The data set needs to be standardized prior to clustering (StandardScaler). Use the predict method to predict cluster membership. The transform method returns the distances to the K cluster centers.

```
from sklearn.cluster import KMeans

almpin = mistat.load_data('ALMPIN.csv')

scaler = StandardScaler()
X = scaler.fit_transform(almpin)
model = KMeans(n_clusters=9, random_state=1).fit(X)
print('Cluster membership (first two data points)')
print(model.predict(X)[:2])
print()
print('Distance to cluster center (first two data points)')
model.transform(X)[:2,:]
```

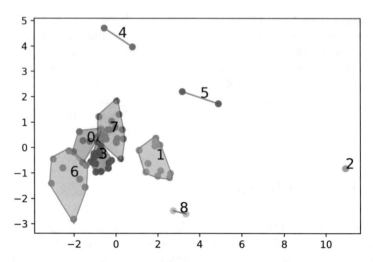

Fig. 7.15 *K*-Means clustering of the six variables in the ALMPIN data with 70 observations

```
Cluster membership (first two data points)
[1 8]

Distance to cluster center (first two data points)

array([[3.33269867, 1.04495283, 9.69783046, 2.48008435, 5.57643458,
         3.98844284, 3.77738831, 2.52221759, 2.4300689 ],
        [5.66992727, 2.7429881 , 7.9467812 , 4.78497545, 7.75144534,
         4.68141833, 5.77196141, 4.89153706, 0.4705252 ]])
```

In Fig. 7.15, the data set is split into nine clusters. Cluster 2 has one observation, and clusters 4, 5, and 8 have two observations each. These clusters include unusual observations that can be characterized by further investigations.

7.9.3   *Cluster Number Selection*

For practical applications, it is necessary to set the number of clusters. This applies to both hierarchical clustering and *K*-means clustering. While this is   often a subjective decision, a large number of methods aim to derive an *optimal* cluster number.

One of these is the elbow method. We determine the overall average within-cluster sum of squares (WSS) as a function of the cluster number. Figure 7.16 demonstrates this for the ALMPIN data set. WSS decreases with the increasing number of clusters. However we can see that the change in WSS gets smaller and smaller. By selecting a cluster number near the *elbow* of the curve, at 3, we find a compromise between complexity and improvement.

`Scikit-learn` provides the following cluster performance metrics: silhouette coefficient, Calinski-Harabasz index, and Davies-Bouldin index. Figure 7.17 shows

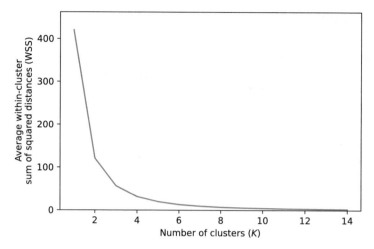

Fig. 7.16 Elbow method for cluster number determination

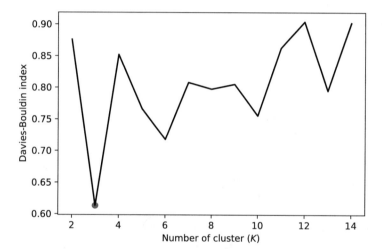

Fig. 7.17 Davies-Bouldin index for cluster number determination

the variation of the Davies-Bouldin index as a function of cluster number. The index combines the average distance of each cluster member to the cluster center and the distance between cluster centers. The *optimal* cluster number is selected based on the minimum of this curve. In this case, we select three clusters.

7.10   Chapter Highlights

This chapter introduced a variety of supervised and unsupervised learning methods. This chapter can only provide a glimpse into modern analytic methods. More in-depth study of these methods will require access to specialized books listed as references.

The topics covered in the chapter include:

- Validation data
- Confusion matrix
- Decision trees
- Boosted tree
- Bootstrap forest
- Random forest
- Bayes' theorem
- Naïve Bayes classifier
- Cluster analysis
- K-Means clusters

7.11   Exercises

Exercise 7.1 Make up a list of supervised and unsupervised applications mentioned in COVID19-related applications.

Exercise 7.2 Create a pruned decision tree model for the `testResult` column in the **SENSORS.csv** data set using `scikit-learn`. Compare the results to the `status` model from Example 7.2.

Exercise 7.3 Fit a gradient boosting model to the sensor data to predict status as the outcome. Use the property `feature_importances_` to identify important predictors and compare to the results from the decision tree model in Sect. 7.5.

Exercise 7.4 Fit a random forest model to the sensor data to predict status as the outcome. Use the property `feature_importances_` to identify important predictors and compare to the results from the decision tree model in Sect. 7.5.

Exercise 7.5 Build decision tree, gradient boosting, and random forest models for the sensor data using status as a target variable.

Use `LabelEncoder` from the `scikit-learn` package to convert the outcome variable into numerical values prior to model building. Split the data set into a 60% training and 40% validation set using `sklearn.model_selection.train_test_split`.

Exercise 7.6 One way of assessing overfitting in models is to assess model performance by repeated randomization of the outcome variable. Build a decision tree model for the sensor data using status as a target variable. Repeat the model training 100 times with randomized outcome.

Exercise 7.7 The data set `DISTILLATION-TOWER.csv` contains a number of sensor data from a distillation tower measured at regular intervals. Use the temperature data measured at different locations in the tower (`TEMP#`) to create a decision tree regressor to predict the resulting vapor pressure (`VapourPressure`):

(i) Split the data set into training and validation set using a 80:20 ratio.
(ii) For each `ccp_alpha` value of the decision tree regressor model, use the test set to estimate the MSE (`mean_squared_error`) of the resulting model. Select a value of `ccp_alpha` to build the final model. The `ccp_alpha` values are returned using the `cost_complexity_pruning_path` method.
(iii) Visualize the final model using any of the available methods.

Exercise 7.8 Create a Naïve Bayes classifier for the sensor data using status as a target. Compare the confusion matrix to the decision tree model (see Fig. 7.9).

Hint: Use the `scikit-learn` method `KBinsDiscretizer` to bin the sensor data, and encode them as `ordinal` data. Try a different number of bins and binning strategies.

Exercise 7.9 Nutritional data from 961 different food items is given in the file **FOOD.csv**. For each food item, there are seven variables: fat (grams), food energy (calories), carbohydrates (grams), protein (grams), cholesterol (milligrams), weight (grams), and saturated fat (grams). Use Ward's distance to construct ten clusters of food items with similarity in the seven recorded variables using cluster analysis of variables.

Exercise 7.10 Repeat Exercise 7.9 with different linkage methods, and compare the results.

Exercise 7.11 Apply the K-means cluster feature to the sensor variables in SENSORS.cvs, and interpret the clusters using the test result and status label.

Exercise 7.12 Develop a procedure based on K-means for quality control using the SENSORS.cvs data. Derive its confusion matrix.

Chapter 8
Modern Analytic Methods: Part II

Preview Chapter 8 includes the tip of the iceberg examples with what we thought were interesting insights, not always available in standard texts. The chapter covers functional data analysis, text analytics, reinforcement learning, Bayesian networks, and causality models.

8.1 Functional Data Analysis

When you collect data from tests or measurements over time or other dimensions, we might want to focus on the functional structure of the data. Examples can be chromatograms from high-performance liquid chromatography (HPLC) systems, dissolution profiles of drug tablets over time, distribution of particle sizes, or measurement of sensors. Functional data is different using individual measurements recorded at different sets of time points. It views functional observations as continuously defined so that an observation is the entire function. With functional data consisting of a set of curves representing repeated measurements, we characterize the main features of the data, for example, with a functional version of principal component analysis (FPCA). The regular version of principal component analysis (PCA) is presented in detail in Chap. 4 (Industrial Statistics book) on Multivariate Statistical Process Control. With this background, let us see an example of functional data analysis (FDA).

Example 8.1 Data set **DISSOLUTION.csv** consists of 12 test and reference tablets measured under dissolution conditions at 5, 10, 15, 20, 30, and 45 s. The level of dissolution recorded at these time instances is the basis for the dissolution functions we will analyze. The test tablets behavior is compared to the reference tablets paths. Ideally the tested generic product is identical to the brand reference.

Supplementary Information The online version contains supplementary material available at https://doi.org/10.1007/978-3-031-07566-7_8.

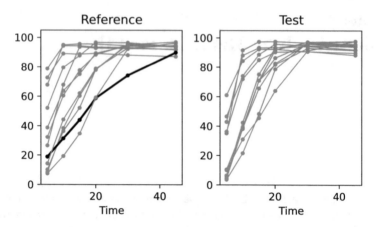

Fig. 8.1 Dissolution paths of reference and tested paths. **T5R** is highlighted

Figure 8.1 shows the dissolution. We can see the curves varying considerably across the different experiments.

```
dissolution = mistat.load_data('DISSOLUTION.csv')

fig, axes = plt.subplots(ncols=2, figsize=(5, 3))
for label, group in dissolution.groupby('Label'):
  ax = axes[0] if label.endswith('R') else axes[1]
  color = 'black' if label == 'T5R' else 'grey'
  lw = 2 if label == 'T5R' else 1
  group.plot(x='Time', y='Data', color=color, ax=ax,
             marker='o', markersize=3, lw=lw)
for ax in axes:
  ax.get_legend().remove()
  ax.set_ylim(0, 105)
axes[0].set_title('Reference')
axes[1].set_title('Test')
plt.tight_layout()
plt.show()
```

To analyze these data using functional data analysis, we use the Python package scikit-fda. It offers a comprehensive set of tools for FDA. The first step is to describe the data using functions. Here, we approximate the function using quadratic splines without smoothing. Periodic functions will be better described using a Fourier basis.

```
from skfda import FDataGrid
from skfda.representation.interpolation import SplineInterpolation

# convert the data to FDataGrid
data = []
labels = []
names = []
for label, group in dissolution.groupby('Label'):
  data.append(group['Data'].values)
  labels.append('Reference' if label.endswith('R') else 'Test')
  names.append(label)
labels = np.array(labels)
grid_points = np.array(sorted(dissolution['Time'].unique()))
```

```
fd = FDataGrid(np.array(data), grid_points,
        dataset_name='Dissolution',
        argument_names=['Time'],
        coordinate_names=['Dissolution'],
        interpolation=SplineInterpolation(2))
```

The functional representation of the data set is stored in `fd` using the specialized class `FDataGrid`. We can use this to calculate the average dissolution curve for the two groups.

```
from skfda.exploratory import stats

mean_ref = stats.mean(fd[labels=='Reference'])
mean_test = stats.mean(fd[labels=='Test'])
means = mean_ref.concatenate(mean_test)
```

Figure 8.2 visualizes the functional representation and the average dissolution curves using the following Python code.

```
group_colors = {'Reference': 'grey', 'Test': 'black'}

fig, axes = plt.subplots(ncols=2)

fd.plot(axes=[axes[0]], group=labels, group_colors=group_colors)
for label, group in dissolution.groupby('Label'):
  color = 'grey' if label.endswith('R') else 'black'
  group.plot.scatter(x='Time', y='Data', c=color, ax=axes[0], alpha=0.2)

means.plot(axes=[axes[1]], group=['Reference', 'Test'],
        group_colors=group_colors)

fig.suptitle('')
axes[0].set_title('Functional representation')
axes[1].set_title('Average dissolution')
for ax in axes:
```

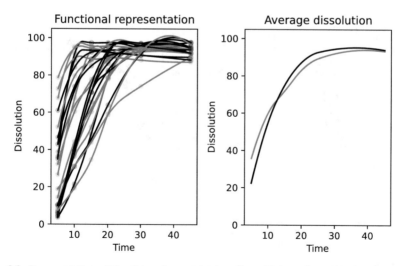

Fig. 8.2 Representation of the data using quadratic splines. The graph on the right shows the average dissolution curve for reference (grey) and test (black)

```
  ax.set_ylim(0, 105)
  ax.set_ylabel('Dissolution')
plt.tight_layout()
plt.show()
```

We see in Fig. 8.2 that the individual curves are shifted to the left and right of the mean curves, most likely due to differences in timing while running the individual dissolution experiments. FDA provides methods to align the individual observations. These approaches are known as registration in FDA. The method we use in our example is **shift registration**.

```
from skfda.preprocessing.registration import ShiftRegistration
shift_registration = ShiftRegistration()
fd_registered = shift_registration.fit_transform(fd)
```

Figure 8.3 shows the effect of shift registration on our data set. The spread is now much tighter, and we can see clearly that the shape of the individual curves is similar. As some curves are shifted considerably, we can see on the right that a few curves indicate decreasing dissolution. This is obviously an artifact of the chosen functional representation. If we use linear splines, the edges are better represented. From here on, we will use the linear spline representation.

Functional data analysis extends the capabilities of traditional statistical techniques in a number of ways. For example, even though an observation is no longer a data point, the concept of outliers still exists. Visual inspection of Fig. 8.1 clearly shows that **T5R** is an outlier. The IQROutlierDetector in scikit-fda implements a generalization of outlier concepts in boxplots and confirms our assumption. It is possible that the dissolution at 30 seconds was misreported too low. A double check of the record should help clarify this.

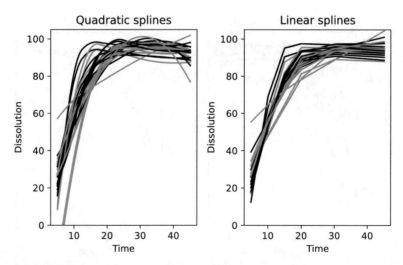

Fig. 8.3 Functional representation after shift registration. Left: result for representing the data using quadratic splines. Right: result for representing the data using linear splines

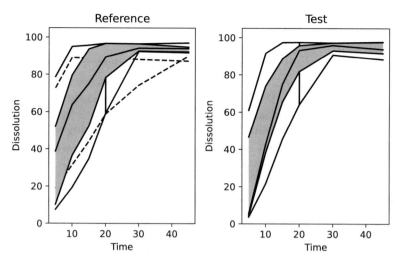

Fig. 8.4 FDA implementation of boxplot. The grey area corresponds to the Q1–Q3 range usual boxplots; the dashed lines are the curves of individual curves that are considered outliers

```
from skfda.exploratory.outliers import IQROutlierDetector

out_detector = IQROutlierDetector()
outliers = out_detector.fit_predict(fd)
print('Outlier:')
for name, outlier in zip(names, outliers):
  if outlier == 1:
    continue
  print('   ', name)
```

```
Outlier:
   T5R
```

Figure 8.4 shows the boxplot representation of the functional data set. A similar representation of the shift-registered data set shows a much tighter grey region in the middle.

```
from skfda.exploratory.visualization import Boxplot
from matplotlib.colors import LinearSegmentedColormap

def addBoxplot(fd, ax):
  cm = LinearSegmentedColormap.from_list('fda', ['grey', 'lightgrey'])
  boxplot = Boxplot(fd)
  boxplot.barcol = 'black'
  boxplot.outliercol = 'black'
  boxplot.colormap = cm
  boxplot.plot(axes=[ax])

fig, axes = plt.subplots(ncols=2)
addBoxplot(fd[labels=='Reference'], axes[0])
addBoxplot(fd[labels=='Test'], axes[1])
fig.suptitle('')
axes[0].set_title('Reference')
axes[1].set_title('Test')
for ax in axes:
```

```
   ax.set_ylim(0, 105)
   ax.set_ylabel('Dissolution')
plt.tight_layout()
plt.show()
```

Functional data analysis methods are considering change over time (or space or some other dimension). Because we are observing curves rather than individual values, the vector-valued observations $X_1, \ldots, X_n$ are replaced by the univariate functions $X_1(t), \ldots, X_n(t)$, where t is a continuous index varying within a closed interval [0, T]. In functional PCA, each sample curve is considered to be an independent realization of a univariate stochastic process $X(t)$ with smooth mean function $EX(t) = \mu(t)$ and covariance function $cov\{X(s), X(t)\} = \sigma(s, t)$.

```
from skfda.preprocessing.dim_reduction.projection import FPCA
fpca = FPCA(n_components=2)
fpca.fit(fd)

df = pd.DataFrame(fpca.transform(fd), columns=['FPCA 1', 'FPCA 2'])
df['labels'] = labels
df['names'] = names

lim1 = (min(df['FPCA 1'])-5, max(df['FPCA 1'])+5)
lim2 = (min(df['FPCA 2'])-5, max(df['FPCA 2'])+5)
fig, axes = plt.subplots(ncols=2, figsize=(5, 3))
for ax, label in zip(axes, ['Reference', 'Test']):
    subset = df[df.labels == label]
    subset.plot.scatter(x='FPCA 1', y='FPCA 2', ax=ax, color='lightgrey')
    ax.set_title(label)
    ax.set_xlim(*lim1)
    ax.set_ylim(*lim2)
outlier = df[df.names == 'T5R']
outlier.plot.scatter(x='FPCA 1', y='FPCA 2', color='black',
    marker='s', ax=axes[0])

plt.tight_layout()
plt.show()
```

Figure 8.5 shows the results of the PCA analysis for the DISSOLV data set with the outlier **T5R** highlighted. In functional PCA we assume that the smooth curves are the completely observed curves. This gives a set of eigenvalues $\{\lambda_j\}$ and (smooth) eigenfunctions $\{Vj(t)\}$ extracted from the sample covariance matrix of the smoothed data. The first and second estimated eigenfunctions are then examined to exhibit location of individual curve variation. Other approaches to functional PCA have been proposed, including the use of roughness penalties and regularization, which optimize the selection of smoothing parameter and choice of the number of principal components simultaneously rather than separately in two stages.

■

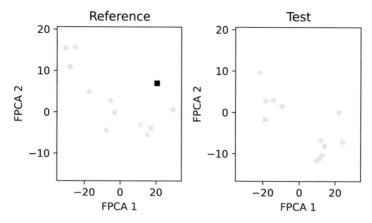

Fig. 8.5 Scatterplot of top two functional principal components; the outlier **T5R** is highlighted by a square marker

8.2 Text Analytics

In this section we discuss methods for analyzing text data, sometimes called unstructured data. Other types of unstructured data include voice recordings and images. The approach we describe is based on a collection of documents or text items that can consist of individual sentences, paragraphs, or a collection of paragraphs. As an example, consider Amazon reviews. You can consider each sentence in the review as a document or the whole review as a single document. A collection of document is called a corpus. We will look at the words, also called tokens, that are included in a document. The analysis we perform is based on a list of terms consisting of tokens included in each document. This approach to text analytics is called "bag of words" where every item has just a collection of individual words. It ignores grammar, word order, sentence structure, and punctuation. Although apparently simplistic, it performs surprisingly well. Before conducting the bag of words analysis, some text preparation is required. This involves tokenizing, phrasing, and terming. The tokenizing stage converts text to lowercase, applies tokenizing method to group characters into tokens, and recodes tokens based on specified recode definitions. For example, to identify dates or other standard formats, we use regular expressions (Regex). Noninformative terms are labeled as stop words and omitted. The phrasing stage collects phrases that occur in the corpus and enables you to specify that individual phrases be treated as terms. The terming stage creates the term list from the tokens and phrases that result from the previous tokenizing and phrasing. For each token, the terming stage checks that minimum and maximum length requirements are met. Tokens that contain only numbers are excluded from this operation. It also checks that a token is qualified and contains at least one alphabetical character. Stemming removes differences such as singular or plural endings. For each phrase added, the terming stage adds the phrase to the term list.

Term frequency shows how frequent a term is in a single document. We also look at how common the term is in the entire corpus. Text processing imposes a small lower limit on the number of items in which a term must occur so that rare terms are ignored. However, terms should also not be too common. A term occurring in every document carries no information. Overly common terms are therefore also eliminated with an upper limit on the number of documents in which a word appears.

In addition to imposing upper and lower limits on term frequency, many systems take into account the distribution of the term over items in a corpus. The fewer documents in which a term occurs, the more significant it is in the documents where it does occur. This sparseness of a term t is measured commonly by an equation called inverse document frequency (IDF). For a given document, d, and term, t, the term frequency is the number of times term t appears in document d: $TF(d;t)$ = # times term t appears in document d. To account for terms that appear frequently in the domain of interest, we compute the inverse document frequency of term t, calculated over the entire corpus and defined as

$$IDF(t) = 1 + log\frac{\text{total number of documents}}{\#\text{ documents containing term t}}. \tag{8.1}$$

The addition of 1 ensures that terms that occur in all document are not ignored. There are variations of calculating TF and IDF described in the literature.

Example 8.2 To demonstrate the data preparation phase of text documents, we use text describing aircraft accidents listed in the National Transportation Board data base: https://www.ntsb.gov/_layouts/ntsb.aviation/Index.aspx.

The **AIRCRAFTINCIDENTS.csv** data was downloaded from http://app.ntsb.gov/aviationquery/Download.ashx?type=csv and is available in the mistat package.

```
incidents = mistat.load_data('AIRCRAFTINCIDENTS.csv')
print(incidents.shape)
```

```
(1906, 27)
```

The data set consists of 1906 incidents in the USA. We will analyze the "Final Narrative" text.

The first step is converting the text into what is known as a bag of words. The following code identifies all words in the text excluding special characters and removes all numbers. The reports also contain a common phrase at the start that we strip here. We also remove "stop words," this means words like "the" or "he" that occur frequently in the English language.

```
import re
from sklearn.feature_extraction.text import CountVectorizer
from sklearn.feature_extraction.text import TfidfTransformer

def preprocessor(text):
```

```
text = text.lower()
text = re.sub(r'\d[\d,]*', '', text)
text = '\n'.join(line for line in text.split('\n')
                 if not line.startswith('ntsb'))
return text

vectorizer = CountVectorizer(preprocessor=preprocessor,
                             stop_words='english')
counts = vectorizer.fit_transform(incidents['Final Narrative'])
print('shape of DTM', counts.shape)
print('total number of terms', np.sum(counts))
```

```
shape of DTM (1906, 8430)
total number of terms 163883
```

The data preparation resulted in 8430 distinct terms, a total of 163,883 terms, an average of 86.0 terms per document. This is translated to a document-term matrix (DTM) with 1906 rows, one for each document, and 8430 columns, one for each term. This matrix contains the number of occurrences of the term in each document but could also be binary, with entries of 1 or 0, depending on the occurrence of a term in a document.

The ten most frequent terms in the data set are listed in Table 8.1. Unsurprisingly, numbers, pilot, airplane, and engine are the most prevalent terms. This is expected for reports that deal with aircraft incidents, and as such, they don't carry much information. Here is where TF-IDF rescoring becomes relevant. We convert the counts into the TF-IDF score to give differential weight of terms depending on their prevalence in the corpus.

```
tfidfTransformer = TfidfTransformer(smooth_idf=False, norm=None)
tfidf = tfidfTransformer.fit_transform(counts)
```

Table 8.2 shows the ten largest TF-IDF scores for the first document. Despite being found 5037 times in the text, "airplane" is not very relevant for this document when considering the TF-IDF score. Words with high TF-IDF scores like Lincoln, Logan, or Illinois capture the information of the document more.

This huge DTM matrix is very sparse. To conduct its analysis, we employ a basic dimension reduction procedure called partial singular value decomposition (SVD).

Table 8.1 Ten most frequently occurring words

Term	Frequency
Pilot	5227
Airplane	5037
Engine	2359
Flight	2194
Landing	2161
Runway	2112
Left	1754
Feet	1609
Fuel	1589
Right	1555

Table 8.2 Terms in the first document that are highly relevant due to TF-IDF rescoring following data preparation (by sentence)

Terms	Counts	TF-IDF	Weight
Lincoln	2	14.908	7.454
Logan	1	8.553	8.553
Said	3	8.234	2.745
Illinois	1	7.166	7.166
Upside	1	6.356	6.356
Came	2	5.671	2.836
Plowed	1	5.557	5.557
Runway	3	5.487	1.829
County	1	5.462	5.462
Feet	3	5.408	1.803

Document = "The pilot said he performed a normal landing to runway 03 (4000 feet by 75 feet, dry asphalt), at the Lincoln-Logan County Airport, Lincoln, Illinois. He said the airplane settled on the runway approximately 1000 feet down from the runway threshold. The pilot raised the flaps and applied full power. The airplane lifted, came back down, and veered to the right. The pilot said the airplane plowed into the snow, nosed over, and came to rest upside down. An examination of the airplane revealed no anomalis"

Partial singular value decomposition approximates the DTM using three matrices: U, S, and V'. The relationship between these matrices is defined as follows:

$$DTM \approx U * S * V \tag{8.2}$$

If k is the number of documents (rows) in the DTM, l is the number of terms (columns) in the DTM and n as a specified number of singular vectors. To achieve data reduction, n must be less than or equal to $\min(k, l)$. It follows that U is an $k \times n$ matrix that contains the left singular vectors of the DTM. S is a diagonal matrix of dimension n. The diagonal entries in S are the singular values of the DTM, and V' is an n by l matrix. The rows in V' (or columns in V) are the right singular vectors.

The right singular vectors capture connections among different terms with similar meanings or topic areas. If three terms tend to appear in the same documents, the SVD is likely to produce a singular vector in V' with large values for those three terms. The U singular vectors represent the documents projected into this new term space.

Principal components, mentioned in Sect. 8.1, are orthogonal linear combinations of variables, and a subset of them can replace the original variables. An analogous dimension reduction method applied to text data is called latent semantic indexing or latent semantic analysis (LSA). LSA is applying partial singular value decomposition (SVD) of the document-term matrix (DTM). This decomposition reduces the text data into a manageable number of dimensions for analysis. For example, we can now perform a topics analysis. The rotated SVD option performs a rotation on the partial singular value decomposition (SVD) of the document-term matrix (DTM). In `scikit-learn` we use the `TruncatedSVD` method for this.

```
from sklearn.decomposition import TruncatedSVD
from sklearn.preprocessing import Normalizer
svd = TruncatedSVD(10)
tfidf = Normalizer().fit_transform(tfidf)
lsa_tfidf = svd.fit_transform(tfidf)
print(lsa_tfidf.shape)
```

| (1906, 10)

You must specify the number of rotated singular vectors, which corresponds to the number of topics that you want to retain from the DTM. As we specified 10 here, the final matrix has 1906 rows and 10 columns.

We can further analyze the terms that contribute most to each of the components. This is called topic analysis. Topic analysis is equivalent to a rotated principal component analysis (PCA). The rotation takes a set of singular vectors and rotates them to make them point more directly in the coordinate directions. This rotation makes the vectors help explain the text as each rotated vector orients toward a set of terms. Negative values indicate a repulsion force. The terms with negative values occur in a topic less frequently compared to the terms with positive values.

Looking at Table 8.3, we identify in Topics 5, 9, and 10 incidents related to weather conditions. Topics 7 and 8 are about incidents involving students and instructors. Topics 1 and 2 are mentioning fuel and engine issues. Topic 3 and 4 involve helicopters, etc.

If we now link the documents to supplementary data such as incident impact, one can link label reports by topic and derive a predictive model that can drive accident prevention initiatives.

◼

8.3 Bayesian Networks

Bayesian networks (BNs) were introduced in Sect. 2.1.6. They implement a graphical model structure known as a directed acyclic graph (DAG) that is popular in statistics, machine learning, and artificial intelligence. BNs enable an effective representation and computation of the joint probability distribution over a set of random variables (Pearl 1985). The structure of a DAG is defined by two sets: the set of nodes and the set of directed arcs; arcs are often also called edges. The nodes represent random variables and are drawn as circles labeled by the variable names. The arcs represent links among the variables and are represented by arrows between nodes. In particular, an arc from node X_i to node X_j represents a relation between the corresponding variables. Thus, an arrow indicates that a value taken by variable X_j depends on the value taken by variable X_i. This property is used to reduce the number of parameters that are required to characterize the joint probability distribution (JPD) of the variables. This reduction provides an efficient way to compute the posterior probabilities given the evidence present in the data

Table 8.3 Topic analysis

Topic 1	Loading 1	Topic 2	Loading 2	Topic 3	Loading 3	Topic 4	Loading 4	Topic 5	Loading 5
Airplane	0.35	Fuel	0.67	Helicopter	0.61	Helicopter	0.36	Fuel	0.34
Pilot	0.28	Engine	0.29	Rotor	0.24	Gear	0.33	Knots	0.22
Runway	0.23	Tank	0.24	Flight	0.14	Landing	0.27	Helicopter	0.20
Landing	0.19	Tanks	0.12	Tail	0.10	Right	0.21	Wind	0.17
Fuel	0.19	Gallons	0.12	Engine	0.10	Main	0.20	Tank	0.17
Engine	0.19	Power	0.12	Feet	0.10	Left	0.17	Aircraft	0.14
Left	0.17	Selector	0.09	Collective	0.10	Rotor	0.15	Runway	0.13
Right	0.15	Forced	0.09	Blades	0.10	Fuel	0.12	Winds	0.12
Reported	0.15	Carburetor	0.08	Ground	0.09	Student	0.12	Degrees	0.11
Flight	0.14	Loss	0.06	Instructor	0.08	Instructor	0.11	Left	0.11

Topic 6	Loading 6	Topic 7	Loading 7	Topic 8	Loading 8	Topic 9	Loading 9	Topic 10	Loading 10
Aircraft	0.70	Gear	0.35	Student	0.53	Knots	0.29	Said	0.38
Gear	0.19	Flight	0.28	Instructor	0.38	Runway	0.28	Runway	0.23
Landing	0.11	Student	0.20	Aircraft	0.26	Engine	0.25	Approach	0.17
Reported	0.11	Landing	0.20	cfi	0.16	Wind	0.20	Feet	0.16
Time	0.07	Instructor	0.19	Flight	0.13	Power	0.17	Aircraft	0.15
Failure	0.06	Weather	0.12	Controls	0.09	Winds	0.16	End	0.13
Control	0.06	Approach	0.12	Runway	0.08	Degrees	0.15	Gear	0.11
Terrain	0.05	Airport	0.12	Dual	0.08	Landing	0.15	Fuel	0.11
Malfunction	0.05	Hours	0.11	Landings	0.07	Carburetor	0.12	Pilot	0.10
Encountered	0.05	Accident	0.10	Solo	0.07	Reported	0.11	Snow	0.09

(Nielsen and Jensen 2007; Pourret et al. 2008; Ben Gal 2008; Pearl 2009; Koski and Noble 2009; Kenett 2016, 2017). In addition to the DAG structure, which is often considered as the "qualitative" part of the model, a BN includes "quantitative" parameters. These parameters are described by applying the Markov property, where the conditional probability distribution (CPD) at each node depends only on its parents. For discrete random variables, this conditional probability is represented by a table, listing the local probability that a child node takes on each of the feasible values—for each combination of values of its parents. The joint distribution of a collection of variables is determined uniquely by these local conditional probability tables (CPT). In learning the network structure, one can include white lists of forced causality links imposed by expert opinion and black lists of links that are not to be included in the network.

To fully specify a BN, and thus represent the joint probability distributions, it is necessary to specify for each node X the probability distribution for X conditional upon X's parents. The distribution of X, conditional upon its parents, may have any form with or without constraints.

These conditional distributions include parameters which are often unknown and must be estimated from data, for example, using maximum likelihood. Direct maximization of the likelihood (or of the posterior probability) is usually based on the expectation-maximization (E-M) algorithm which alternates computing expected values of the unobserved variables conditional on observed data, with maximizing the complete likelihood assuming that previously computed expected values are correct. Under mild regularity conditions, this process converges to maximum likelihood (or maximum posterior) values of parameters (Heckerman 1995).

A Bayesian approach treats parameters as additional unobserved variables and computes a full posterior distribution over all nodes conditional upon observed data and then integrates out the parameters. This, however, can be expensive and leads to large dimension models, and in practice classical parameter-setting approaches are more common.

Bayesian networks (BNs) can be specified by expert knowledge (using white lists and black lists) or learned from data or in combinations of both (Kenett 2016). The parameters of the local distributions are learned from data, priors elicited from experts, or both. Learning the graph structure of a BN requires a scoring function and a search strategy. Common scoring functions include the posterior probability of the structure given the training data, the Bayesian information criterion (BIC), or Akaike information criterion (AIC). When fitting models, adding parameters increases the likelihood, which may result in overfitting. Both BIC and AIC resolve this problem by introducing a penalty term for the number of parameters in the model with the penalty term being larger in BIC than in AIC. The time requirement of an exhaustive search, returning back a structure that maximizes the score, is super-exponential in the number of variables. A local search strategy makes incremental changes aimed at improving the score of the structure. A global search algorithm like Markov chain Monte Carlo (MCMC) can avoid getting trapped in local minima. A partial list of structure learning algorithms includes hill-climbing with score functions BIC and AIC grow-shrink, incremental association, fast incremental association, interleaved incremental association, hybrid algorithms, and phase-restricted maximization.

Example 8.3 The data set ABC2.csv contains data from an electronic product company's annual customer satisfaction survey collected from 266 companies (customers) (Kenett and Salini 2009). The data set contains for each company its location (country) and feedback summarized responses on:

- Equipment
- SalesSup (sales support)
- TechnicalSup (technical support)
- Suppliers
- AdministrativeSup (administrative support)
- TermsCondPrices (terms, conditions, and prices)

Additional information:

- Satisfaction: overall satisfaction
- Recommendation: recommending the product to others
- Repurchase: intent to repurchase

The response data are ordinal data ranging from 1 (very low satisfaction, very unlikely) to 5 (very high satisfaction, very likely).

The HillClimbSearch method from the pgmpy is used to derive a structure of the Bayesian network from the data. The Bayesian network is shown in Fig. 8.7.

```
from pgmpy.estimators import HillClimbSearch

abc = mistat.load_data('ABC2.csv')
abc = abc.drop(columns=['ID'])

est = HillClimbSearch(data=abc)
model = est.estimate(max_indegree=4, max_iter=int(1e4), show_progress=False,
                     scoring_method='k2score')
```

```
import pydotplus

def layoutGraph(dot_data, pdfFile):
    graph = pydotplus.graph_from_dot_data(dot_data)
    with open(pdfFile, 'wb') as f:
      f.write(graph.create_pdf())

def createGraph(G, pdfFile):
    sortedNodes = list(nx.topological_sort(G))
    commonSettings = """
    edge [ fontsize=11, color=gray55 ];
    # size="10,10"
    graph [ranksep="0.2", dpi=300];
    """
    def makeNode(label):
        return f'{label} [ label="{label}", fontsize=11, color=white ];'
    def makeEdge(edge):
        fromNode, toNode = edge
        return f'{fromNode} -> {toNode};'

    allNodes = '\n'.join(makeNode(node) for node in sortedNodes)
    allEdges = '\n'.join(makeEdge(edge) for edge in G.edges)
```

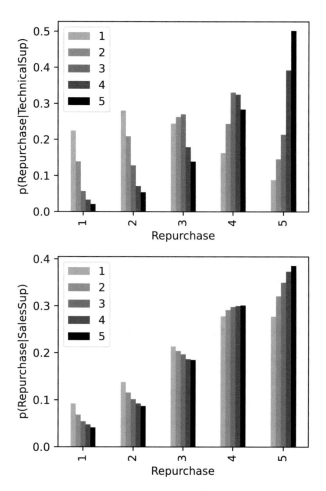

Fig. 8.6 Inferred conditional probability of repurchasing decision on quality of technical support

```
s = f"""
digraph ethane {{
{ commonSettings }
{ allNodes }
{ allEdges }
}}
"""
return layoutGraph(s, pdfFile)

createGraph(model, 'compiled/figures/Chap008_abcBNmodel.pdf')
```

Using the derived network structure, we can now fit the data to the Bayesian network to deduce the CPD:

```
from pgmpy.models import BayesianNetwork
from pgmpy.estimators import MaximumLikelihoodEstimator
```

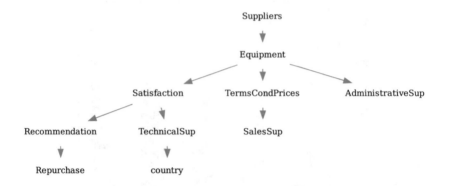

Fig. 8.7 Bayesian network estimated from ABC data set

```
# convert to BayesianNetwork and fit data
model = BayesianNetwork(model.edges())
model.fit(data=abc, estimator=MaximumLikelihoodEstimator)
```

As an example, here is the CPD for the arc Equipment-AdministrativeSup:

```
for cpd in model.get_cpds():
    df = pd.DataFrame(cpd.values)
    v0 = cpd.variables[0]
    df.index = pd.MultiIndex.from_tuples([(v0, state) for state in cpd.state_names[v0]
    if len(cpd.variables) > 1:
        v1 = cpd.variables[1]
        df.columns = pd.MultiIndex.from_tuples([(v1, state) for state in cpd.state_names
    print(df.round(3))
    break
```

		Equipment				
		1.0	2.0	3.0	4.0	5.0
Satisfaction	1.0	0.8	0.167	0.033	0.008	0.000
	2.0	0.0	0.444	0.100	0.042	0.154
	3.0	0.0	0.278	0.533	0.158	0.077
	4.0	0.2	0.111	0.283	0.583	0.385
	5.0	0.0	0.000	0.050	0.208	0.385

We can also infer probabilities between nodes that are not directly connected. The Python code for inferring the influence of technical support experience on repurchasing decision is shown below and the result visualized in Fig. 8.6.

```
from pgmpy.inference import VariableElimination, BeliefPropagation
infer = BeliefPropagation(model)
results = {i: infer.query(variables=['Repurchase'],
                          evidence={'TechnicalSup': i}).values
           for i in range(1, 6)}
```

The Bayesian network model indicates a strong positive correlation between quality of technical support and repurchasing decision. Sales support on the other hand has only little influence (see Fig. 8.6).

We can further query the BN model using belief propagation. The estimated probability distribution for all variables is shown in Fig. 8.8. Belief propagation also allows to study the effect of additional evidence on these probability distributions. In Fig. 8.9 *Recommendation* is set to very high. The largest changes are unsurprisingly seen for *Satisfaction* and *Repurchase*.

Above we've seen that good technical support experience has a positive influence on repurchasing. We can use the BN to identify countries where we should improve technical support. Figure 8.10 shows the estimated distribution of technical support scores by country. The data tell us that technical support should be improved in the Benelux countries.

■

8.4 Causality Models

Causality analysis has been studied from two main different points of view, the "probabilistic" view and the "mechanistic" view. Under the probabilistic view, the causal effect of an intervention is judged by comparing the evolution of the system when the intervention is and when it is not present. The mechanistic point of view focuses on understanding the mechanisms determining how specific effects come about. The interventionist and mechanistic viewpoints are not mutually exclusive. For example, when studying biological systems, scientists carry out experiments where they intervene on the system by adding a substance or by knocking out genes. However, the effect of a drug product on the human body cannot be decided only

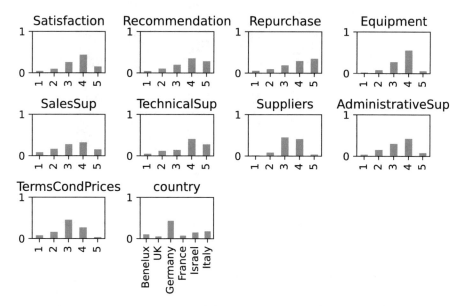

Fig. 8.8 Estimated probability distribution derived using belief propagation

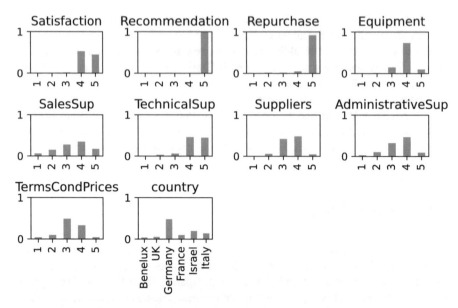

Fig. 8.9 Estimated probability distribution if recommendation is very high

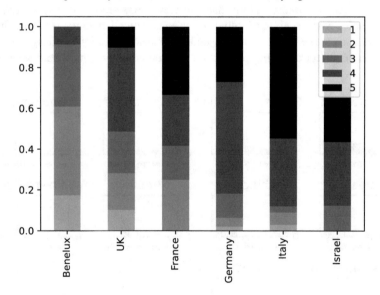

Fig. 8.10 Distribution of estimated technical support score conditioned by country. Countries are sorted by increasing expected technical score

in the laboratory. A mechanistic understanding based on pharmacometrics models is a preliminary condition for determining if a certain medicinal treatment should be studied in order to elucidate biological mechanisms used to intervene and either prevent or cure a disease. The concept of potential outcomes is present in the work on randomized experiments by Fisher and Neyman in the 1920s (Fisher 1935; Neyman and Pearson 1967) and was extended by Rubin in the 1970s (Dempster et al. 1977) to non-randomized studies and different modes of inference (Mealli et al. 2012). In their work, causal effects are viewed as comparisons of potential outcomes, each corresponding to a level of the treatment, and each observable had the treatment taken on the corresponding level with at most one outcome actually observed, the one corresponding to the treatment level realized. In addition, the assignment mechanism needs to be explicitly defined as a probability model for how units receive the different treatment levels. With this perspective, a causal inference problem is viewed as a problem of missing data, where the assignment mechanism is explicitly modeled as a process for revealing the observed data. The assumptions on the assignment mechanism are crucial for identifying and deriving methods to estimate causal effects (Frosini 2006).

Imai et al. (2013) study how to design randomized experiments to identify causal mechanisms. They study designs that are useful in situations where researchers can directly manipulate the intermediate variable that lies on the causal path from the treatment to the outcome. Such a variable is often referred to as a "mediator." Under the parallel design, each subject is randomly assigned to one of two experiments. In one experiment only the treatment variable is randomized, whereas in the other, both the treatment and the mediator are randomized. Under the crossover design, each experimental unit is sequentially assigned to two experiments where the first assignment is conducted randomly and the subsequent assignment is determined without randomization on the basis of the treatment and mediator values in the previous experiment. They propose designs that permit the use of indirect and subtle manipulation. Under the parallel encouragement design, experimental subjects who are assigned to the second experiment are randomly encouraged to take (rather than assigned to) certain values of the mediator after the treatment has been randomized. Similarly, the crossover encouragement design employs randomized encouragement rather than the direct manipulation in the second experiment. These two designs generalize the classical parallel and crossover designs in clinical trials, allowing for imperfect manipulation, thus providing informative inferences about causal mechanisms by focusing on a subset of the population.

Causal Bayesian networks are BNs where the effect of any intervention can be defined by a "do" operator that separates intervention from conditioning. The basic idea is that intervention breaks the influence of a confounder so that one can make a true causal assessment. The established counterfactual definitions of direct and indirect effects depend on an ability to manipulate mediators. A BN graphical representations, based on local independence graphs and dynamic path analysis, can be used to provide an overview of dynamic relations (Aalen et al. 2012). As an alternative approach, the econometric approach develops explicit models of outcomes, where the causes of effects are investigated and the mechanisms governing the choice of treatment are analyzed. In such investigations, counterfactuals are studied (counterfactuals are possible outcomes in different hypothetical states of the world). The study of causality in studies of economic policies involves:

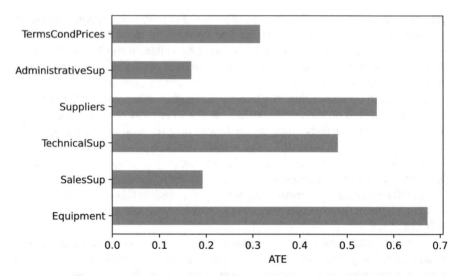

Fig. 8.11 Average treatment effect (ATE) for different interventions on *Satisfaction*

(a) Defining counterfactuals
(b) Identifying causal models from idealized data of population distributions
(c) Identifying causal models from actual data, where sampling variability is an
     issue (Heckman 2008)

Pearl developed BNs as the method of choice for reasoning in artificial intelligence and expert systems, replacing earlier ad hoc rule-based systems. His extensive work covers topics such as causal calculus, counterfactuals, do-calculus, transportability, missingness graphs, causal mediation, graph mutilation, and external validity (Pearl 1988).

Example 8.4 Continuing with the previous example, we can determine the average treatment effect (ATE) of changes on *Satisfaction*. The results are shown in Fig. 8.11. The ATE is defined as

$$\text{ATE} = E(Y_1) - E(Y_0)$$

where Y defines the outcome variable (here *Satisfaction*). $E(Y_1)$ is the average outcome for the case of the treatment and $E(Y_0)$ for the case of not having the treatment. We can see that changes to the *Equipment*, *Suppliers*, and *TechnicalSup* have the largest ATE.   ∎

Granger (1969) developed an approach to test whether one time series is forecasting another. With this approach that leverages the temporal dimension, causality is tested by measuring the ability to predict future values of a time series using prior values of another time series. For two time series X_t and Y_t, the *Granger causality test* is comparing two auto-regression models:

$$Y_t = a_0 + \sum_{j=1}^{p} a_j Y_{t-j} + e_t$$

$$Y_t = a_0 + \sum_{j=1}^{p} a_j Y_{t-j} + \sum_{j=1}^{p} b_j X_{t-j} + e_t.$$

In the second model, the contributed lagged terms b_j of X_{t-j} are tested for significance. If some of the b_j are significant and the addition of the X_{t-j} adds explanatory power, X_t Granger causes Y_t. The assumptions here are that:

1. The cause happens prior to its effect.
2. The cause has unique information about the future values of its effect.

Any lagged value of one of the variables is retained in the regression if it is significant according to a t-test, and it and the other lagged values of the variable jointly add explanatory power to the model according to an F-test. The null hypothesis of no Granger causality is not rejected if and only if no lagged values of an explanatory variable are retained in the regression. For more on time series, see Chap. 6.

This basic idea was further extended to multivariate problems using regularization (Lozano et al. 2009). Tank et al. (2021) extended the Granger causality test to allow nonlinear relationships in the time series regression.

Example 8.5 The data set **DISTILLATION-TOWER.csv** contains snapshot measurements for 27 variables from a distillation tower. Using the Granger causality test available in statsmodels, we can test if the temperature columns have a causal effect on vapor pressure.

```
distTower = mistat.load_data('DISTILLATION-TOWER.csv')
distTower = distTower.set_index('Date')
subset = ['VapourPressure', 'Temp1', 'Temp2', 'Temp3', 'Temp4', 'Temp5',
          'Temp6', 'Temp7', 'Temp8', 'Temp9', 'Temp10','Temp11','Temp12']
distTower = distTower[subset]
results = []
for c in subset[1:]:
    # use pct_change to make time series stationary
    data = distTower[["VapourPressure", c]].pct_change().dropna()
    gc_res = grangercausalitytests(data, 10, verbose=False)
    results.append({f'lag {i}': gc[0]['ssr_ftest'][1]
                    for i, gc in gc_res.items()})
df = pd.DataFrame(results, index=subset[1:])
df['id'] = df.index
```

The resulting p-values for the various measurements and lag times are shown in Fig. 8.12. The Granger causality test for Temp2 shows a causal relationship between this measurements and vapor pressure. While the process location of the temperature measurements is not known for this data set, it is highly likely that Temp2 is measured at the top of the distillation tower. ∎

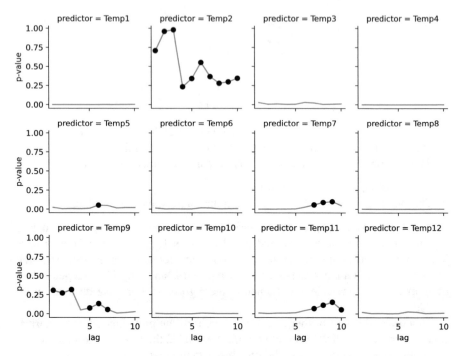

Fig. 8.12 Granger causality test for the causal effect of temperature measurements on the vapor pressure

8.5   Chapter Highlights

This last chapter extends modern analytic methods covered in Chap. 7. It is designed as an introductory chapter with examples of Python applications and real-life case studies. The companion text on Industrial Statistics builds on the methods presented in this book.

The topics covered in the chapter include:

- Functional data analysis
- Functional principal components
- Text analytics
- Bag of words
- Topic analysis
- Bayesian networks
- Causality

8.6   Exercises

Exercise 8.1 Use functional data analysis to analyze the dissolution data of reference and test tablets. Use shift registration with split interpolation of order 1, 2, and 3 to align the curves. Determine the mean dissolution curves for the reference and test tablets, and compare the result for the different interpolation methods. Compare the curves and discuss the differences.

Exercise 8.2 The **Pinch** data set contains measurements of pinch force for 20 replications from the start of measurement. The pinch force is measured every 2 ms over a 300 ms of interval:

  (i) Load the data. The data are available in the `fda` R-package as data sets `pinchraw` and `pinchtime`. Load the two data sets using the command `fetch_cran` command and combine in a `FDataGrid`.
       `(skfda.datasets.fetch_cran(name, package_name))`
  (ii) Plot the data set and discuss the graph.
 (iii) Use smoothing to focus on the shape of the curve. You can use
       `skfda.preprocessing.smoothing.kernel_smoothers.NadarayaWatson`
       `Smoother`.
       Explore various values for the `smoothing_parameter`, and discuss its effect. Select a suitable `smoothing_parameter` to create a smoothed version of the data set for further processing.
  (iv) Use landmark registration to align the smoothed measurements by their maximum value. As a first step, identify the times at which each measurement had it maximum (use `fd.data_matrix.argmax(axis=1)` to identify the index of the measurement and use `pinchtime` to get the time to get the landmark values). Next use `skfda.preprocessing.registration.` `landmark_shift` to register the smoothed curves.
  (v) Plot the registered curves and discuss the graph.

Exercise 8.3 The **Moisture** data set contains near-infrared reflectance spectra of 100 wheat samples together with the samples' moisture content. Convert the moisture values into two classes, and develop a classification model to predict the moisture content of the sample:

  (i) Load the data. The data are available in the `fds` R-package as data sets `Moisturespectrum` and `Moisturevalues`. Load the two data sets using the `skfda.datasets.fetch_cran(name, package_name)` command.
  (ii) Determine a threshold value to split the moisture values in `high` and `low` moisture content.
 (iii) Convert the spectrum information into the `FDataGrid` representation of the `scikit-fda` package, and plot the spectra. What do you observe?

(iv) Normalize the sample spectra so that the differences in intensities are less influential. This is in general achieved using the standard normal variate (SVN) method. For each spectrum, subtract the mean of the intensities and divide by their standard deviation. As before plot the spectra and discuss the observed difference.

(v) Create k-nearest neighbor classification models to predict the moisture content class from the raw and normalized spectra.
(use `skfda.ml.classification.KNeighborsClassifier`)

Exercise 8.4 Repeat the previous Exercise 8.3 creating K-nearest neighbor regression models to predict the moisture content of the samples:

(i) Load and preprocess the **Moisture** data as described in Exercise 8.3.
(ii) Create k-nearest neighbor regression models to predict the moisture content from the raw and normalized spectra (use `skfda.ml.regression.KNeighborsRegressor`). Discuss the results.
(iii) Using one of the regression models based on the normalized spectra, plot predicted versus actual moisture content. Discuss the result. Does a regression model add additional information compared to a classification model?

Exercise 8.5 In this exercise, we look at the result of a functional PCA using the **Moisture** data set from Exercise 8.3:

(i) Load and preprocess the **Moisture** data as described in Exercise 8.3.
(ii) Carry out a functional principal component analysis of the raw and normalized spectra with two components. Plot the projection of the spectra onto the two components and color by moisture class. Discuss the results (use `skfda.preprocessing.dim_reduction.projection.FPCA`).

Exercise 8.6 Pick articles on global warming from two journals on the web. Use the same procedure for identifying stop words, phrases, and other data preparation steps. Compare the topics in these two articles using five topics. Repeat the analysis using ten topics. Report on the differences:

(i) Convert the two documents into a list of paragraphs and labels.
(ii) Treating each paragraph as an individual document, create a a document-term matrix (DTM). Ignore numerical values as terms. Which terms occur most frequently in the two articles?
(iii) Use TF-IDF to convert the DTM.
(iv) Use latent semantic analysis (LSA) to find five topics.

Exercise 8.7 Pick three articles on COVID-19 economic impact from the same author. Use the same procedure for identifying stop words, phrases, and other data preparation steps. Compare the topics in these three articles using ten topics.

Exercise 8.8 Use the **LAPTOP_REVIEWS.csv** data set to analyze reviews, and build a model to predict positive and negative reviews:

(i) Load the **LAPTOP_REVIEWS** data using the `mistat` package. Preprocess the data set by combining the values of the columns `Review title` and `Review content` into a new column `Review`, and remove missing rows with missing values in these two columns.

(ii) Convert the `Reviews` into a document-term matrix (DTM) using a count vectorizer. Split the reviews into words and remove English stop words. Use a custom preprocessor to remove numbers from each word.

(iii) Convert the counts in the DTM into TF-IDF scores.

(iv) Normalize the TF-IDF scores, and apply partial singular value decomposition (SVD) to convert the sparse document representation into a dense representation. Keep 20 components from the SVD.

(v) Build a logistic regression model to predict positive and negative reviews. A review is positive if the `User rating` is 5. Determine the predictive accuracy of the model by splitting the data set into 60% training and 40% test sets.

Appendix A
Introduction to Python

There are many excellent books and online resources that can introduce you to Python. Python itself comes with an excellent tutorial that you can find at https://docs.python.org/3/tutorial/. Instead of duplicating here what has been improved over many years, we suggest the reader to follow the Python tutorial. In particular, we recommend reading the following chapters in the tutorial:

- An informal introduction to python
- More control flow tools
- Data structures

In the following, we will point out a selection of more specialized topics that we use in the code examples throughout the book.

A.1 List, Set, and Dictionary Comprehensions

Many data handling tasks require the creation of lists or dictionaries. We can use a `for` loop in this case:

```
the_list = []
for i in range(10):
  the_list.append(2 * i)
the_list
```

```
[0, 2, 4, 6, 8, 10, 12, 14, 16, 18]
```

Instead of using the `for` loop, Python has a more concise way of achieving the same outcome using what is called a list comprehension:

```
the_list = [2 * i for i in range(10)]
the_list
```

```
[0, 2, 4, 6, 8, 10, 12, 14, 16, 18]
```

List comprehensions can also be used if the addition to the list is conditional. In the following example, we create a list of numbers divisible by 3.

```
the_list = []
for i in range(20):
  if i % 3 == 0:
    the_list.append(i)

the_list = [i for i in range(20) if i % 3 == 0]
the_list
```

```
[0, 3, 6, 9, 12, 15, 18]
```

The list comprehension is easier to read.

A similar construct can also be used to create sets:

```
letters = ['a', 'y', 'x', 'a', 'y', 'z']
unique_letters = {c for c in letters}
unique_letters
```

```
{'a', 'x', 'y', 'z'}
```

The set comprehension uses curly brackets instead of the square brackets in list comprehensions.

Dictionary comprehensions create dictionaries. The following example creates a dictionary that maps a number to its square. We show first the implementation using a for loop and then the dictionary comprehension:

```
squares = {}
for i in range(10):
  squares[i] = i * i

squares = {i: i * i for i in range(10)}
squares
```

```
{0: 0, 1: 1, 2: 4, 3: 9, 4: 16, 5: 25, 6: 36, 7: 49, 8: 64, 9: 81}
```

A.2   Pandas Data Frames

Most of the data sets used in this book are either in list form or tabular. The pandas package (https://pandas.pydata.org/) implements these data structures. The mistat package returns the data either as pandas DataFrame or Series objects:

```
import mistat

almpin = mistat.load_data('ALMPIN')
print('ALMPIN', type(almpin))

steelrod = mistat.load_data('STEELROD')
print('STEELROD', type(steelrod))
```

```
ALMPIN <class 'pandas.core.frame.DataFrame'>
STEELROD <class 'pandas.core.series.Series'>
```

The DataFrame and Series objects offer additional functionality to use these them in an efficient and fast manner. As an example, here is the calculation of the column means:

```
almpin.mean()
```

```
diam1          9.992857
diam2          9.987286
diam3          9.983571
capDiam       14.984571
lenNocp       49.907857
lenWcp        60.027857
dtype: float64
```

The `describe` method returns basic statistics for each column in a DataFrame:

```
almpin.describe().round(3)
```

	diam1	diam2	diam3	capDiam	lenNocp	lenWcp
count	70.000	70.000	70.000	70.000	70.000	70.000
mean	9.993	9.987	9.984	14.985	49.908	60.028
std	0.016	0.018	0.017	0.019	0.044	0.048
min	9.900	9.890	9.910	14.880	49.810	59.910
25%	9.990	9.982	9.980	14.980	49.890	60.000
50%	10.000	9.990	9.990	14.990	49.910	60.020
75%	10.000	10.000	9.990	14.990	49.928	60.050
max	10.010	10.010	10.010	15.010	50.070	60.150

As the `pandas` package is used frequently in many machine learning packages, we recommend that you make yourself familiar by reading the documentation.

A.3   Data Visualization Using `Pandas` and `Matplotlib`

Packages like `pandas` or `seaborn` support a variety of visualizations that are often sufficient for exploratory data analysis. However there may be cases where you want to customize the graph further to highlight aspects of your analysis. As these packages often use the `matplotlib` package (https://matplotlib.org/) as their foundation, we can achieve this customization using basic `matplotlib` commands.

This is demonstrated in Fig. A.1. Here, we use the `matplotlib` axis object that is returned from the `pandas` plot function to add additional lines to the graph.

There are many more examples that can be found in the accompanying source code repository at https://gedeck.github.io/mistat-code-solutions/ModernStatistics/.

(a)

```
import matplotlib.pyplot as plt

steelrod = mistat.load_data('STEELROD')
steelrod.plot(style='.', color='black',
              xlabel='Index', ylabel='Steel rod Length')
plt.show()
```

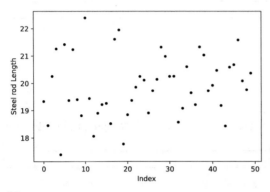

(b)

```
ax = steelrod.plot(style='.', color='black',
                   xlabel='Index', ylabel='Steel rod Length')
ax.hlines(y=steelrod[:26].mean(), xmin=0, xmax=26)
ax.hlines(y=steelrod[26:].mean(), xmin=26, xmax=len(steelrod))
plt.show()
```

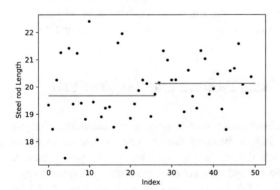

Fig. A.1 Data visualization using pandas and customization. (**a**) Default graph created using pandas. (**b**) Customization of (b) using `matplotlib` commands

Appendix B
List of Python Packages

`bootstrapped`:
>   *Implementations of the percentile based bootstrap*
>   https://pypi.org/project/bootstrapped/
>   https://github.com/facebookincubator/bootstrapped

`dtreeviz`:
>   *A Python 3 library for scikit-learn, XGBoost, LightGBM, and Spark decision tree visualization*
>   https://pypi.org/project/dtreeviz/
>   https://github.com/parrt/dtreeviz

`KDEpy`:
>   *Kernel Density Estimation in Python.* https://pypi.org/project/KDEpy/
>   https://github.com/tommyod/KDEpy

`matplotlib`:
>   *Python plotting package*
>   https://pypi.org/project/matplotlib/
>   https://matplotlib.org/

`mistat`:
>   *Modern Statistics/Industrial Statistics: A Computer Based Approach with Python*
>   https://pypi.org/project/mistat/

`mpl_toolkits`:
>   *Package distributed with matplotlib*
>   https://matplotlib.org/api/toolkits/index.html

`networkx`:
>   *Python package for creating and manipulating graphs and networks*
>   https://pypi.org/project/networkx/
>   https://networkx.org/

`numpy`:
>   *NumPy is the fundamental package for array computing with Python*
>   https://pypi.org/project/numpy/

https://numpy.org/

`pandas`:

Powerful data structures for data analysis, time series, and statistics
https://pypi.org/project/pandas/
https://pandas.pydata.org/

`pgmpy`:

A library for Probabilistic Graphical Models
https://pypi.org/project/pgmpy/
https://github.com/pgmpy/pgmpy

`pingouin`

Pingouin: statistical package for Python
https://pypi.org/project/pingouin/
https://pingouin-stats.org/

`scipy`:

SciPy: Scientific Library for Python
https://pypi.org/project/scipy/
https://www.scipy.org/

`seaborn`:

seaborn: statistical data visualization
https://pypi.org/project/seaborn/
https://seaborn.pydata.org/

`scikit-fda` (skfda):

Functional Data Analysis Python package
https://pypi.org/project/scikit-fda/
https://fda.readthedocs.io/

`scikit-learn` (sklearn):

A set of python modules for machine learning and data mining
https://pypi.org/project/scikit-learn/
https://scikit-learn.org/

`statsmodels`:

Statistical computations and models for Python
https://pypi.org/project/statsmodels/
https://www.statsmodels.org/

`xgboost`:

XGBoost is an optimized distributed gradient boosting library designed to be highly efficient, flexible and portable
https://pypi.org/project/xgboost/
https://github.com/dmlc/xgboost
https://xgboost.readthedocs.io/en/latest/

Appendix C
Code Repository and Solution Manual

The source code used in this book both shown as code examples and used to create the figures in this book is available from the GitHub repository https://gedeck.github.io/mistat-code-solutions/ModernStatistics/.

The repository also contains the solutions for the exercises in the book.

Bibliography

Aalen O, Røysland K, Gran J, Ledergerber B (2012) Causality, mediation and time: A dynamic viewpoint. J R Stat Soc A Stat Soc 175:831–861. https://doi.org/10.1111/j.1467-985X.2011.01030.x

Ben Gal I (2008) Bayesian Networks. In: Ruggeri F, Kenett RS, Faltin FW (eds) Encyclopedia of statistics in quality and reliability. Wiley-Interscience, Chichester, pp 175–185

Box GEP, Jenkins GM, Reinsel GC, Ljung GM (2015) Time series analysis: forecasting and control, 5th edn. Wiley, New York

Bratley P, Fox BL, Schrage LE (1983) A guide to simulation. Springer, New York. https://doi.org/10.1007/978-1-4684-0167-7

Breiman L (2001a) Random forests. Mach Learn 45(1):5–32. https://doi.org/10.1023/A:1010933404324

Breiman L (2001b) Statistical modeling: the two cultures (with comments and a rejoinder by the author). Stat Sci 16(3):199–231. https://doi.org/10.1214/ss/1009213726

Breiman L, Friedman JH, Olshen RA, Stone CJ (1984) Classification and regression trees. Chapman & Hall, London

Cochran WG (1977) Sampling Techniques, 3rd edn. Wiley, New York

Craven P, Wahba G (1978) Smoothing noisy data with spline functions. Numer. Math. 31(4):377–403. https://doi.org/10.1007/BF01404567

Daniel C, Wood FS (1999) Fitting equations to data: computer analysis of multifactor data, 2nd edn. Wiley, New York

del Rosario Z, Iaccarino G (2022) All Models are Uncertain: Case Studies with a Python Grammar of Model Analysis. https://zdelrosario.github.io/uq-book-preview/00_frontmatter/landing-page.html

Dempster AP, Laird NM, Rubin DB (1977) Maximum likelihood from incomplete data via the EM algorithm. J R Stat Soc Ser B Methodol 39(1):1–38

Draper NR, Smith H (1998) Applied regression analysis, 3rd edn. Wiley, New York

Efron B, Hastie T (2016) Computer age statistical inference, 1st edn. Cambridge University, New York

Fisher RA (1935) The design of experiments. Oliver and Boyd, Ltd., Edinburgh

Frosini BV (2006) Causality and causal models: a conceptual perspective*. Int Stat Rev 74(3):305–334. https://doi.org/10.1111/j.1751-5823.2006.tb00298.x

Granger CWJ (1969) Investigating causal relations by econometric models and cross-spectral methods. Econometrica 37(3):424–438. https://doi.org/10.2307/1912791

Hastie T, Tibshirani R, Friedman J (2009) The elements of statistical learning: data mining, inference, and prediction, 2nd edn. Springer Series in Statistics, Springer, New York. https://doi.org/10.1007/978-0-387-84858-7

Heckerman D (1995) A tutorial on learning with bayesian networks. Tech. Rep. MSR-TR-95-06, Microsoft

Heckman JJ (2008) Econometric causality. Int. Stat. Rev. **76**(1):1–27. https://doi.org/10.1111/j.1751-5823.2007.00024.x

Imai K, Tingley D, Yamamoto T (2013) Experimental designs for identifying causal mechanisms. J R Stat Soc A Stat Soc 176(1):5–51. https://doi.org/10.1111/j.1467-985X.2012.01032.x

Kenett RS (1983) On an exploratory analysis of contingency tables. J R Stat Soc Ser D (The Statistician) 32(4):395–403. https://doi.org/10.2307/2987541

Kenett RS (2016) On generating high InfoQ with Bayesian networks. Quality Technology and Quantitative Management 13(3):309–332. https://doi.org/10.1080/16843703.2016.1189182

Kenett RS (2017) Bayesian networks: Theory, applications and sensitivity issues. Encyclopedia with Semantic Computing and Robotic Intelligence 01(01):1630014. https://doi.org/10.1142/S2425038416300147

Kenett RS, Redman TC (2019) The real work of data science: turning data into information, better decisions, and stronger organizations, 1st edn. Wiley, Hoboken

Kenett RS, Rubinstein A (2021) Generalizing research findings for enhanced reproducibility: An approach based on verbal alternative representations. Scientometrics 126(5):4137–4151. https://doi.org/10.1007/s11192-021-03914-1

Kenett RS, Salini S (2008) Relative Linkage Disequilibrium applications to aircraft accidents and operational risks. Transactions on Machine Learning and Data Mining 1(2):83–96

Kenett R, Salini S (2009) New Frontiers: Bayesian networks give insight into survey-data analysis. Qual Prog 42:30–36

Kenett RS, Salini S (eds) (2012) Modern analysis of customer surveys: with applications using R. Wiley, London

Kenett RS, Shmueli G (2016) Information quality: the potential of data and analytics to generate knowledge, 1st edn. Wiley, Chichester

Koski T, Noble J (2009) Bayesian networks: an introduction, 1st edn. Wiley, Chichester

Kotz S, Johnson NL, Read CB (eds) (1988) Encyclopedia of statistical sciences, vol. 9. Set, 1st edn. Wiley-Interscience, New York

Lawrence ND (2017) Data Readiness Levels. arXiv:170502245 [cs] 1705.02245

Lozano AC, Abe N, Liu Y, Rosset S (2009) Grouped graphical Granger modeling for gene expression regulatory networks discovery. Bioinformatics 25(12):i110–i118. https://doi.org/10.1093/bioinformatics/btp199

Mealli F, Pacini B, Rubin DB (2012) Statistical inference for causal effects. In: Kenett RS, Salini S (eds) Modern analysis of customer surveys: with applications using R. Wiley, London

Neyman J, Pearson ES (1967) On the problem of two samples. In: Joint statistical papers. University of California, California. chap Joint Statistical Papers, pp 99–115

Nielsen TD, Jensen FV (2007) Bayesian networks and decision graphs, 2nd edn. Springer, New York

Pearl J (1985) Bayesian networks: a model of self-activated memory for evidential reasoning. Technical Report CSD-850021, University of California, Los Angeles. Computer Science Department

Pearl J (1988) Probabilistic reasoning in intelligent systems. Elsevier, Amsterdam. https://doi.org/10.1016/C2009-0-27609-4

Pearl J (2009) Causality, 2nd edn. Cambridge University Press, Cambridge

Pourret O, Naïm P, Marcot B (2008) Bayesian networks: a practical guide to applications, 1st edn. Wiley, Chichester

Price PN, Nero AV, Gelman A (1996) Bayesian prediction of mean indoor radon concentrations for Minnesota counties. Health Phys. 71(6):922–936. https://doi.org/10.1097/00004032-199612000-00009

Ruggeri F, Kenett RS, Faltin FW (eds) (2008) Encyclopedia of statistics in quality and reliability. Wiley, New York

Ryan BF, Joiner BL (2000) MINITAB handbook, 4th edn. Duxbury Press, Pacific Grove

Scheffé H (1999) The analysis of variance, 1st edn. Wiley-Interscience, New York

Seeger MW (2008) Cross-validation optimization for large scale structured classification kernel methods. J. Mach. Learn. Res. 9(39):1147–1178

Shumway RH, Stoffer DS (2010) Time series analysis and its applications: with R examples, 3rd edn. Springer, New York

Tank A, Covert I, Foti N, Shojaie A, Fox E (2021) Neural Granger Causality. In: IEEE Transactions on Pattern Analysis and Machine Intelligence, pp 1–1. https://doi.org/10.1109/TPAMI.2021.3065601, 1802.05842

Wang W, Gelman A (2015) Difficulty of selecting among multilevel models using predictive accuracy. Statistics and Its Interface 8(2):153–160. https://doi.org/10.4310/SII.2015.v8.n2.a3

Zacks S (2009) Stage-wise adaptive designs, 1st edn. Wiley, Hoboken

Index

Symbols
3D-scatterplots, 228, 290

A

Acceptance, 9, 149, 160, 214
Acceptance region, 149, 160, 214
Acceptance sampling, 9
Accuracy, 6–7, 34, 35, 362, 365, 369, 375, 376, 386
Activation function, 382
Adjusted coefficient of determination, 365
Akaike information criterion (AIC), 365
Alternative hypothesis, 149, 158
Analysis of variance (ANOVA), 201, 214, 246, 271, 290
ANOVA Table, 272, 275
Arcsin transformation, 289, 291
ARIMA model, 358
ARMA model, 345, 358
Attained significance level, 152
Auto regressive, 358
Average linkage, 388
Average treatment effect (ATE), 414

B

Bag of words, 401, 402, 416
Balanced accuracy, 365
Balanced sample, 323
Bar diagram, 10
Batches, 39, 205, 230
Bayes decision function, 182
Bayes estimator, 184
Bayesian decisions, 176, 215

Bayesian information criterion (BIC), 365, 407
Bayesian network (BN), 53, 405, 407, 408, 410, 413, 416
Bayes risk, 182, 184
Bayes' theorem, 51, 53, 125, 179, 378, 392
Bernoulli trials, 66, 164, 204
Beta distribution, 92, 93, 178, 180
Beta function, 92
Binomial distribution, 63, 67, 69, 72, 75–78, 119, 148, 158, 159, 164, 178, 186, 204, 209, 216, 217, 288
Binomial experiments, 288
Bivariate frequency distribution, 233
Bivariate normal distribution, 105, 107
Boosted tree, 369, 376, 392
Bootstrap, 9, 144, 191, 192, 208, 214, 301, 305, 366, 377
Bootstrap ANOVA, 215, 271
Bootstrap confidence intervals, 190, 198, 215
Bootstrap confidence limits, 189, 190, 198
Bootstrap distribution, 189, 192
Bootstrap estimate, 189
Bootstrap forest, 369, 376, 392
Bootstrap method, 175, 189, 190, 192, 196, 201, 206, 215
Bootstrap sampling, 189, 191, 193
Bootstrap testing, 192
Bootstrap tolerance interval, 204, 205, 210, 215
Box and whiskers plot, 25, 28, 34, 233

C

Categorical data analysis, 288, 291
Cauchy distribution, 60

© The Author(s), under exclusive license to Springer Nature Switzerland AG 2022   433
R. S. Kenett et al., *Modern Statistics*, Statistics for Industry, Technology, and
Engineering, https://doi.org/10.1007/978-3-031-07566-7

Causal Bayesian networks, 413
Causality, 53, 407, 411, 413, 414, 416
c.d.f., *see* Cumulative distribution function (c.d.f.)
Central limit theorem (CLT), 117, 126, 193
Central moments, 59, 63, 85, 125, 309
Characteristic polynomials, 339, 358
Chebyshev's inequality, 24
Chi-square automatic interaction detector (CHAID), 369
Chi-squared statistic, 285
Chi-squared test, 175, 215, 291
Chi-squared test for contingency tables, 291
Classification and regression trees (CART), 368
Class intervals, 14, 30
CLT, *see* Central limit theorem (CLT)
Cluster analysis, 392
Clustering, 227, 366, 386, 388, 390
Cluster number selection, 390
Code variables, 229, 230, 290
Coding, 7
Coefficient of determination, 242, 290, 365
Coefficient of variation, 23, 322, 323
Complementary event, 43, 50
Complete linkage, 388, 389
Computerized numerically controlled (CNC), 231
Conditional distribution, 93, 99–101, 103, 105, 106, 125, 234, 407
Conditional expectation, 99, 100
Conditional frequency distribution, 284, 290
Conditional independence, 53, 125
Conditional probabilities, 49, 51, 52, 125
Conditional variance, 100, 101
Confidence intervals, 160–165, 170, 171, 192–193, 196–199, 204, 205, 214, 275–279, 303, 346
Confidence level, 25, 160, 162, 163, 167, 190, 205–208, 275, 308, 314
Confusion matrix, 369, 374, 386, 392
Consistent estimator, 141, 214, 269
Contingency tables, 279–281, 283, 286–289, 291
Contingency tables analysis, 286, 291
Continuous random variable, 13, 34, 54, 93, 105, 125
Continuous variable, 9, 205, 378
Contrasts, 20, 25, 266, 275, 278, 291, 362
Convergence in probability, 112, 126
Cook distance, 267, 268, 290
Correlation, 96–99, 170–172, 236–245, 247, 253, 262–264, 283, 284, 288, 292, 329, 330, 410

Covariance, 96–99, 236–237, 246, 248, 330–331, 338, 346, 357, 400
Covariance matrix, 347, 354
Covariance stationary, 336, 338, 339, 343, 346, 357, 358
Covariates, 320, 325, 368, 375
Cramer's index, 287
Credibility intervals, 185
Critical region, 149–152, 154, 156, 158, 159
Cross-validation, 26, 189, 361, 366, 375
Cumulative distribution function (c.d.f.), 55, 125
Cycle time, 2, 19, 54, 174, 176, 206, 209

D

Decision trees, 363, 365, 366, 368–370, 372–374, 376, 378, 381, 392
Deep learning, 383, 385
Degrees of freedom, 121, 122, 124, 162, 164, 174, 175, 258, 259, 272, 285
De Morgan rule, 42
Descriptive analysis, 9, 384
Design of experiments (DoE), 2, 367
Deterministic component, 237
Directed acyclic graph (DAG), 53, 405
Discrete random variables, 9, 53, 54, 93, 98, 125, 146, 178
Discrete variable, 9, 56
Disjoint events, 44, 51, 125
Distribution free tolerance limits, 206
Distribution median, 212
Document-term matrix (DTM), 403, 404
Dynamic linear model (DLM), 352–355, 358

E

Elementary events, 40, 125
Empirical bootstrap distribution (EBD), 189
Ensemble models, 376, 377
Entropy, 371
Estimator, 139–145, 160, 161, 163, 184, 185, 191, 192, 205, 214, 255, 269, 270, 300, 305
Event, 39, 40, 42, 43, 45, 47–49, 52, 54, 55, 66, 72, 74, 76, 87, 100, 110, 125, 284, 318
Expected frequency, 173, 175, 284
Expected loss, 176, 184
Expected value, 60, 62, 67, 70, 75, 77, 78, 83, 96, 97, 99, 101, 109, 111, 117, 121, 145, 191, 300, 305, 306, 407
Experiment, 2, 3, 14, 39, 40, 43, 46, 48, 49, 52, 54, 55, 60, 199, 204, 229, 240, 268,

269, 271, 274, 288, 302, 367, 396, 398,
411, 413
Explainable, 240, 247, 257, 261
Exponential distribution, 85, 87, 89

F
Factor levels, 199
Failure rate, 281
Failures per million (FPM), 279
FDA, *see* Functional data analysis (FDA)
Finite population, 8, 46, 54, 190, 299, 301,
309, 312, 320, 325
Finite population multiplier, 312, 325
FPM, *see* Failures per million (FPM)
Frequency distribution, 9–16, 18, 34, 140, 173,
187–189, 227, 230–235, 284, 290, 304
Functional data analysis (FDA), 395, 396, 398,
400, 416
Functional principal components, 401, 416

G
Gage repeatability and reproducibility (GRR),
7
Gamma distribution, 88, 116, 119, 185
Gaussian distribution, 24, 79
Generators, 43, 301
Geometric distribution, 75, 76
Geometric mean, 23
Gini impurity, 371
GRR, *see* Gage repeatability and
reproducibility (GRR)

H
Hierarchical clustering, 386, 390
Histograms, 14, 16, 20, 25, 26, 30, 31, 140,
141, 172, 190, 197, 202, 203, 228, 236,
303, 304, 324
Homogeneous groups, 277, 368
Hypergeometric distribution, 69, 104, 312

I
Inclusion relationship, 40
Incomplete beta function ratio, 92
Independent events, 50, 51, 125
Independent random variables, 98, 115, 116
Independent trials, 66, 75, 102, 125, 163, 288
Indices of association, 282, 284
Information quality (InfoQ), 363
Inspection, 67, 299, 398
Interactions, 369

Interquartile range, 23, 28, 33
Intersection of events, 41, 51
Inverse document frequency (IDF), 402

J
Joint distribution, 53, 96, 105, 125, 407
Joint frequency distribution, 231, 232

K
Kolmogorov-Smirnov test (KS), 175
Kurtosis, 21, 24, 34, 62, 64, 65, 80, 87, 125

L
Lag-correlation, 336–338, 340, 341, 358
Lagrangian, 319
Laplace transform, 66
Latent semantic analysis (LSA), 404
Law of iterated expectation, 125
Law of large numbers (LLN), 117, 126, 142,
184
Law of total variance, 101, 125
Least squares, 144, 145, 214, 239, 240, 242,
246, 249, 251, 331, 362
Least squares estimator, 144, 145, 214, 246
Level of significance, 149, 161, 162, 200, 202,
209, 273, 289
Life length, 274
Likelihood function, 146–148, 214
Likelihood statistic, 147
Linear combination, 111, 145, 310, 337, 344,
383, 404
Linearity, 7
Linear model, 239, 245, 246, 265, 332,
351–355, 358
Linear predictors, 346–349, 358
Log-normal distribution, 171
Loss function, 177, 181, 184
Lot, 8, 9, 67, 72, 105, 119, 230, 273, 300, 318
Lower tolerance limit, 167, 168

M
MAE, *see* Mean absolute error (MAE)
MA model, 358
Marginal distribution, 93, 100, 101, 103, 105,
106, 125, 234, 305, 378
Marginal frequencies, 231, 232, 235
Matrix scatterplot, 290
Maximum likelihood estimator (MLE), 147
Mean absolute error (MAE), 365, 373
Mean squared contingency, 285

Mean squared error (MSE), 323, 346, 365, 373
Mean vector, 388
Measurements, 1, 3, 4, 6–8, 11, 19, 20, 23, 24,
    29, 39–41, 230, 231, 237, 238, 247,
    363, 369, 384, 386, 395, 416
m.g.f., *see* Moment generating function
    (m.g.f.)
Mixing, 39
Mode, 413
Model, 1, 39–138, 173, 206, 209, 225–297,
    321, 322, 332–336, 353, 355, 361–368,
    370, 376–386, 405, 407, 410, 412–414
Moment equation estimator, 142, 214
Moment generating function (m.g.f.), 65, 125
Moments, 59, 63, 65, 66, 79, 80, 85, 92, 125,
    142, 143, 214, 309
Moving averages, 337–339, 343, 344, 351, 358
Moving average smoother, 351, 358
MSE, *see* Mean squared error (MSE)
Multi-hypergeometric distribution, 104
Multinomial distribution, 102
Multiple boxplot, 229
Multiple comparisons, 275, 291
Multiple regression, 240, 245–246, 251, 257,
    258, 261, 263, 264, 266, 271, 290
Multiple squared correlation, 290
Mutual independence, 50, 125

N
Negative-binomial distribution, 77
Neural network, 381–386
Non-parametric test, 208, 215
Normal approximation, 119, 120, 165, 193,
    313
Normal distribution, 21, 24, 25, 33, 79–84,
    119, 122, 142, 152–158, 161, 163,
    166–173, 176, 178, 183, 206
Normal equations, 254, 340
Normal probability plot, 172, 173, 215, 335
Normal scores, 215
Null event, 40
Null hypothesis, 149, 150, 152, 158, 163, 175,
    192, 208, 209, 213, 275, 288, 290

O
Objectives, 1, 186, 205, 266, 366, 384
Observed frequency, 173, 175
OC curve, 151, 214
OC function, 150, 153–156, 159, 214
Operating characteristic (OC), 150, 153–156,
    158, 159, 214
Optimal allocation, 316, 319, 325

Order statistics, 16, 23, 24, 93, 108, 112, 126,
    175, 205, 206, 208
Outliers, 23, 28, 29, 31, 33, 237, 362, 368, 388,
    398–400

P
Parameters, 9, 25, 26, 43, 54, 67, 72, 77, 80,
    81, 88, 92, 103, 119, 120, 130, 139,
    141, 142, 149, 152, 160, 161, 176–178,
    184–186, 189, 212, 214, 217, 220, 268,
    299, 300, 303, 318, 345, 349, 352, 354,
    362, 366, 384, 400, 405, 407
Parameter space, 142, 143, 147, 161, 178, 181,
    214
Parametric family, 141, 146
Partial correlation, 251, 252, 262, 264, 290,
    340
Partial lag-correlation, 341, 358
Partial regression, 251, 290
p.d.f., *see* Probability density function (p.d.f.)
Piston, 1, 2, 19, 174, 176, 206, 209
Point estimator, 142, 214
Poisson distribution, 72, 73, 87, 88, 115, 119,
    120, 143, 157, 178, 185
Polynomial predictors, 358
Population, 8, 17, 24, 34, 40, 46–48, 54, 69,
    117, 123, 139–142, 149, 160, 163, 166,
    167, 169, 187, 188, 190–192, 195, 198,
    199, 201, 204–208, 214, 215, 234, 271,
    273, 299–327, 362, 364, 369, 384, 413,
    414
Population mean, 117, 167, 195, 204, 214, 271,
    299, 300, 304, 307, 314
Population quantiles, 324
Population quantities, 300, 302, 305, 309
Posterior distribution, 177, 184, 185, 353, 407
Posterior expectation, 184, 353
Posterior probability, 52, 53, 125, 183, 405,
    407
Posterior risk, 183, 184
Power function, 152, 155, 157, 160, 214
Precision, 6, 34, 141, 300, 305, 320, 325, 375
Precision of an estimator, 325
Predicted values, 243, 247, 257, 266, 268, 290,
    374
Prediction errors, 366, 367
Prediction intervals, 23, 25, 34, 204, 244
Prediction model, 320, 323, 325
Prediction MSE, 325
Predictors, 145, 240, 245, 247, 256, 262, 263,
    266, 269, 321, 322, 324, 325, 327, 329,
    330, 345–352, 354, 358, 366, 368, 369,
    371, 377, 378, 380, 381, 384–386

Principal component analysis (PCA), 395, 405
Principal components, 395, 401, 404, 405, 416
Principle of least squares, 246
Prior distribution, 353
Prior probability, 125, 178, 181, 184
Prior risk, 182
Probability density function, 57, 178
Probability distribution function (p.d.f), 55, 72, 125, 146
Probability function, 44, 46, 54, 75, 102
Probability of events, 125
Process control, 2
Proportional allocation, 316–319, 325
Protocol, 40
P-value, 215

Q

Quadratic model, 367
Quantile plot, 25, 29, 34
Quantiles, 19, 34, 62–65, 68, 121, 130, 142, 153, 158, 162, 164, 166, 167, 170, 177, 187–189, 198, 200, 202, 205, 207–209, 261

R

Random component, 5, 6, 34, 237, 246, 321
Random forest, 369, 377, 381, 392
Randomization, 192, 208, 210, 214, 413
Randomization test, 208, 210, 214
Random measurements, 39
Randomness, 40, 234
Random numbers, 3, 42, 46, 171, 172, 301
Random sample, 8, 46, 67, 105, 111, 117, 139, 140, 142, 152, 163, 167, 185–187, 189, 192, 195, 198, 204, 208, 209, 212, 271, 299–303, 305–314, 367
Random sample without replacement (RSWOR), 9, 301
Random sample with replacement (RSWR), 9, 301
Random variable, 3, 9, 13, 17, 34, 54–67, 69, 75, 77, 78, 81, 83, 88, 96, 98, 99, 101, 111–119, 121, 122, 139, 141–142, 145, 149, 156, 177, 178, 191, 206, 207, 268, 271, 294, 305, 306, 310, 315, 330, 337, 357, 405, 407
Ratio predictor, 321, 322, 324, 325
Reference distribution, 186–188, 192, 193, 210, 213–215
Regression coefficients, 145, 242, 246, 248, 251, 254–256, 267–269, 320
Regression diagnostics, 265, 290

Regularization, 384, 385
Rejection region, 149, 151, 214
Reliable, 273, 332, 368
Repeatability, 6, 7
Reproducibility, 7
Residuals around the regression, 243, 247, 262, 290
Resolution, 364
Response surface, 383
Response variable, 382
Robust statistics, 25, 31, 33, 34
Runs, 76, 259, 286, 384

S

Sample, 8–25, 35, 40, 46–49, 54, 55, 67, 72, 105, 117–125, 139–145, 149–163, 167, 169–176, 187–196, 198–209, 211–213, 231, 234, 240, 265, 271, 273, 274, 284, 286, 288, 300–314, 318, 321, 322, 324, 366, 367, 373, 384, 400
Sample allocation, 316, 325
Sample correlation, 236–238, 247, 283, 290
Sample covariance, 236, 237, 400
Sample kurtosis, 21
Sample maximum, 17, 108
Sample mean, 19, 20, 22, 24, 25, 28, 31, 34, 111, 117, 120, 140, 141, 145, 152, 156, 161, 162, 167, 187, 188, 191, 195, 201, 225, 236, 239, 277, 302, 304, 305, 307, 321
Sample median, 17–19, 209
Sample minimum, 17, 23
Sample percentile, 19
Sample quantiles, 19, 29
Sample quartiles, 17
Sample range, 17
Sample skewness, 21, 24
Sample space, 40, 42, 43, 46–49, 52, 54, 55, 125
Sample standard deviation, 20, 31, 34, 122, 156, 157, 163, 198
Sample variance, 20, 34, 111, 121, 141, 162, 191, 198, 200, 236, 240, 247, 255, 305
Sampling distribution, 117, 140, 141, 160, 187, 189, 214, 215, 289, 302, 303, 306, 313
Sampling distribution of an estimate, 215
Sampling with replacement, 34, 125, 301
Scatterplots, 226, 227, 229, 236, 238, 244, 249, 250, 252, 257, 290, 336
Schwarz inequality, 236
SE, *see* Standard error (SE)
Sequential SS, 260
Shift, 4, 176, 301, 350, 398

Significance level, 150, 151, 153, 156, 160, 214
Sign test, 208, 209, 211, 214
Simple linear regression, 145, 214, 239, 240, 251, 252, 260, 266, 290
Simple random sample, 140, 300, 305, 324
Simulation, 2, 43, 160, 170, 187, 214, 313, 314
Simultaneous confidence interval, 275, 291
Single linkage, 388
Singular value decomposition (SVD), 403, 404
Skewness, 21, 24, 28, 30, 34, 62, 69, 85, 87, 125
Slope coefficient, 239, 251
SST, *see* Total sum of squares (SST)
Stability, 7, 376
Standard deviation (STD), 23–25, 31, 33, 34, 62–65, 81, 83, 96, 97, 130, 141, 155, 162, 163, 166, 167, 176, 189, 196, 234, 267, 302, 304, 306, 375
Standard error (SE), 141, 214, 290, 305, 308, 319
Standard error of predicted value, 290
Standardized residual, 267, 290
Statistical hypotheses, 149, 170, 214
Statistical inference, 25, 93, 120, 189, 190, 214, 215, 366
Statistical model, 141, 271, 299, 361, 366
Statistical process control (SPC), 2
Statistic of central tendency, 17
Statistics, 1–38, 112, 120–125, 176, 189, 198, 199, 202, 206, 208, 212, 236, 261, 271, 302, 303, 305, 306, 361–362, 364, 375, 405, 423
Stem and leaf diagram, 34
Step-wise regression, 263, 290
Stratified random sample, 324
Stratified sampling, 364, 367, 368
Studentized test for the mean, 193
Studentized test statistic, 193
Sufficient statistic, 147
Sum of squares of deviations (SSD), 272
Sure event, 40
Symmetric matrix, 255

T
Term frequency (TF), 402
Testing hypotheses, 158, 181, 192
Testing statistical hypotheses, 170, 214
Text analytics, 401, 416

Time till failure (TTF), 109, 110, 120
Toeplitz matrix, 340, 348, 358
Tolerance interval, 25, 166, 167, 204, 206–209, 214, 215
Topic analysis, 404, 416
Total sum of squares (SST), 257, 272
Treatment combinations, 271, 288, 290
Treatments, 1, 271, 272, 275–277, 288, 290, 412
Trend function, 348, 358
Trial, 3, 39, 47, 66, 75, 102, 110, 125, 158, 163, 164, 186, 204, 279, 288, 413
Two-sided test, 154
Type I error, 150, 152, 214
Type II error, 150, 152, 214

U
Unbiased estimator, 141, 143, 163, 191, 214, 249, 255, 307, 310, 314, 318, 325
Unbiased predictor, 322, 323
Uniform distribution, 30, 78, 95, 100, 110, 112, 140, 174, 175, 206
Upper tolerance limit, 167

V
Validation data, 384, 386, 392
Validity, 414
Variance, 60, 67, 70, 75, 77, 78, 81, 83, 96, 100, 101, 111, 112, 117, 121, 123, 130, 139, 140, 145, 149, 156, 163, 183, 191, 192, 198–204, 208, 213, 214, 236, 237, 239, 240, 242, 248, 249, 255, 256, 260, 270, 288, 289, 299, 300, 307, 310, 314, 315, 318, 319, 321, 322, 324, 337, 388

W
Weibull distribution, 88, 90
White noise, 335, 358
Wilcoxon signed rank test, 211

Y
Yule-Walker equations, 339, 358

Z
Z-transform, 338, 358

Printed in the United States
by Baker & Taylor Publisher Services